COMPUTER-AIDED CONTROL SYSTEMS ENGINEERING

edited by

M. JAMSHIDI
CAD Laboratory for Systems
Department of Electrical and Computer Engineering
University of New Mexico
Albuquerque, NM
U.S.A.

and

C. J. HERGET
Lawrence Livermore National Laboratory
Livermore, CA
U.S.A.

1985

NORTH-HOLLAND
AMSTERDAM · NEW YORK · OXFORD

ISBN: 0 444 87779 7

Publishers:

ELSEVIER SCIENCE PUBLISHERS B.V.
P.O. Box 1991
1000 BZ Amsterdam
The Netherlands

Sole distributors for the U.S.A. and Canada:

ELSEVIER SCIENCE PUBLISHING COMPANY, INC.
52 Vanderbilt Avenue
New York, N.Y. 10017
U.S.A.

Library of Congress Cataloging in Publication Data
Main entry under title:

Computer-aided control systems engineering.

1. Automatic control--Data processing--Addresses, essays, lectures. 2. Computer-aided design--Addresses, essays, lectures. I. Jamshidi, Mohammad. II. Herget, Charles J.
TJ213.7.C64 1985 629.8'95 85-10138
ISBN 0-444-87779-7

PRINTED IN THE NETHERLANDS

COMPUTER-AIDED CONTROL SYSTEMS ENGINEERING

PREFACE

It was only a few years ago that the only tools for analysis and synthesis available to the control engineer were paper, pencil, slide rule, spirule, and analog computer. The tools and the methods were simple enough that an engineer could easily master them in a relatively short time. However, over the past twenty-five years, control theory has evolved to a state where the digital computer has become a requirement for the control system engineer, and Computer-Aided Control System Design (CACSD) has emerged as an indispensable tool.

A good CACSD system draws on expertise from many disciplines including aspects of computer engineering, computer science, applied mathematics (e.g. numerical analysis and optimization), as well as control systems engineering and theory. The need for such a breadth is partially responsible for the paucity of high quality CACSD software today, and indeed, CACSD must be considered in its infancy. Professional societies, such as the IEEE Control Systems Society and IFAC, have recognized the need for increased professional activities in this field and have sponsored a number of Workshops and Symposia on this subject, the IEEE's being held in the United States and IFAC's being held in Europe and the United Kingdom. Most of these meetings did not have formal proceedings for the outside user to read, and papers on CACSD have not been readily available to the professional community. In recognition of the relevance of this topic, the IEEE Control Systems Society published a Special Issue on CACSD in the *Control Systems Magazine* (December 1982) and the IEEE has published a Special Section on CACSD in the *Proceedings of the IEEE* (December 1984).

It is the object of this volume, therefore, to make available a thorough and up-to-date account of Computer-Aided Control Systems Engineering by collecting together in one place some of the various papers presented at these workshops and symposia as well selected papers from some of the professional journals. We point out that we have used the term computer-aided control systems engineering in the title of this book instead of computer-aided control system design to emphasize the much broader facets of the problem which the control engineer faces. *I.e.*, aspects of modeling, analysis and simulation must always be considered in addition to design. In addition to the selected technical papers, there is a collection of software summaries describing CACSD packages from around the world.

The book consists of nineteen individual contributions from thirty-eight authors and the Software Summaries describe thirty-seven different packages. The papers fall into three general categories: (*i*) computer-aided control system design packages and languages, (*ii*) perspectives and expository looks at computer-aided control system design, and (*iii*) algorithms and techniques for CACSD.

The editors would like to take this opportunity and thank every one of the authors for their worthy contributions to CACSD in general and to this volume in particular. We wish to express our gratitude to Ms. Fran McFarland at Lawrence Livermore National Laboratory and

Ms. Gladys Ericksen at the University of New Mexico for their tireless assistance to the editors. We want to express our appreciation to Dr. Gerard Wanrooy and his staff at North Holland in Amsterdam for their collaboration. We also want to thank IEEE and Springer-Verlag for permitting us to use some articles on which they hold the copyright in this book. The first editor would also like to express his appreciation for the support he received from Dr. R. H. Seacat, Chairman of the Department of Electrical and Computer Engineering, and Dr. G. W. May, Dean of Engineering, at the University of New Mexico.

M. Jamshidi
Albuquerque, NM
U.S.A.

C. J. Herget
Livermore, CA
U.S.A.

CONTRIBUTOR LIST

William F. Arnold, III
Naval Weapons Center
China Lake, California, USA

K. J. Åström
Lund Institute of Technology
Lund, Sweden

S. N. Bangert
Systems Control Technology, Inc.
Palo Alto, California, USA

S. P. Bingulac,
University of Belgrade
Belgrade, Yugoslavia

François E. Cellier
University of Arizona
Tucson, Arizona, USA

Michael J. Denham
Kingston Polytechnic
Kingston-upon-Thames, England

Howard Elliott
University of Massachusetts
Amherst, Massachusetts, USA

A. Emami-Naeini
Systems Control Technology, Inc.
Palo Alto, California, USA

Michel A. Floyd
Integrated Systems, Inc.
Palo Alto, California, USA

G. F. Franklin
Stanford University
Stanford, California, USA

Dean K. Frederick
Rensselaer Polytechnic Institute
Troy, New York, USA

Charles J. Herget
Lawrence Livermore National Laboratory
Livermore, California, USA

M. Jamshidi
University of New Mexico
Albuquerque, New Mexico, USA

Russel P. Kraft
Mechanical Technologies, Inc.
Latham, New York, USA

Alan J. Laub
University of California
Santa Barbara, California, USA

Eugene A. Lee
The Aerospace Corporation
El Segundo, California, USA

Gordon K. F. Lee
Colorado State University
Fort Collins, Colorado, USA

Larry L. Lehman
Integrated Systems, Inc.
Palo Alto, California, USA

J. N. Little
Systems Control Technology, Inc.
Palo Alto, California, USA

A. G. J. MacFarlane
Cambridge University
Cambridge, England

J. M. Maciejowski
Cambridge University
Cambridge, England

D. Q. Mayne
Imperial College
London, England

R. Morel
Los Alamos National Laboratory
Los Alamos, New Mexico, USA

W. T. Nye
University of California
Berkeley, California, USA

R. V. Patel
Concordia University
Montreal, Canada

W. R. Perkins
University of Illinois
Urbana, Illinois, USA

E. Polak
University of California
Berkeley, California, USA

Magnus Rimvall
Swiss Federal Institute of Technology
Zurich, Switzerland

Tahm Sadeghi
Fairchild Republic Company
Farmingdale, New York, USA

Chr. Schmid
Ruhr University
Bochum, Federal Republic of Germany

J. Schotik
University of New Mexico
Albuquerque, New Mexico, USA

Sunil C. Shah
Integrated Systems, Inc.
Palo Alto, California, USA

P. Siegel
University of California
Berkeley, California, USA

H. Austin Spang, III
General Electric R&D Center
Schenectady, New York, USA

Diane M. Tilly
Lawrence Livermore National Laboratory
Livermore, California, USA

P. P. J. van den Bosch
Delft University of Technology
Delft, The Netherlands

P. J. West
University of Illinois
Urbana, Illinois, USA

T. Wuu
University of California
Berkeley, California, USA

T. C. Yenn
University of New Mexico
Albuquerque, New Mexico, USA

TABLE OF CONTENTS

Section 1. PACKAGES

Section 1. PACKAGES

COMPUTER-AIDED CONTROL SYSTEMS ENGINEERING
M. Jamshidi and C.J. Herget (Editors)

COMPUTER AIDED TOOLS FOR CONTROL SYSTEM DESIGN

K.J. Aström

Department of Automatic Control
Lund Institute of Technology
Lund, Sweden

The paper describes experiences of development and use of interactive software for computer aided design of control systems. The experiences are drawn from a comprehensive set of packages for modeling, identification, analysis, simulation and design, which have been in use for about a decade. Problems associated with structuring, portability, maintainability, and extensibility are discussed. Experiences from uses of the packages in teaching and industrial environments are discussed. Views on future development of CAD for control systems are also given.

1. INTRODUCTION

Thirty years ago pencil, paper, slide rules and analog computers were the major tools for analysis and synthesis of control systems. The methods and the tools were so simple that an engineer could master both problems and tools. Many new methods for analysis and design of control systems have emerged during the last 30 years. These methods differ from the classical techniques. They are more sophisticated analytically and their use require extensive calculations. An extensive subroutine library is required to apply these methods to a practical problem. Even if such a library is available it is a major effort to write the software necessary to solve a particular problem. This means that modern control theory is costly to use. Another drawback is that the problem solver interacts with his tools (the computer) via intermediaries (programmers). This easily leads to confusion and mistakes. The intensive interaction between problem formulation and solution is also lost.

Based on experience from industrial application of modern control theory in the early sixties it was clear to me that modern control theory could be used very successfully in a research laboratory or at a university. It was, however, equally clear that the methods would not be widely used in normal engineering practice unless the proper tools were developed. A number of projects were therefore carried out in order to efficiently develop the proper tools for using control theory cost. This paper summarizes some of the software developed in the project and experiences from its use.

The projects were based on the idea of combining an engineer's intuition and overview with digital computing power. The approach included development of design techniques and man-machine interfaces, for interactive use of the computer. Graphics is of course a major ingredient.

The paper is organized as follows. A brief overview of the projects is given in Section 2. Interaction principles are discussed in Section 3. Data structures are discussed in Section 4. The comprehensive set of program packages which is one result of the projects is described in Section 5. Some special problems associated with large systems are discussed in Section 6. Experiences from use of the packages in university and industrial environments are presented in Section 7. Sections 8 and 9 give suggestions for future work and conclusions.

2. THE PROJECTS

The objectives of the projects were to make advanced methods for modeling, analysis and design of control systems easily accessible to engineers, researchers and students and to explore the potentials of interactive computing for control system design, see Åström and Wieslander (1981). When the projects were initiated around 1970 we had extensive experience of analog simulation, programming in Fortran, Basic, and APL. There was common consensus about the power of digital computation and the superiority of the man-machine interaction in analog simulation. We were familiar with the ease of debugging and running programs in an interactive implementation like APL. But we were also aware of the limited portability of such programs and of the difficulties of extending such systems. The software was developed in close interaction with the users. A system outline was sketched. The ideas were discussed in seminars. A system of moderate size was implemented and tested by several users. The system was then modified. In the initial phases we were also quite willing to scrap a system and start all over again. As the projects progressed we got a much better feel for what could be done and how it should be done. It also became clear that a fairly comprehensive package was necessary to evaluate the ideas. Such packages were also developed. They went through many revisions to improve portability, modularity and efficiency. The packages matured and reached a stable state around 1979. From then on we have only made moderate maintenance and updating. Uses of the packages have increased rapidly since then.

What can be done with interactive computing depends much on the available hardware. Since the hardware has undergone a revolutionary development over the past ten years it is useful to describe what was available in the projects. When the activity was started, in 1971, we had access to a DEC PDP 15 with 32 kbytes of core memory, a 256 kbytes disk and a storage oscilloscope. After a few years the activity was moved to large mainframe computers. We are currently using a DEC Vax-11/780 with 2 Mbyte of fast memory and a 600 Mbyte disc for most of the work. Our sponsoring agency (STU) also introduced constraints by insisting that the programs should be portable and useful to industry. One way to achieve this was to use standard Fortran.

The projects have resulted in a comprehensive set of program packages for modeling, identification, analysis, simulation and design of control systems. We have several years experience of using these packages in different university and industrial environments. Ideas on the use of graphics and interactive computing in future systems have also been developed.

3. INTERACTION PRINCIPLES

When designing a system for man-machine interaction it is important to realize that there is a wide range of users, from <u>novices</u> to <u>experts</u>, with different abilities and demands. For a novice who needs a lot of guidance it is natural to have a system where the computer has the initiative and the user is gently led towards a solution of his problem. For an expert user it is much better to have a system where the user keeps the initiative and where he gets advice and and help on request only. Attempts of guidance and control by the computer can easily lead to frustration and inefficiency. It is highly desirable to design a system so that it will accomodate a wide range of users. This makes it more universal. It also makes it possible to grow with the system and to gradually shift the initiative from the computer to the user as he becomes more proficient. Many aspects on interaction principles and their implementation are found in Barstow et al. (1984).

To obtain an efficient man-machine interface it is desirable to have hardware with a high communication rate and a communication language with a good expression power. When our projects were started we were limited to a teletype and a storage oscilloscope. There were also limited experiences of design of man-machine interfaces. The predominant approach was a question-and-answer dialog, see e.g. Rosenbrock (1974).

In our projects it was discovered at an early stage that the simple question-and-answer dialog was too rigid and very frustrating for an experienced user. The main disadvantage is that the computer is in command of the work rather than the user. This was even more pronounced because of the slow input-output device (teletype) which was used initially.

Our primary design goal was to develop tools for the expert. A secondary goal was to make the tools useful also for a novice. To make sure that the initiative would remain with the user it was decided to make the interaction command oriented. This was also inspired by experiences from programming in APL. Use of a command dialog also had the unexpected effect that it was easy to create new user defined commands. The packages could thus be used in ways which were not anticipated when they were designed. The decision to use commands instead of a question and answer dialog had far reaching consequences. A more detailed discussion of the different types of dialogs and of our experiences of them is given in Wieslander (1979a,b). Today there is a wide range of experiences of designing man-machine interfaces in many different fields. Our own conclusions agree well with those found in Newman and Sproull (1979), and Foley and van Dam (1983) although their conclusions are based on different hardware.

Examples of commands

The structure of the commands we introduced will now be described. The general form of a command is

```
NAME LARG1 LARG2...  ←  RARG1 RARG2...
```

A command has a name. It may also have left arguments and right arguments. The arguments may be numbers or names of objects in a data base. In our packages the objects are implemented as files because this is a simple way to deal with objects having different types. A few examples of commands are given to further illustrate the notion of a command. The command

```
MATOP S  ←  A * B + C
```

simply performs the matrix operation expressed to the right of the arrow. The command

```
POLOP S  ←  A * B + C
```

performs the same operation on polynomials. The command

```
INSI U 100
   >PRBS 4 7
   >X
```

generates an input signal of length 100 called U. The command has options to generate several input signals. The options are selected by additional subcommands. PRBS is a subcommand which selects a PRBS signal. The optional arguments 4 and 7 indicate that the PRBS signal should change at most every fourth sampling period and that its period should be 2^7-1. The subcommand EXIT denotes the end of the subcommands. The command

```
DETER Y  ←  SYST U
```

generates the response of the linear system called SYST to the input signal U. The command

```
ML PAR  ←  DAT N
```

fits an ARMAX model of order N to the data in the file called DAT and stores the parameters in a file called PAR. The command

```
OPTFB L S  ←  LOSS SYS
```

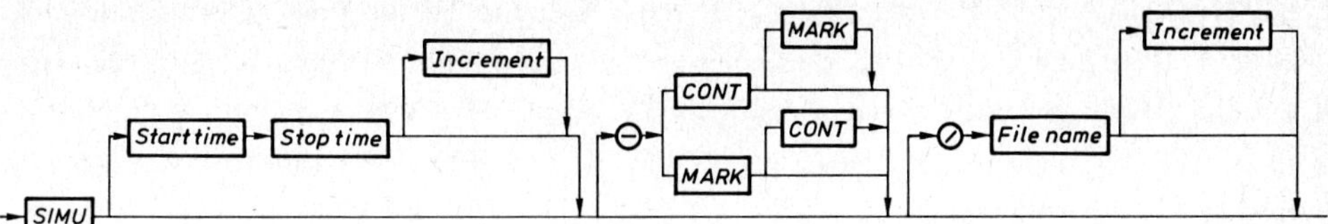

Figure 1
Syntax diagram for the command SIMU.

computes the optimal feedback gain L and the solution S to the steady state Riccati equation for the system SYS and the loss function LOSS.

<u>Short form commands and default values</u>

In a command dialog it is highly desirable to have simple commands. This is in conflict with the requirement that commands should be explicit and that it may sometimes be desirable to have variants of the commands. These opposite requirements may be resolved by allowing short forms of the commands. The standard form for the simulation command is SIMU. If no other command starts with the letter S it is, however, sufficient to type S alone. Another interesting possibility is to correlate a given command with the command list and to choose the command from the list which is closest to the given one. This automatically gives short form commands. The scheme will also be insensitive to spelling errors. It is, however, also dangerous because totally unexpected commands will be obtained. It is also useful to have a simple way of renaming the commands. Such mechanisms are now available in many systems. We have experimented with short form commands, command correlation, and renaming mechanisms. These functions are, however, not implemented in our standard packages.

Similar mechanisms may be used for commands with arguments by introducing a default mechanism so that previous values of the arguments are used unless new values are specified explicitly. The concept is illustrated by an example.

The syntax diagram for the command SIMU is shown in Fig. 1. The diagram implies that any form of the command which is obtained by traversing the graph in the directions of the arrows is allowed. For example the command

```
SIMU 0 100
```

simulates a system from time 0 to time 100. If we want to repeat the simulation a second time with different parameters it suffices to write

```
SIMU
```

The arguments 0 and 100 are then taken as the previously used values.

It follows from Fig. 1 that start and stop times and the initial time increment may be specified. It is also possible to mark curves by the argument MARK. A simulation may also be continued by using the end conditions of a previous simulation as initial values. This is done by the command extension CONT. The results of a simulation may also be stored in a file by appending /filename.

Macros and user defined procedures

Commands are normally read from a terminal in a command driven system. It is, however, useful to have the option of reading a sequence of commands from a file instead. Since this is analogous to a macro facility in an ordinary programming language the same nomenclature is adopted. See e.g. Wegner (1968). The construction

```
MACRO NAME
   Command 1
   Command 2
   Command 3
END
```

thus indicates that the commands 1, 2 and 3 are not executed but stored in memory. The command sequence is then activated simply by typing NAME.

Macros are convenient for simplification of a dialog. Command sequences that are commonly used may be defined as macros. A simple macro call will then activate a whole sequence of commands. The macro facility is also useful in order to generate new commands. Macros may also be used to rename commands. This is useful in order to tailor a system to the needs of a particular user. By introducing commands for reading the keyboard and for writing on the terminal it is also possible to implement menu driven dialogs using macros.

The usefulness of macros may be extended considerably by introducing commands to control the program flow in a macro, facilities for handling local and global variables and by allowing macros to have arguments. We will then have a mechanism for making user defined procedures.

Error checking

It is important in interactive systems to have test for avoiding errors. It is thus useful to check data types and to test problems for consistency whenever possible.

Implementation

It is straightforward to implement a command driven interactive program. The structure used in our packages is shown in Fig. 2. The main loop reads a command, decodes it and performs the required actions. All parts of Fig. 2 except the action routines are implemented as a package of subroutines called Intrac. These subroutines perform

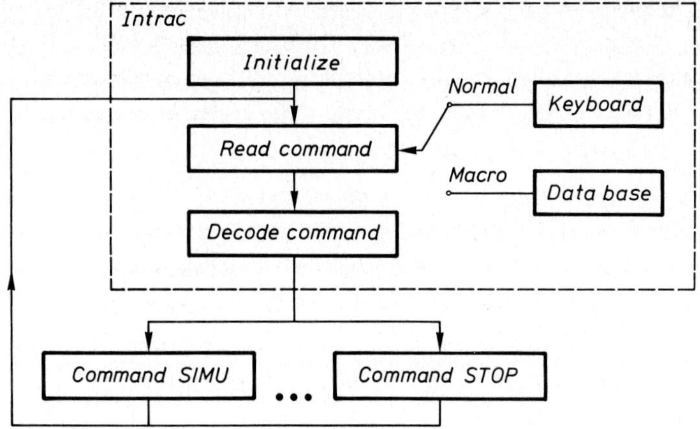

Figure 2
Skeleton flow chart for a command driven program.

command decoding, file handling and plotting. Intrac also contains facilities for creating macros and user defined functions. Macros may have formal arguments, local and global variables. They permit conditional and repeated execution of commands as well as nested use of macros. There are read and write commands, which can be used to implement menu dialogs. It is possible to mix command mode and question mode, since the execution of a macro may be suspended and resumed later. A description of Intrac is given in Wieslander and Elmqvist (1978) and Wieslander (1980a). The commands available in Intrac are listed in Appendix F.

To build a package using Intrac it is necessary to write the action routines i.e. the subroutines that performs the desired tasks. The commands are then entered in the command table of the command decoder. It is also easy to add a command to a package, to move commands between packages and to create special purpose packages. Intrac may thus be viewed as a tool for converting a collection of Fortran subroutines into an interactive package. Intrac has also been used to implement other packages by other groups.

The structure with a common user interface for all packages is advantageous for the user because the interaction and the macro commands are the same in all packages. This simplifies learning and use of the packages. The advantages of interactive computing environments for problem solving are now widely recognized. They are built into systems like Basic, Apl, Lisp and Smalltalk. There are also several systems, which provide interactive features in static computing environments. Speakeasy, CTS are typical examples which are similar to Intrac.

Problem solving languages

The design of a command driven package may be viewed as a construction of a high level language for solving control problems. From this viewpoint design of a CAD system involves the selection of vocabulary, grammar and semantics. The vocabulary defines the basic language elements i.e. the data structures and the operators. The grammar tells how the basic language elements may be combined into new language elements and the semantics tells how the language elements should be interpreted. The language should be rich enough to solve many problems. It should also be simple so that it is easily learned. A CAD program is simply an interpreter for the design language.

The selection of commands is one of the major issues when designing a CAD package. The commands determine how useful a package is and how easy it is to learn. It is important that commands are complete in the sense that they allow use of a wide range of techniques in an area. Otherwise the designer will only try those approaches for which commands are available. Commands should also have a considerable expression power so that a control system designer can do what he wants with a few commands. The commands should also reflect the natural concepts from a theoretical point of view. This would make it easy for a user well versed in control theory to use a package. The commands should also be few and simple so that they are easy to learn and remember. This is of course in conflict with requirements on completeness and expression power. Selection of commands is thus a good exercise in engineering design.

Based on experiences from our projects we have arrived at some design principles. A set of basic commands which correspond to the elements of the theory and which allow coverage of a certain problem area is first determined. Simplifications and extensions are then generated using the macro facility.

4. DATA STRUCTURES

Accepting the viewpoint that a CAE program may be viewed as a high level problem solving language the design of a vocabulary is one of the key issues. A wide range of data structures are required to deal naturally with control problems. Apart from common mathematical objects like integers, real and complex numbers, it is desirable to have polynomials, rational functions, matrices and matrix fractions. Such objects are conveniently viewed as abstract data types. There is also a need to have signals and systems. Signals are conveniently represented as arrays. When working with experimental data there is also a need to tag the data with verbal information about experimental conditions. This can be handled by making signal a record where the pure signal is an array which is part of a record.

Descriptions of control systems problems require flexible data structures. Many problems may be characterized in terms of arrays only. Arrays will go a long way to describe linear systems in state space form and to describe signals. Many problems can be solved using a matrix language like MATLAB, Moler (1981) or its derivatives Matrix_X, Walker et al. (1982), CTRL-C, Little et al. (1984), or Impact, Rimvall and Cellier (1984). It is, however, clear that it is not sufficient to only have matrices. A detailed discussion of this is found in Åström (1984).

For simple systems with only one data type, like matrices, all data may be stored in a stack or in a simple array. A more sophisticated data structure was used in the Lund packages. Our experiences indicate that it would be very useful to have a more flexible system. It is probably a good idea to build a system around some general database system. The need for multiple descriptions of a system is one special problem which is conveniently solved using databases. A typical example is when a system is represented both as a transfer function and as a state equation. Small systems are not much of a problem because it is easy to transform from one form to another. Such computations may however be extensive for large systems. To obtain a reasonable efficiency it is then necessary to store the different descriptions. It may also be desirable to have models of different complexity for the same physical object as well as linearized models for different operating conditions. Since it is very difficult to visualize all possible combinations a priori, it is a useful to have a database system which admits modifications of the structure of the data.

System descriptions

Since dynamical systems is a fundamental notion, its representation becomes a key issue. Many different representations of systems are used in control theory. The ordinary differential equation model

$$\begin{cases} \dfrac{dx}{dt} = f(x,u,t) \\ y = g(x,u,t) \end{cases} \tag{1}$$

where x is the state vector, u the input vector and y the output vector, is a common case. Often the fundamental form of the equations is not (1), where the derivative is solved explicitly but rather

$$\begin{cases} F\left(\dfrac{dx}{dt}, x, u, t\right) = 0 \\ G(x,y,u,t) = 0 \end{cases} \tag{2}$$

The following discussion is restricted to systems of type (1). Other issues arise when operating with models of type (2). This is treated in depth in Elmqvist (1978, 1979a, 1979b). Partial differential equations and differential equations with time delay are also common. In this paper the discussion is, however, restricted to differential equation models.

Linear systems where the functions f and g take the form

$$\begin{cases} f = A(t)x + B(t)u \\ g = C(t)x + D(t)u \end{cases} \tag{3}$$

is an important special case. For linear systems it is also possible to use other representations like input-output models of the form

$$\frac{d^n y}{dt^n} + A_1 \frac{d^{n-1} y}{dt^{n-1}} + \ldots + A_n y = B_1 \frac{d^{n-1} y}{dt} + B_2 \frac{d^{n-2} y}{dt^{n-2}} + \ldots + B_n u \tag{4}$$

which also can be represented by the matrix fraction

$$G(s) = A^{-1}(s)\, B(s) \tag{5}$$

where A and B are the polynomials

$$\begin{cases} A(s) = s^n + A_1 s^{n-1} + \ldots + A_n \\ B(s) = B_1 s^{n-1} + B_2 s^{n-2} + \ldots + B_n \end{cases}$$

The discrete time versions of (1), (2), (3), (4) and (5) are also needed. It is a key issue to find suitable computer representations of systems. Some operations to be performed on systems will be discussed before treating this.

System interconnections

Interconnections of systems is a fundamental issue. The elementary connections of systems are series, parallel and feedback connections. They can be represented graphically as is shown in Fig. 3 or algebraically as

$S_P = S_A + S_B$ "Parallel connection

$S_c = S_A \cdot S_B$ "Series connection

$S_c = [I+S_B \cdot S_A]^{-1} S_A = [I+S_B \cdot S_A] \backslash S_A$ "Feedback connection

where "\" is the notation introduced in MATLAB to denote the solution X = A\B of the linear equation

$$AX = B.$$

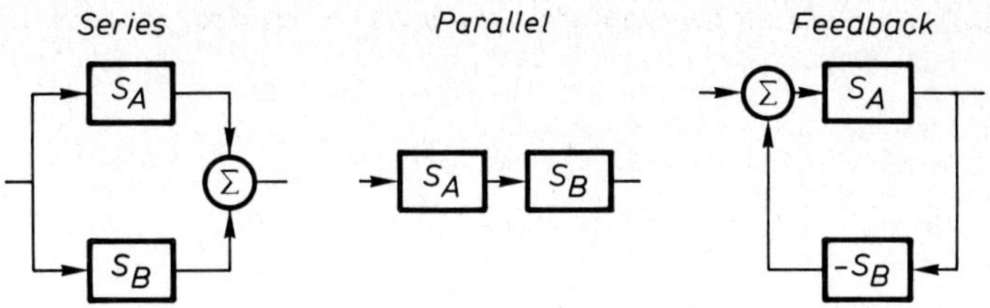

Figure 3
The basic system interconnections.

For more complex systems it is desirable to have appropriate notations for interconnected hierarchical systems. These notations should be such that details of the subsystem can be hidden and that signals and variables at the lower levels can be accessed in a well controlled fashion.

The system description introduced by Elmqvist (1977) in the simulation language Simnon has been very easy to operate with and very easy to teach. Elmqvist introduced the classes of continuous and discrete time systems defined as follows.

```
CONTINUOUS SYSTEM   <system identifier>
[INPUT    <simple variable>*]
[OUTPUT   <simple variable>*]
[STATE    <simple variable>*]
[DER      <simple variable>*]
[TIME     <simple variable>*]

[INITIAL                                            ]
[Computation of initial values for state variables  ]

[Computation of auxiliary variables]
[Computation of output variables]
[Computation of derivatives]
[Parameter assignment]
[Initial value assignment]
END

DISCRETE SYSTEM  <system identifier>
[INPUT    <simple variable>*]
[OUTPUT   <simple variable>*]
[STATE    <simple variable>*]
[NEW      <simple variable>*]
[TIME     <simple variable>*]
TSAMP     <simple variable>*]

[INITIAL                                                  ]
[[Computation of initial values for state variables]      ]
[[Computation of initial values for output variables]     ]
[[Computation of initial values for the TSAMP-variable]   ]
[SORT                                                     ]
```

```
[Computation of auxiliary variables]
[Computation of output variables]
[Computation of new values of the states]
Updating of the TSAMP-variable
[Modification of states in continuous subsystems]
[Parameter assignment]
[Initial value assignment]
END
```

where the standard BNF notations of <...> as a syntactic element, * as repetition, and [ ] an optimal element are used. Elmqvist allowed connections of systems at one level using connecting systems defined as

```
CONNECTING SYSTEM  <system identifier>
[TIME   <simple variable>]
[Computation of auxiliary variables]
[Computation of input variables]
[Parameter assignment]
END
```

The variables in each system description are local. The notation <variable name>[<system name>] is used in the connecting system and at the interaction level to separate variables with the same names.

Elmqvists notation is very natural. Long experience of using it has shown that it is very easy to teach and use. It therefore seems attractive to make marginal extensions of it to allow hierarchical interconnections with controlled access to parameters and variables.

Hierarchical system connections can be obtained by adding definitions of inputs and outputs in the connecting system and adding the new language element INCLUDE to give the names of the subsystems that are connected. The syntax of the connecting system thus becomes

```
CONNECTING SYSTEM  <system identifier>
[INPUT     <variable>*]
[OUTPUT    <variable>*]
[INCLUDE   <system identifier>*]
[TIME      <simple variable>]
[Computation of input variables]
[Parameter assignment]
END
```

The variables in Simnon have only one type reals. This should be extended to allow other datatypes like arrays, matrices and polynomials. When using hierarchically connected systems it is also useful to replace the notation for referencing variables to the common dot notation. A variable would thus be referenced as

```
<system name>.<system name>...<variable name>
```

Controlled access to variables and parameters can be obtained by introducing a heading

```
EXPORT  <variable>*
```

to list those variables which are exported up to the next level. In many teaching situations it is desirable to have access to all variables. This can be achieved by using EXPORT ALL or some similar construction.

Since system interconnections are often visualized graphically there should be facilities for representing and manipulating system interconnections graphically as well as textually. Interesting ideas in this direction have been proposed by Elmqvist (1982).

It would also be desirable to have the notion of system type and facilities for creating instances of the type. This would give a simple way of generating special classes of systems. Linear systems can then be defined as

```
type LINEAR_STATE_SPACE_SYSTEM
INPUT   u: vector
OUTPUT y: vector
STATE   x: vector
DERIVATIVE  dx: vector
A B C D : matrix
y  = C*x + D*y
dx = A*x + B*u
END
```

A similar construction can be used for linear polynomial systems. Instance of linear systems can then be created by

```
S: LINEAR_STATE_SPACE_SYSTEM
```

The parameters can be accessed as

```
S.A = matrix (1 2 ; 3 4)
```

It is a nontrivial design issue to decide when and how dimension compatibility should be checked. This has to do with how arrays are implemented. From the user point of view it would, however, be desirable to define a linear system as was done above without a need for specifying the dimensions.

In some cases it is also desirable to be able to hide a system description so that a user of the system can only make operations like simulation. An example from teaching is in courses on system identification, where it is desirable for students to find the properties of an unknown system, or in courses on adaptive control, when it is desirable to check that an algorithm works on an unknown system. The possibility to hide details of a system description would also be a possibility to get controlled access to industrial models. This can be achieved by using the mechanisms introduced in Ada, where the declarations and a body of a procedure are separated, see DOD (1983).

System operations

Apart from interconnections there are many other operations that are desirable to perform on systems, e.g. computation of equilibrium values, simulation, linearization,

Table 1
Examples of Program Sizes

	Number of commands	Source code lines	Program size kbytes
Intrac	17	7 000	90
Idpac	39	37 000	470
Modpac	37	41 000	570
Simnon	24	25 000	360
Synpac	46	43 000	630
Polpac	32	32 000	460

system inversion. For linear systems it is also natural to be able to transform coordinates, compute poles and zeros, determine observability and controllability, and perform Kalman decomposition. Some of these operations are conveniently done numerically. Others require formula manipulation.

5. PROGRAM PACKAGES

A description of our packages is given in this section. To work with programs of reasonable size a family of interactive program packages for modeling, identification, simulation, analysis and design of control systems were developed. The packages are all based on the common user interface Intrac, which was discussed in Section 3. The different packages are listed in Table 1, which also summarizes some data about them. Brief descriptions of the different packages are given below. The commands used in the packages are listed in the appendices. There are also a large number of macros available for all packages.

Idpac

Idpac is a package for data analysis and identification of linear systems having one output and many inputs. Time series analysis of ARMA and ARIMA models is a special case. The package has commands for manipulation and plotting of data, correlation analysis, spectral analysis and parametric system identification. There are also commands for model validation and simulation. The basic techniques used for parameter estimation are the least squares method and the maximum likelihood method. By using the macro facility it is, however, possible to generate commands for most of the parameter estimation methods which are proposed in literature. It was actually in the development of Idpac that the power of the macro concept became apparent. In the early Idpac versions there were many commands necessary to cover the available identifications methods. It was, however, discovered that almost all methods could be obtained by combinations of correlation analysis, spectral analysis, least squares and maximum likelihood estimation. Commands were thus constructed to give primitives for these operations and the special methods were then implemented as macros which used the primitive commands. This approach is also a pedagogical way to structure the problem area.

Idpac can be viewed as a convenient way of packaging the research in systems identification that has been done at our department for a period of 15 years. Idpac has gone through several steps of development. It grew out of the software described in Åström et al. (1965). The latest version is described in Wieslander (1980b). The paper Åström (1980) gives the relevant theory for the parametric identification methods. It also contains a comprehensive set of examples of using Idpac. A summary of the commands are given in Appendix A. Descriptions of some of the Idpac macros are given in Gustavsson (1979). Typical examples of using Idpac are given in Gustavsson and Nilsson (1979).

Modpac

There are many ways to describe a control system. Nonparametric methods in the time and frequency domain can be used. Parametric descriptions like state equations, rational transfer functions and fractions of matrix polynomials may also be used. There are also many ways in which state equations can be transformed. For digital control it is necessary to go between continuous time and discrete time representations. All these problems can be handled by Modpac. The package also has facilities for finding the Kalman decomposition of a system and for calculating observers. Modpac is described in Wieslander (1980c). A list of the commands in Modpac is given in Appendix B.

Simnon

Simnon is a package for interactive simulation of nonlinear continuous time systems with discrete time regulators. The package also includes noise generators, time-delays, a facility for using data files from Idpac as inputs to the system and an optimizer.

Simnon allows a system to be described as an interconnection of subsystems. There are two types of subsystems, continuous time systems and discrete time systems. These were described in detail in Section 4. This makes Simnon well suited for simulation of digital control systems. Simnon has two abstract datatypes continuous system and discrete systems to describe the subsystems and a third type connecting system to describe the interconnenctions. The characteristics of Simnon are illustrated by an example.

Listing 1 gives a description of a feedback loop consisting of a continuous time process called PROC and a digital PI regulator called REG. The process is an integrator with input saturation. The interconnections are described by the connecting system CON.

```
-----------------------------------------------------------------

CONTINUOUS SYSTEM PROC
"Integrator with input saturation
Input u
Output y
State x
Der dx
upr=if u<-0.1 then -0.1 else if u<0.1 then u else 0.1
dx=upr
END


DISCRETE SYSTEM REG
"PI regulator with anti-windup
Input yr y
Output u
State i
New ni
Time t
Tsamp ts
e=yr-y
v=k*e+i
u=if v<ulow then ulow else if v<uhigh then v else uhigh
ni=i+k*h*e/ti+h/tt*(u-v)
ts=t+h
k:1
ti:1
tt:0.5
h:0.5
ulow:-1
uhigh:1
END


CONNECTING SYSTEM CON
"Connecting system for simulation of process PROC
"with PI regulation by system REG
yr[REG]=1
y[REG]=y[PROC]
u[PROC]=u[REG]
END
-----------------------------------------------------------------
```

Listing 1
Simnon description of a simple control loop consisting of a continuous time process and a discrete PI regulator.

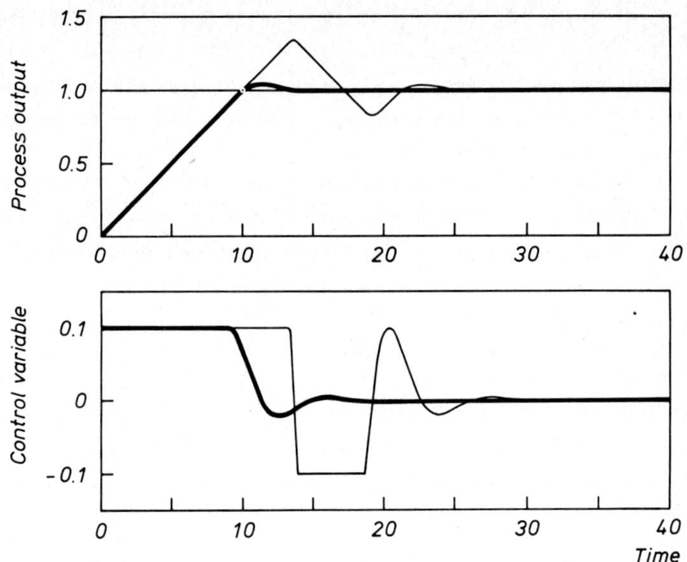

Figure 4
Results of simulation of process with a PI regulator.
Thin lines show results with ordinary regulator and thick lines show results for regulator with anti-windup.

The following annotated dialog illustrates how Simnon is used.

Command	Action
SYST PROC REG CON	Activate the systems.
AXES H 0 100 V -1 1	Draw axes.
PLOT yr y[proc] u[reg]	Determine variables to be plotted.
STORE yr y[proc] u[reg]	Select variable to be stored.
SIMU 0 100	Simulate.
SPLIT 2 1	Form two screen windows.
ASHOW y SHOW yr	Draw y with automatic scaling and yr with the same scales in first window.
ASHOW u	Draw u with automatic scaling in second window.

The result is shown by the curves in thin lines shown in Fig. 4. These curves show that there is a considerable overshoot due to integral windup. The regulator REG has anti-windup. The state of the regulator is reset when its output is equal to ulow or uhigh. The limits were set to ulow = -1 and uhigh = 1 in the simulations shown with thin lines in Fig. 4. These values are so large that the integral is never reset. The simulation, shows in thin lines in Fig. 4, thus correspond to a regulator without wind-up. The actual actuator limitations correspond to ulow = -0.1 and uhigh = 0.1. The commands

```
PAR   ulow:-0.1
PAR   uhigh:0.1
```

change the parameters and the command SIMU now generates the curves shown in thick lines in Fig. 3. Notice the drastic improvements due to the nonlinearity in the regulator.

The first version of Simnon was implemented in an MS project. Simnon has gone through several stages of development, see Elmqvist (1975, 1977). A list of the commands in Simnon is given in Appendix C. Uses of Simnon are described further in the tutorial Åström (1982) and the manuals Elmqvist (1975) and Frederick (1982b).

Synpac

Synpac is a state space oriented design package. It includes facilities for calculating state feedback and Kalman filters for continuous and discrete time LQG problems. It also has facilities for transforming continuous time problems into discrete time problems.

An example from Åström (1979) illustrates some features of Synpac. Consider the standard LQG problem

$$dx = Axdt + Budt + dv$$
$$dy = Cxdt + de$$

where $\{v\}$ and $\{e\}$ are Wiener processes with joint incremental covariances

$$\operatorname{cov} \begin{bmatrix} dv \\ de \end{bmatrix} \begin{bmatrix} dv \\ de \end{bmatrix}^T = \begin{bmatrix} R_1 & R_{12} \\ R_{12}^T & R_2 \end{bmatrix}.$$

Let the control problem be to minimize

$$J = \lim_{T\to\infty} \frac{1}{T} E \int_0^T \left[x^T(t)Q_1x(t) + 2x^T(t)Q_{12}u(t) + u^T(t)Q_2u(t)\right]dt.$$

Furthermore assume that a digital regulator will be used and that sampling periods from 0.5 to 5 s are of interest.

Assume that a system description which contains the matrices A, B, C, R_1, R_{12}, R_2, Q_1, Q_{12} and Q_2 has been introduced in a file called CSYS, and that the design parameter i.e. the parameter which will be modified in the design, is the 3,3 element of the matrix Q_1. The following macro then executes the design

```
 1:   MACRO DESIGN ALPHA
 2:     ALTER Q1 3 3 ALPHA
 3:     FOR H = 0.5 TO 5 STEP 0.5
 4:       SAMP DSYS ← CSYS H
 5:       TRANS Q DSYS ← CSYS H
 6:       TRANS R DSYS ← CSYS H
 7:       OPTFB L ← DSYS
 8:       KALFI K ← DSYS
 9:       CONNECT CLSYS ← DSYS K L
10:       SIMU Y X ← CLSYS UREF
11:       PLOT X(1) X(7) X(8) XE(1) U
12:     NEXT H
13:   END {MACRO}
```

Line numbers have been introduced only to be able to describe the algorithm. A macro with the name DESIGN and the parameter ALPHA is defined on line 1. The macro definition ends on line 13. The 3,3 element of the matrix Q1 is assigned the value ALPHA on line 2. Line 3 is a repetition statement which repeats the commands 4 through 11 for sampling periods 0.5 to 5 with an increment of 0.5. The system description is sampled on line 4 and the criterion and the covariances are transformed on lines 5 and 6. The command on line 7 computes the optimal state feedback matrix L and the command on line 8 computes the Kalman filter gain. The command on line 9 forms a closed loop system composed of the original system the Kalman filter and the state feedback. The command on line 10 simulates the closed loop system with a reference input UREF. The command on line 11 plots state variables 1, 7 and 8, the estimate of the first state and the control signal. The following dialog illustrates how the macro may be used

```
EDIT FILE CSYS
INPUT UREF ← STEP
DESIGN 3
DESIGN 8
```

The system file is first edited. A step is generated as a command signal and the macro design is executed with parameters 3 and 8.

The example illustrates some Synpac commands. It also shows how a macro may be used to create a special purpose command.

Synpac was the first package that was implemented. It was based on the Fortran programs described in Åström (1963). A test version was made as an MS project. The current version is described in Wieslander (1980d). An introduction is presented in Gutman et al. (1984). A list of the commands in Synpac is given in Appendix D.

Polpac

Polpac is a polynomial oriented design package for multi-output single-input systems. It includes algorithms for pole placement, minimum variance control, and LQG control. The package allows classical design using root loci and Bode plots. Root loci may be drawn with respect to arbitrary parameters. A list of the commands in Polpac is given in Appendix E.

Portability

The programs were initially written in FORTRAN for a minicomputer PDP-15. A considerable effort was also devoted to development of subroutine libraries and programming standards, see Elmqvist et al. (1976), Wieslander (1977) and Cowell (1977). The advantage of using a large main frame computer for program development was soon apparent. The program development was therefore moved to a Univac 1108 at the Lund University Computing Center. More powerful program development tools like Pfort could then be used, see Ryder (1975). Since there was a considerable interest from external groups to use the programs, a substantial effort went into making the software portable. This included development of Fortran routines for file, character and string handling. A plotting library in Fortran was also developed. These routines are interfaced with a well-defined small set of installation dependent routines. A result of the efforts is that the packages are indeed portable. Packages are currently running on the following computers: PDP-15, PDP-11, DEC-10, VAX-11/780, NOVA-3, Nord-100, ECLIPSE, IBM-1800, IBM-360, CDC-1700, CDC-6400, HP-3000, Honeywell, SEL-32, Univac 1108, PRIME-750.

6. LARGE SYSTEMS

When working with the projects it was found that there were certain problems where interactive computing is not feasible. These problems typically involve large systems where it can easily happen that the computing time required is so large that it does not make sense to wait for the results at the terminal. We experienced this in connection with identification and simulation of large systems.

Lispid

For identification of large systems we found that it was better to use a batch program which allows an interactive start up and an interactive inspection of the results. LISPID is an example of such a program. This program allows estimation of parameters in linear stochastic systems with arbitrary parameterization and in special types of nonlinear systems, see Källström et al. (1976).

Dymola

Another problem was also encountered in connection with modeling of large systems. It is straightforward to write down and check the balance equations. It is however a major effort to reduce the equations obtained to forms that are suitable for simulation and control design. The language Dymola which admits a simple description of a large hierarchical system was therefore developed, see Elmqvist (1978, 1979a,b). Experimental software, which operates on the basic system description and generates simulation programs e.g. in Simnon, and linearized system equations have also been developed. We believe that this is an important step towards effective methods for dealing with large systems.

7. EXPERIENCES OF USING THE PACKAGES

The packages have been used at our department, at other universities, and in industry. The early use of the packages provided very good feedback to their development. There has been a continuous dialog between users and implementors at all stages of the development. Very valuable input was provided by visitors to the department and from industrial use of the packages. They often had different ideas on how to use the programs.

All staff members of the department and a large number of the students who have done MS or PhD dissertations have used the packages. The programs have been used in some of our advanced courses and in courses for industrial audiences. They are now being introduced also in elementary courses. The bottleneck for this has been the availability of a sufficient number of graphic terminals. By using the packages it has been possible to focus on concepts and ideas in the lectures and to work with realistic examples with considerable detail in exercises and projects.

The simulation language Simnon is used as a standard language for documenting models. The availability of a library of realistic models of different complexity is of course very beneficial in teaching. Simnon has been used in an interesting way in a recent book on computer control, Åström and Wittenmark (1984), which makes extensive use of simulation. All simulation results are implemented as Macros in Simnon which are accessable from the student terminals. This means that the students may conveniently check the results and also look into effects of variations of data. Simnon has also been used in many applied projects at the institute. A typical example is a study on modeling and simulation of a wind turbine, Bergman et al. (1983).

We have found Idpac to be a very good tool to teach system identification. It is possible for the students to gain a lot of experience by working with real data. Idpac has also been used in many industrial projects. Typical examples are given in Åström and Källström (1976) and Källström and Åström (1981). Trouble shooting in the paper industry has e.g. been another interesting application area, see Lundqvist and Nordström (1980) and Johansson et al. (1980).

Similarly we have found that Synpac is an excellent tool for teaching LQG design. The students can work with realistic problems with reasonable effort. Synpac has also been used to design control laws for digital flight control systems. These control laws have also been flight tested, see Åström and Elgcrona (1976) and Folkesson et al. (1982). Synpac was the major tool in the development of the control laws for the flight control system of the new Swedish multirole combat aircraft JAS.

The programs are now used by a number of industries. Their use has resulted in a significant increase in engineering productivity. A reduction of engineering effort from four man-months to four man-weeks compared with traditional practices is quite typical. An important factor for this reduction is that one person now handles the whole problem instead of using programmers and intermediaries. An increase of the quality of the designs has also been observed. This is partly due to the fact that it is so easy to repeat a design or a simulation with slightly modified conditions. Improved documentation is another factor which increases productivity.

8. FUTURE WORK

Computer aided design of control systems is still in its early stages. There are a number of packages like ours. An overview of some packages are found in Armstrong (1980), Atherton (1981), Edgar (1981), Edmunds (1979), Frederick (1982a), Furuta and Kajiwara (1979) Hashimoto and Takamatsu (1981), Lemmens and van den Boom (1979), Munro (1979), Rosenbrock (1974), Tyssø (1979,1981) and Wieslander (1973, 1979a,b). More references are also found in these papers. Special workshops and symposia devoted to CAD for control systems have been organized by IFAC, GE-RPI, and IEEE CSS, see Mansour (1979), Leininger (1982), Spang and Gerhart (1981), Herget and Laub (1982). Computer aided tools are also popular in many other fields e.g. mechanical design and VLSI design. The seminal work on computer graphics by Newman and Sproull (1979) and the text Foley and van Dam (1983) contain much material and many references.

The field is in a state of rapid development due to an increased understanding of the technology and the drastic development of computer and graphics hardware. It is safe to predict that future computer aided design tools will be much more powerful than the packages described in this paper. Some speculations on future development are given in this section.

Computer hardware

The virtual memory requirement for each package is less than 1 Mbyte. A package can run on a computer having a primary storage of 128 kbytes. They require fast floating point operations for reasonable efficiency. The packages can be implemented on a computer like the IBM PC XT with an Intel 8087 floating point processor. Prototype implementation indicates that the performance of such implementation is surprisingly good.

The personal computers which are projected to appear within a few years have specifications like: a primary memory of 2 Mbytes, a secondary memory 100 Mbytes, a computing speed of one megaflop/s and a price less than 20k$, see Dertouzos and Moses (1980). These computers are also expected to have a high resolution bit mapped color graphics display. With computers like this it is possible to have single user work-stations with packages which are much more sophisticated than all our current packages. The existence of computers like Apollo, Lisa, Mackintosh, Sun, and Iris make the predictions quite credible.

There has been a drastic development of the computer output devices. A teletype is capable of writing at a speed of 10 ch/s (110 Baud). A regular terminal connected to a 19.2 kBaud channel can write a screen i.e. 80 x 24 ch in a second. A good vector graphics terminal can refresh up to 100 000 long vectors or a million short vectors per second. A high resolution bit mapped display may refresh 512 x 512 pixel frames at rates of 60 frames/s (15 Mbit/s).

The input devices have unfortunately not developed at the same rate. We still have ordinary keyboards, see Montgomery (1982). A very good typist may type at a rate of 8 ch/s. A normal engineer types considerably slower. Pointing devices like roll balls, mouses and touch panels have been invented. These devices may perhaps be used to increase the input rate indirectly by combining the rapid output rate with feedback via the picking device (dynamic menus). Speech input is another possibility. There are, however, no indications of a more drastic increase in the input rate.

The renaissance of graphics

Graphics has played a major role in engineering. The first books used in engineering education were books of drawings of machines by Leonardo da Vinci. Graphical representations have been used extensively ever since. Graphics in the forms of Bode diagrams, Nichols charts, root loci, block diagrams and signal flow diagrams are important tools in classical control theory. Modern control theory has, however, not been much influenced by graphics. This can partly be explained by lack of proper tools for graphics. The situation may change drastically in the future because good graphics hardware will be available at a reasonable cost.

The man-machine interface

A high bandwidth information transmission is required for an efficient man-machine communication. This implies a high rate of transmission of symbols and a high information content in each symbol. The user interfaces in our packages were designed for teletypes combined with graphic terminals having storage screens and data rates of 4800 Baud. These were the only tools available at reasonable cost when our design was frozen. A storage scope is very limited. Curves may be shown but they can not be erased individually. Bit mapped graphics is faster and much more flexible. Individual picture elements may be changed instantaneously. This makes it possible to zoom, scroll and pan a picture. Color and animation add extra dimensions. Imaginative use of color graphics is still in its infancy in CAD packages for control systems. Interesting ideas have been proposed by Polak (1981) in connection with applications of optimization techniques. Animation has not been used much. It is clear that we have a lot to learn from designers of video games, see Perry et al. (1982).

The information content of each symbol is related to the expression power of the commands. Our experience indicate clearly the need for having a CAD language with considerable expression power. It would be nice to describe all operations using notations similar to those used in system theory.

Short form commands and default values are two simple techniques for increasing the efficiency of the man-machine dialog. More sophisticated techniques like conceptual dependency are found in semantically oriented programs for processing natural language, see Schank (1975) and Schank and Abelson (1977). Ideas in this direction have been persued by Gale and Pregibon (1983) who have tried to construct an expert interface called REX (Regression Expert) to a program for regression analysis. It would be interesting to explore how these ideas can be incorporated in CAD systems. Another possibility is to have an operator communication which is more oriented towards graphics. Interesting ideas in this direction are demonstrated in Elmqvist (1982).

Numerical algorithms and design tools

The numerical algorithms for the design primitives are key elements in the software. There have been major advances in numerical software for linear algebra over the past years. A substantial effort has gone into subroutine packages such as Eispack, Wilkinson and Reinsch (1971), Garbow et al. (1977) and Smith et al. (1976), and Linpack, Dongarra et al. (1979), which are now available in the public domain. A similar effort has not yet been devoted to the numerical calculations required for analysis and design of control systems, although libraries are maintained at many departments. The numerical problems that arise in automatic control are, however, starting to receive attention from numerical analysts, see van Doren (1981), Hammarling (1982) and Laub (1980). This is crucial for the future development.

Most data processing in current packages is inspired from numerical analysis. The powers of non-numeric data processing have not been exploited. It would be highly desirable to have facilities for symbolic manipulation. This can e.g. be used for model simplification, generation of code for computing equilibrium points, generation of simulation code, linearization, etc. If symbolic manipulations are included it is also possible to generate code for realization of the control laws. Symbolic calculations were not used in our packages because of the limited computing facilities available. It is, however, feasible in future packages.

When transfering our packages we have noticed that their power increases considerably if an experienced user is around. The possibilities of providing the packages with a rule based expert system or an advisory system, Barr and Feigenbaum (1982), is therefore very appealing. It is an interesting research problem to find out if expert knowledge in identification, analysis and design of control systems can be incorporated in the packages.

Implementation languages

The source code for the smallest package described in this paper is about 30 000 lines of source code. A future package may be an order of magnitude larger. A good programming environment and efficient software tools are necessary to develop and maintain such systems. Fortran was used in our packages to make them portable. It is however unlikely that future systems will be written only in Fortran.

Fortran libraries like Eispack, Linpack, (hopefully also a control package), and some graphics package will probably be used. Although Intrac was written in Fortran it is not convenient to do so. Pascal would be much more convenient, particularly if we want to include formula manipulation and the other features that we may expect in a future system. An expression parser is needed. Macros and user defined procedures are very useful in order to increase the efficiency of the man-machine dialog. More flexible control structures and more powerful commands than those used in Intrac would be desirable. One possible extension is the system Delight which is based on the language RATTLE developed by Nye et al (1981). Other possibilities are to replace Intrac by languages with an interactive implementation like Apl, Lisp or Logo or an interpretive threaded language like Forth, see Winston and Horn (1981), Abelson, (1982) and Kogge, (1982). Systems based on Lisp will be extendable automatically, symbolic manipulations are also easy to implement. There are good programming environments for Lisp which have been used to implement very large systems. Natural language interfaces and expert systems are also often written in Lisp.

Programs like Idpac and Delight, which handle the interpretations of the commands and the man-machine interaction, have many features in common with operating systems like UNIX, Kernighan and Pike (1983). They may be viewed as an interpretative programming language whose data types are files. Such programs can be implemented as extensions to an operating system. The software Honey-X developed by Honeywell is such a system which is based on Multics, see Anon. (1982). The advantage of such an approach is that the major part of the code is the ordinary operating system. Only a small portion has to be added. Another advantage is that the user does not have to learn a new environment. A drawback is, however, that the software becomes less portable, since it is tied to a specific operating system. Unless great care is taken in the design it will also be necessary to update the system when new versions of the operating system are issued.

The programming language Ada, DOD (1983), which will be available in a few years time is another interesting alternative. The basic subroutine libraries can conveniently be implemented as packages in Ada. A wide range of libraries can be expected to be available for Ada. Since Ada supports the concept of tasks it will also be possible to apply ideas from concurrent programming. The overloading facility is also useful in order to give readable code. A good programming environment which will be a substantial help in software development is also planned for Ada. The deciding factor will probably depend on how well Ada will be accepted.

Some computations, such as simulation and identification, are quite demanding computationally. The problem solving would be more efficient and convenient if the the user could perform several tasks like plotting, editing and report writing in parallel. This mode of operation is particularly useful for a system with windowing. See Goldberg et al. (1983). It is thus likely that future systems will be implemented using several programming languages. One indication of this is the design package ISER-CSD which is written in Fortran and Pascal, see Suleyman (1981). This system is, however, restricted to one computer system.

Implementation tools

Our packages were developed from scratch. Future packages may be expected to use ready made modules to a much larger extent. Descriptions of control systems problems require flexible data structures. Many problems may be characterized in terms of arrays only. Arrays will go a long way to describe linear systems in state space form and to describe signals. Many problems can be solved using a matrix language like Matlab, Moler (1981) and one of its extension Matrix_X, Walker et al. (1982). It is, however, clear, that it is very useful to also have polynomials, rational functions and good general structures for linear and nonlinear systems as was discussed in Section 4. In some cases it is also valuable to describe systems as hierarchical interconnections of subsystems.

In our packages we had to develop our own graphics interface. A few simple routines which were compatible with Tektronix 4010 systems were used. The situation will be much better when standards like the Graphical Kernel System (GKS) or raster graphics extensions of SIGGRAPH Core materialize, see Foley and van Dam (1983) and GKS (1982).

9. CONCLUSIONS

Interactive computing is a powerful tool for problem solving. An engineer can come to the work station with a problem and he can leave with a complete solution after a few hours. The results are well documented in terms of listings, text and graphs. The problem solver can obtain the solution by himself without relying on programmers as intermediaries. Our projects have shown that the productivity in analysing and designing control systems can be increased substantially by using these tools. We believe that interactive computer aided design tools is one possibility to make modern control theory cost effective.

Computer aided design of control systems is still in its infancy. A small number of systems have been implemented in a few places. There are many possible future developments which are mainly driven by the computer development. Packages of the type we have been experimenting with can easily be fitted into the personal computers or work stations that will be available in a few years time. The bit mapped high resolution color displays that will be available on these computers offer new possibilities for an efficient man-machine dialog. With the drastic increase in computer capacity, that is forth coming, it is also possible to make much more ambitious projects. Applications of computer aided design also appear in many other branches of engineering. Cross fertilization between the fields will most likely lead to a rapid development.

10. ACKNOWLEDGEMENTS

The work reported in this paper has been supported by the Swedish Board of Technical Development for many years. This support is gratefully acknowledged. The projects, which the paper draws upon, have been true team efforts. Many members of the department have contributed to discussions of command structures, implementation, testing and evaluation. Particular thanks are due to Johan Wieslander and Hilding Elmqvist, who generated many of the important ideas, and to Tommy Essebo and Thomas Schönthal, who did a major part of the programming of the packages.

REFERENCES

Abelson, H (1982): Logo for the Apple II. Byte/McGraw-Hill, Peterborough, New Hampshire.

Anon. (1982): HONEY-X Users manual. Honeywell, Ridgeway Park, Minneapolis.

Armstrong, E S (1980): ORACLS - A design system for linear multivariable control. Marcel Dekker, New York.

Åström, K J (1963): On the choice of sampling rates in optimal linear systems. IBM Research Report RJ-243.

Åström, K J (1979): Reflections on theory and practice of automatic control. Plenary lecture 17th CDC, San Diego, 1979, also Dept of Automatic Control, Lund Institute of Technology, Lund, Sweden, Report CODEN: LUTFD2/(TFRT-7178)/1-26/(1979).

Åström, K J (1980): Maximum likelihood and prediction error methods. Automatica 16, 551-574.

Åström, K J (1982): A Simnon tutorial. Dept of Automatic Control, Lund Institute of Technology, Report CODEN: LUTFD2/(TFRT-3168)/1-52/(1982).

Åström, K J (1984): Computer aided design of control systems. In Bensoussan and Lions (Eds.) Analysis and Optimization of Systems. Springer Lecture Notes in Control and Information Sciences, Springer, Berlin.

Åström, K J, Bohlin, T, and Wensmark, S (1965): Automatic construction of linear stochastic dynamic models for stationary industrial processes with random disturbances using operating records. Report TP 18.150, IBM Nordic Laboratory, Sweden.

Åström, K J, and Elgcrona, P O (1976): Use of LQG theory and Synpac in design of flight control systems. Reports SAAB, Linköping, Sweden.

Åström, K J, and Källström, C G (1976): Identification of ship steering dynamics. Automatica 12, 9-22.

Åström, K J, and Wieslander, J (1981): Computer aided design of control systems - Final Report STU Projects 73-3553, 75-2776 and 77-3548. Dept of Automatic Control, Lund Institute of Technology, Lund, Sweden, Report CODEN: LUTFD2/(TFRT-3160)/1-23/(1981).

Åström, K J, and Wittenmark, B (1984): Computer Control Theory. Prentice Hall, Englewood Cliffs, N J.

Atherton, D P (1981): The role of CAD in education and research. IFAC Congress VIII, Kyoto, Japan.

Barr, A, and Feigenbaum, E A (1982): The Handbook of Artificial Intelligence. Vol II. W. Kaufmann Inc., Los Altos, Calif.

Barstow, D R, Shrobe, H E, and Sandewall, E (1984): Interactive Programming Environments. McGraw-Hill, New York.

Bergman, S, Mattson, S E, and Östberg, A B (1983): A modular simulation model for a wind turbine system. Journal of Energy 7, 319-324.

Cowell, W (Ed.) (1977): Portability of numerical software. Lecture Notes in Computer Science, Vol. 57, Springer-Verlag, New York.

Dertouzos, M L, and Moses, J (1980): The Computer Age: A twenty year view. MIT Press, Cambridge, Mass.

DOD (1983): Reference Manual for the Ada Programming Language. ANSI/MIL-STD-1815A--1983, United States Department of Defense, Washington, DC.

Dongarra, J J, Moler, C B, Bunch, J R, and Stewart, G W (1979): LINPACK - Users' guide. SIAM, Philadelphia.

Edgar, T F (1981): New results and the status of computer-aided process control system design in North America. Engineering Foundation Conference on Chemical Process Control-II, Sea Island, Georgia.

Edmunds, J M (1979): Cambridge linear analysis and design programs. IFAC Symposium on Computer Aided Design of Control Systems, Zurich, 253-258.

Elmqvist, H (1975): SIMNON, an interactive simulation program for nonlinear systems. Dept of Automatic Control, Lund Institute of Technology, Lund, Sweden, Report CODEN: LUTFD2/(TFRT-7502).

Elmqvist, H, Tyssø, A, and Wieslander, J (1976): Scandinavian control library. Programming. Dept of Automatic Control, Lund Institute of Technology, Lund, Sweden, Report CODEN: LUTFD2/(TFRT-3139)/(1976).

Elmqvist, H (1977): SIMNON - An Interactive Simulation Program for Nonlinear Systems. Simulation '77, Montreux, Switzerland, June 1977.

Elmqvist, H (1978): A Structured Model Language for Large Continuous Systems. Ph.D. Thesis. Dept of Automatic Control, Lund Institute of Technology, Lund, Sweden, Report CODEN: LUTFD2/(TFRT-1015)/1-226/(1978).

Elmqvist, H (1979a): Dymola - A Structured Model Language for Large Continuous Systems. Summer Computer Simulation Conference, Toronto, Canada, July 1979.

Elmqvist, H (1979b): Manipulation of Continuous Models Based on Equations to Assignment Statements. Simulation of Systems '79. Sorrento, Italy, September 1979.

Elmqvist, H (1982): A graphical approach to documentation and implementation of control systems. Proc 3rd IFAC/IFIP Symposium on Software for Computer Control, SOCOCO 82, Madrid, Spain.

Folkesson K, Elgcrona, P O, and Haglund, R (1980): Design and experience with a low-cost digital fly-by-wire system in the SAAB JA37 Viggen A/C. Proc 13th International Council of the Aeronautical Sciences, Seattle, WA.

Foley, J D, and van Dam, A (1983): Fundamentals of interactive computer graphics. Addison Wesley, Reading, Mass.

Frederick, D K (1982a): Computer packages for the simulation and design of control systems. Lecture notes, Arab school on science and technology.

Frederick, D K (1982b): Simnon reference manual. Automation and Control Laboratory, Corporate Research and Development, General Electric Company, Schenectedy, New York.

Furuta, K, and Kajiwara, H (1979): CAD system for control system design. J of the Society of Instrument and Control Engineers, Japan, 18 (9). (In Japanese).

Gale, W A, and Pregibon, D (1983): Using expert systems for developing statistical strategy. Proc Joint Statistical Meetings, Toronto.

Garbow, B S., et al. (1977): Matrix eigensystem routines - Eispack Guide Extension. Lecture Notes in Computer Science, Vol. 51, Springer-Verlag, New York.

GKS (1982): Graphical Kernel System (GKS) - Functional Description. Draft International Standard ISO/DIS 7942 Version 7.02, August 9, 1982. Available through American National Standards Institute Inc., New York, N.Y.

Goldberg, A J, Robson, D, and Ingalls, D H H (1983): Smalltalk-80: The Language and its Implementation. Addison-Wesley, Reading, MA.

Gustavsson, I (1979): Processidentifiering - overheadbilder (Process identification - transparencies). Dept of Automatic Control, Lund Institute of Technology, Lund, Sweden, CODEN: LUTFD2/(TFRT-7166)/1-248/(1979).

Gustavsson, I, and Nilsson, A-B (1979): Övningar för Idpac (Exercises for Idpac). Dept of Automatic Control, Lund Institute of Technology, Lund, Sweden, Report CODEN: LUTFD2/(TFRT-7169)/1-55/(1979).

Gutman, P-O, Hagander, P, and Lilja, M (1984): Introduction to Synpac. Dept of Automatic Control, Lund Institute of Technology, Lund, Sweden, Report CODEN: LUTFD2/(TFRT--3172)/(1984).

Hammarling, S (1982): Some notes on the use of orthogonal similarity transformations in control. NPL Report DITC.

Hashimoto, I, and Takamatsu, Y (1981): New results and the status of computer aided process control systems design in Japan. Engineering Foundation Conference on Chemical Process Control-II, Sea Island, Georgia.

Herget, C J, and Laub, A J (Eds.) (1982): Proc IEEE CSS Workshop on Computer Aided Control System Design. Berkeley, Calif. IEEE CSM 2:4. Special Issue on Computer--Aided Design of Control Systems.

Johansson, B L, Karlsson, H, and Ljung, E (1980): Experiences with computer control based on optical sensors for pulp quality of a two state TMP plant. Preprints Process Control Conf. Canadian Pulp and paper Ass., Halifax, Canada.

Källström, C G, Essebo, T, and Åström, K J (1976): A computer program for maximum likelihood identification of linear multivariable stochastic systems. Proc 4th IFAC Symposium on Identification and System Parameter Estimation, Tblisi, USSR.

Källström, C G, and Åström, K J (1981): Experiences of system identification applied to ship steering. Automatica 17, 187-198.

Kernighan, B W, and Pike, R (1983): The Unix Programming Environment. Prentice-Hall, Englewood Cliffs, NJ.

Kogge, P M (1982): An Architectural trail to threaded-code systems. Computer 15:3, 22-32.

Laub, A J (1980): Survey of computational methods in control theory. In Erisman, A M, et al. (Eds.), Electric power problems. The mathematical challenge, SIAM, Philadelphia, pp 231-260.

Leininger, G (Ed.) (1982): Computer aided design of multivariable technological systems. Preprints second IFAC symposium on Computer Aided Design of Multivariable Technological systems, West Lafayette, Indiana, USA.

Lemmens, W J M, and Van den Boom, A J W (1979): Interactive computer programs for education and research: a survey. Automatica 15, 113-121.

Little, J N, Emami-Noeini, A, and Bangert, S N (1984): CTRL-C and matrix environments for the computer-aided design of control systems. In Bensoussan and Lions (Eds.): Analysis and Optimization of Systems. Springer Lecture Notes in Control and Information Sciences, Springer, Berlin.

Lundqvist, S O, and Nordström, H (1980): The development of a control system for a pulp washing plant through the use of dynamic simulation. Preprints IFAC Conf. on Instrumentation and automation in the paper, rubber, plastics and polymerisation industries. Gent, Belgium.

Mansour, M editor (1979): Preprints first IFAC Symposium on CAD of Control systems, Zurich. Pergamon Press.

Moler, C B (1981): MATLAB user's guide. Report Department of Computer Science, University of New Mexico.

Montgomery, E B (1982): New keyboard concepts. IEEE Computer 15:3, 11-18.

Munro, N (1979): The UMIST control system design and synthesis suites. IFAC Symposium on Computer Aided Design of Control Systems, Zurich, 343-348.

Newman, W M, and Sproull, R F (1979): Principles of interactive computer graphics. McGraw-Hill, New York.

Nye, W, Polak, E, Sangiovanni-Vincentilli, A, and Tits, A (1981): An optimization-based computer-aided-design system. Proc ISCAS, April 24-27.

Perry, T, Truxal, C, and Wallich, P (1982): Video games: the electronic big bang. IEEE Spectrum 19:12, 20-33.

Polak, E (1981): Interactive software for computer-aided-design of control systems via optimization. Proc. 20th IEEE Conf. on Decision and Control, San Diego, CA, December 16-18, pp 408-412.

Rimvall, M, and Cellier, F (1984): IMPACT Interactive mathematical program for automatic control theory. In Bensoussan and Lions (Eds.): Analysis and Optimization of Systems. Springer Lecture Notes in Control and Information Sciences, Springer, Berlin.

Rosenbrock, H H (1974): Computer-aided control system design. Academic Press, New York.

Ryder, B G (1975): The Pfort verifier: User's guide. CS Tech. Rept. 12, Bell Labs.

Schank, R C (1975): Conceptual Information Processing. North Holland. Amsterdam.

Schank, R C, and Abelson, R P (1977): Scripts, plans, goals and understanding. Lawrence Erlbaum Associates, Hillsdale NJ.

Smith, B T, et al. (1976): Matrix eigensystem routines - Eispack guide. 2nd ed., Lecture Notes in Computer Science, Vol. 6, Springer-Verlag, New York.

Spang, H A, III, and Gerhart, L (Eds.) (1981): Preprints GE-RPI, Workshop on control design, Schenectady, N.Y.

Suleyman, C (1981): Interactive system for education and research in control system design. IEEE International Conference on Cybernetics and Society, Atlanta, Georgia.

Tyssø, A (1979): CYPROS: Cybernetic program packages. IFAC Symposium on Computer Aided Design of Control Systems, Zurich, 383-389.

Tyssø, A (1981): New results and the status of computer aided process control systems design in Europe. Engineering Foundation Conference on Chemical Process Control-II,

Sea Island, Georgia.

Van Doren, P (1981): A generalized eigenvalue approach for solving Riccati equations. SIAM J Sci. Stat. Comput. 2, 121-135.

Walker, R, Gregory, C, and Shah, S (1982): $Matrix_X$ A data analysis, system identification, control design and simulation package. IEEE CSM 2:4, 30-37.

Wegner, P (1968): Programming Languages, Information Structures, and Machine organization. McGraw-Hill, New York.

Wieslander, J (1973): Computer aided control system design. IEE, Publ. no 96.

Wieslander, J (1977): Scandinavian control library. A subroutine library in the field of automatic control. Report, Dept of Automatic Control, Lund Institute of Technology, Lund, Sweden, CODEN: LUTFD2/(TFRT-3146)/1-38/(1977).

Wieslander, J (1979a): Interaction in computer aided analysis and design of control systems. PhD thesis, Dept of Automatic Control, Lund Institute of Technology, Lund, Sweden, Report CODEN: LUTFD2/(TFRT-1019)/1-222/(1979).

Wieslander, J (1979b): Design principles for computer aided design software. Preprints, IFAC Symposium on CAD of Control Systems, Zurich, 493.

Wieslander, J (1980a): Interactive programs - General guide. Dept of Automatic Control, Lund Institute of Technology, Lund, Sweden, Report CODEN: LUTFD2/(TFRT-3156)/1--30/(1980).

Wieslander, J (1980b): IDPAC commands - User's guide. Dept of Automatic Control, Lund Institute of Technology, Lund, Sweden, Report CODEN: LUTFD2/(TFRT-3157)/1-108/(1980).

Wieslander, J (1980c): MODPAC commands - User's guide. Dept of Automatic Control, Lund Institute of Technology, Lund, Sweden, Report CODEN: LUTFD2/(TFRT-3158)/1-81/(1980).

Wieslander, J (1980d): Synpac commands - User's guide. Dept of Automatic Control, Lund Institute of Technology, Lund, Sweden, Report CODEN: LUTFD2/(TFRT-3159)/1-130/(1980).

Wieslander, J, and Elmqvist, H (1978): INTRAC, A communication module for interactive programs. Language manual. Dept of Automatic Control, Lund Institute of Technology, Lund, Sweden, Report CODEN: LUTFD2/(TFRT-3149)/1-60/(1978).

Wilkinson, J H, and Reinsch, C (1971): Linear algebra. Springer-Verlag, Berlin.

Winston, P H, and Horn, B K P (1981): Lisp. Addison-Wesley, Reading, Mass.

APPENDIX A - IDPAC COMMANDS

1. Utilities

CONV	- Conversion of data to internal standard format
DELET	- Delete a file
EDIT	- Edit system description
FHEAD	- Inspect and change file parameters
FORMAT	- Conversion of data to symbolic external form
FTEST	- Check existence of a file
LIST	- List files
MOVE	- Move data in database
TURN	- Change program switches

2. Graphic output

BODE	- Plot Bode diagrams
HCOPY	- Make hard copy
PLMAG	- Magnify plot and allow changes of data
PLOT	- Plot curves with linear scales

3. Time series operations

ACOF	- Compute autocorrelation function
CCOF	- Compute cross correlation function
CONC	- Concatenate time series
CUT	- Extract a part of a time series
INSI	- Generate time series
PICK	- Pick equidistant time points
SCLOP	- Do scalar operations on a time series
SLIDE	- Introduce relative delays between time series
STAT	- Compute statistical characteristics
TREND	- Remove a trend
VECOP	- Do vector operations on a time series

4. Frequency response operations

ASPEC	- Compute an auto spectrum
CSPEC	- Compute a cross spectrum
DFT	- Discrete Fourier Transform
FROP	- Operate on frequency responses
IDFT	- Inverse Discrete Fourier Transform

5. Simulation and model analysis

DETER	- Deterministic Simulation
DSIM	- Simulation with noise
FILT	- Compute a filter system
RANPA	- Pick parameters from a random distribution
RESID	- Compute residuals with statistical test
SPTRF	- Compute the frequency response of a transfer function

6. Identification

LS	- Least Squares identification
ML	- Maximum Likelihood identification
SQR	- Least Squares data reduction
STRUC	- Least Squares structure definition

APPENDIX B - MODPAC COMMANDS

1. Utilities

AGR	- Edit an aggregate file
CONV	- Conversion of data to internal standard format
DELET	- Delete a file
EDIT	- Edit system description
FHEAD	- Inspect and change file parameters
FORMAT	- Conversion of data to symbolic external form
FTEST	- Check existence of a file
LIST	- List files
MOVE	- Move data in database
TURN	- Change program switches

2. Graphic output

BODE	- Plot Bode diagrams
HCOPY	- Make hard copy
NIC	- Display a frequency response in a Nichols diagram
NYQ	- Display a frequency response in a Nyquist diagram
PLEV	- Display eigenvalues, etc. in the complex plane
PLOT	- Plot curves with linear scales

3. Matrix operations

ALTER	- Alter elements in a matrix
EIGEN	- Compute eigenvalues of a matrix
ENTER	- Enter a matrix element by element
EXPAN	- Generate a matrix from sub-matrices
MATOP	- Perform matrix operations
REDUC	- Extract a submatrix
UNITM	- Generate a unit matrix
ZEROM	- Generate a zero matrix

4. Polynomial operations

POCONV	- Polynomial image - polynomial file conversion
POLY	- Generate or edit a polynomial
POLZ	- Compute and plot the zeros of a polynomial
ZERPOL	- Create a polynomial from its zeros

5. System operations

CONT	- Convert to continuous time form
KALD	- Do a Kalman decomposition
SAMP	- Convert to discrete time form
SPSS	- Compute the frequency response
SSTRF1	- Convert from state space to transfer function
SYST	- Generate a system description
SYSTR	- Do a general coordinate transformation
TBALAN	- Transform to balanced form
TCON	- Transform to controllable form
TDIAG	- Transform to diagonal form
THESS	- Transform to Hessenberg form
TOBS	- Transform to observable form
TRFSS1	- Convert from transfer function to state space

APPENDIX C - SIMNON COMMANDS

1. Utilities

EDIT	- Edit system description
GET	- Get parameters and initial values
LIST	- List files
PRINT	- Print files
SAVE	- Save parameter values and initial values in a file
STOP	- Stop

2. Graphic output

AREA	- Select window on screen
ASHOW	- Plot stored variables with automatic scaling
AXES	- Draw axes
HCOPY	- Make hard copy
SHOW	- Plot stored variables
SPLIT	- Split screen into windows
TEXT	- Transfer text string to graph

3. Simulation Commands

ALGOR	- Select integration algorithm
DISP	- Display parameters
ERROR	- Choose error bound for integration routine
INIT	- Change initial values of state variables
PAR	- Change parameters
PLOT	- Choose variables to be plotted
SIMU	- Simulate a system
STORE	- Choose variables to be stored
SYST	- Activate systems

APPENDIX D - SYNPAC COMMANDS

1. Utilities

CONV	- Conversion of data to internal standard format
DELET	- Edit system description
EDIT	- Symbolic text editor
FHEAD	- Inspect and change file parameters
FORMAT	- Conversion of data to symbolic external form
FTEST	- Check existence of a file
LIST	- List files
MOVE	- Move data in database
TURN	- Change program switches

2. Graphic output

BODE	- Plot Bode diagrams
HCOPY	- Make hard copy
NIC	- Display a frequency response in a Nichols diagram
NYQ	- Display a frequency response in a Nyquist diagram
PLEV	- Display eigenvalues etc in the complex plane
PLOT	- Plot curves with linear scales

3. Time series operations

CONC	- Concatenate two time series
CORNO	- Generate a correlated noise time series
CUT	- Extract a part of a time series
INSI	- Generate time series
PICK	- Pick equidistant time points
SCLOP	- Do scalar operations on a time series
STAT	- Compute statistical characteristics
VECOP	- Do vector operations on a time series

4. Matrix operations

ALTER	- Alter elements in a matrix
EIGEN	- Compute eigenvalues of a matrix
ENTER	- Enter a matrix element by element
EXPAN	- Generate a matrix from sub-matrices
MATOP	- Perform matrix operations
REDUC	- Extract a submatrix
UNITM	- Generate a unit matrix
ZEROM	- Generate a zero matrix

5. System conversion and analysis

CONT	- Convert to continuous time form
POLES	- Compute the poles of a system
SAMP	- Convert to discrete time form
SIMU	- Simulate the time response of a system
SPSS	- Compute the frequency response of a system
SYSOP	- Generate a system from its subsystems
SYST	- Generate and edit system descriptions
TRANS	- Convert a criterion from continuous time to discrete time form

6. Design

FEEDF	- Design feedforward control
KALFI	- Compute a Kalman filter gain
LUEN	- Compute a Luenberg observer
OPTFB	- Compute a linear quadratic state feedback
PENLT	- Reduce a penalty function to standard form
PPLAC	- Pole placement for single input systems
RECON	- State reconstruction for single input systems
REDFB	- Compute an output feedback

APPENDIX E - COMMANDS IN POLPAC

1. Utilities

CONV - Conversion of data to internal standard format
DELET - Delete a file
EDOT - Edit system description
FHEAD - Inspect and change file parameters
FORMAT - Conversion of data to symbolic external form
FTEST - Check existence of a file
LIST - List files
MOVE - Move data in database
TURN - Change program switches

2. Graphics

BODE - Plot Bode diagrams
HCOPY - Make hard copy
LOCPLOT - Plot root locus diagrams
NIC - Plot Nichols diagrams
NYQ - Plot Nyquist diagrams
PLEV - Plot eigenvalues and allow editing
PLOT - Plot curves with linear scales

3. System and polynomial operations

INSI - Generate a data file
POLOP - Evaluate algebraic polynomial expressions
POLSYS - Create a system file or a polynomial file
POLY - Generate or edit a polynomial
POLZ - Compute and plot the zeros of a polynomial
SIMU - Simulate a system
SYSOP - Build a system from subsystems

4. Analysis

PROP - Compute bandwidth, rise time, error coefficients
ROTLOC - Compute the root locus
ROUTH - Compute and display Routh's tableau
TRFFR - Compute frequency response of a transfer function
TRFSIM - Simulate

5. Synthesis

DEADBE - Compute dead-beat strategy
MIVRE - Compute minimum variance control
POLPLA - Make a pole placement design

APPENDIX F - INTRAC COMMANDS

1. Input and output

READ	- Read string or variable from keyboard
SWITCH	- Utility command
WRITE	- Write string or variable on terminal

2. Assignment

DEFAULT	- Assign default values
FREE	- Release assigned global variables
LET	- Assignment of variables and global parameters
STOP	- Stop execution and return to OS

3. Control of program flow

FOR..TO	- Loop
NEXT V	
LABEL L	- Declaration of label
GOTO L	- Transfer control
IF..GOTO	- Transfer control

4. Macro

END	- End of macro definition
FORMAL	- Declaration of formal arguments
MACRO	- Macro definition
RESUME	- Resume execution of macro
SUSPEND	- Suspend execution of macro

COMPUTER-AIDED CONTROL SYSTEMS ENGINEERIN(
M. Jamshidi and C.J. Herget (Editors)
Elsevier Science Publishers B.V. (North-Holland), 1985

INTERACTIVE COMPUTER-AIDED DESIGN OF CONTROL SYSTEMS

A. Emami-Naeini
Systems Control Technology, Inc.
Palo Alto, California 94304
and
G.F. Franklin
Information System Laboratory
Stanford University, Stanford, CA 94305

We describe four computer packages for interactive computer-aided design of control systems. DIGICON (digital control) is a computer aid to the design of digital controls for single-input/single-output (SISO) systems using the state-space techniques of pole and zero assignment. CONCON (continuous control) is a similar package for the design of SISO continuous systems using the same techniques. The program DOPTICON (discrete optimal control) is a computer aid for the design of discrete optimal control systems using the linear quadratic Gaussian (LQG) theory and OPTICON (optimal control) is its continuous-time counterpart. All four packages contain algorithms for computation of the poles and zeros of the resulting designs as well as the evaluation of the transient responses.

I. PURPOSE

The programs DIGICON and CONCON are computer aids for the design of digital and continuous controls for scalar systems, respectively. They employ the state-space techniques of pole and zero assignment and provision is made to implement the designs using an estimator for the unmeasured states. Evaluation of the designs is by means of step function transient response. The program DOPTICON has been written for designing optimal controls for discrete-time systems using the linear quadratic Gaussian (LQG) theory and solves the Hamiltonian equations by eigenvector decomposition technique. OPTICON is the continuous-time analog of DOPTICON. We refer the reader to Refs. 1 through 4 for further reading. All four packages contain algorithms for the computation of poles and zeros of the resulting systems, as well as programs for evaluating their transient responses. In the sequel, we describe only two of the packages, namely DIGICON and DOPTICON, as the two other packages have very similar descriptions.

DIGICON [1] assumes that the dynamic object to be controlled is the plant, described by the equations

$$\begin{aligned} \dot{x}(t) &= Fx(t) + Gu(t-\lambda) \\ y(t) &= Hx(t) + Ju(t) \end{aligned} \tag{1}$$

where

$x =$ state vector, n_s dimensions;

$u =$ control vector, n_c dimensions;

$y =$ output, $n_o \times 1$;

(If $\lambda \neq 0$, x is not really the state, but rather the state has infinite dimensions in this case.)

λ = time delay;

F = system matrix;

G = input matrix;

H = output matrix;

J = direct transmission matrix.

Since we are interested in digital controls, we require a sampled-data or discrete model of the plant. Usually we will assume that u is piecewise constant as occurs with a controller acting via a zero-order hold. In any event, the discrete evolution of (1) is given by

$$\begin{aligned} x(k+1) &= \Phi x(k) + \Gamma u(k) \\ y(k) &= H_d x(k) + J_d u(k) \end{aligned} \tag{2}$$

where x, u, and y are the state, input, and output, respectively, and

Φ = discrete system matrix

Γ = discrete input matrix

H_d = discrete output matrix

J_d = discrete direct transmission matrix.

Notice that if $\lambda > 0$, then the dimension of $x(k)$ and Φ will be greater than n_s, the dimension of $x(t)$ and F.

The control for the discrete plant is implemented by a discrete dynamical system called the controller described by the equations

$$\begin{aligned} x_c(k+1) &= Ax_c(k) + By(k) + Mr(k) \\ u(k) &= Cx_c(k) + Dy(k) + Nr(k) \end{aligned} \tag{3}$$

In Eq. (3), x_c is the controller state and r is the system reference input. The vectors y and u are the plant output and input as in Eq. (2). Also, we define

A = controller system matrix

B = controller input matrix

C = controller output matrix

D = controller direct-transmission matrix

M = controller reference input matrix

N = controller reference signal direct transmission matrix

The controller described by Eq. (3) is the result of a control law, $u = -K\hat{x}$ and an estimator which approximates the plant (discrete) state, $x(k)$, by $\hat{x}(k)$. The estimator, in turn, has an error equation

$$\tilde{x}(k+1) = [\Phi - LH_d]\tilde{x}(k) \tag{4}$$

In Eq. (4), the matrix L is the estimator gain matrix.

The control gain K is selected to make the closed-loop poles of the plant be located at the roots of the characteristic polynomial $\alpha_c(z)$. The estimator gain L is selected to make the characteristic roots of the error equation (4) be located at the roots of the estimator characteristic polynomial $\alpha_e(z)$. The reference input matrices M and N are selected to give desired locations to the variable zeros in the system transfer function. These zeros may be arbitrarily chosen, may be chosen to guarantee that the error equation (4) is independent of r, or finally, may be chosen so that the only system feedback is via system error, r-y [1].

The DOPTICON program provides computer aids to solve the steady-state, discrete-time optimal control problems. The user specifies the system matrices, the weighting matrices in the performance index, and the covariances of the noise sources. Depending on the options chosen, the optimal controller gains, filter gains, the RMS state, and control responses are computed as well as the related eigensystems. DOPTICON assumes that the system under consideration is the constant coefficient discrete-time linear system of the form

$$\begin{aligned} x(k+1) &= \Phi x(k) + \Gamma u(k) + \Gamma 1\, w(k) \\ y(k) &= H_d x(k) + v(k) \end{aligned} \tag{5}$$

where w and v are the process and sensor noises with covariance matrices R_w and R_v, respectively. DOPTICON assumes a quadratic cost function and minimizes the following loss function in the statistical steady state,

$$\mathcal{J} = E\left\{ \frac{1}{2} \sum_{k=0}^{\infty} x^T(k)\, Q_1\, x(k) + u^T(k)\, Q_2\, u(k) \right\} \tag{6}$$

where Q_1 and Q_2 are the symmetric non-negative definite state and control weighting matrices, respectively. The program uses the certainty equivalence principle which states that the optimal feedback system is the optimal state feedback control law applied to an optimal estimate of the state.

It is known that the steady-state optimal control of Eq. (5) with respect to Eq. (6) with known states is given by

$$u = -Kx \tag{7}$$

where the matrix K is the control gain. K is computed from

$$K = (Q_2 + \Gamma^T S\Gamma)^{-1}\, \Gamma^T S\Phi \tag{8}$$

where S is the solution to the discrete algebraic Riccati equation and can be found from the solution of the control Hamiltonian by eigenvector decomposition. If x is not available but control must be based on the noisy observations, y, then it is known that the optimal control is given by

$$u = -K\hat{x} \tag{9}$$

where $\hat{x}$ is the least square estimate of x given y and satisfies the corrected model (or Kalman filter) equation:

$$\hat{x}(k) = \bar{x}(k) + L(y(k) - H_d\,\bar{x}(k)) \qquad \text{observation update}$$
$$\bar{x}(k) = \Phi\hat{x}(k) + \Gamma u(k) \qquad \text{state update} \tag{10}$$

In Eq. (10) the steady-state estimator gain L is given by a solution dual to that for K in Eq. (7). If we combine Eqs. (5), (8), and (10), and define the state error

$$e(k) \triangleq \bar{x}(k) - x(k) \; ,$$

we can write the solution in the form

$$\begin{aligned} e(k+1) &= \Phi(I - L\,H_d)\,e(k) + \Phi L\,v(k) - \Gamma 1\,w(k) \\ x(k+1) &= (\Phi - \Gamma k)(x(k) - L\,H_d\,e(k) + L\,v(k)) \end{aligned} \tag{11}$$

We define the following mean square quantities:

$$\begin{aligned} P_m &\triangleq E\left\{ e(k)\,e^T(k) \right\} \\ R_{\bar{x}} &\triangleq E\left\{ \bar{x}(k)\,\bar{x}^T(k) \right\} \\ R_x &\triangleq E\left\{ x(k)\,x^T(k) \right\} = R_{\bar{x}} + P_m \\ P &\triangleq E\left\{ (\hat{x}(k) - x(k))(\hat{x}(k) - x(k))^T \right\} \\ R_{\hat{x}} &\triangleq E\left\{ \hat{x}(k)\,\hat{x}^T(k) \right\} = R_x + P_m - P \end{aligned} \tag{12}$$

P_m is found from the solution of the filter Hamiltonian by eigenvector decomposition and then P is computed as follows:

$$P = \Phi^{-1}(P_m - \Gamma 1\,R_w\,\Gamma 1^T)\Phi^{-T} \tag{13}$$

The estimator gain is given by:

$$L = P_m\,H_d^T(H_d\,P_m\,H_d^T + R_v)^{-1} \tag{14}$$

As an indication of the control effort required by a given design, we frequently compute the covariance of the control. When $e \equiv 0$ (perfect state measurement), this is

$$E\left\{ uu^T \right\} = K\,R_x\,K^T \tag{15}$$

(If Φ^{-1} does not exist, the problem can be solved via a generalized eigenvalue formulation [8].)

The state covariance R_x in Eq. (15) is the solution to the discrete Lyapunov equation:

$$R_x = \Phi_c R_x \Phi_c^T + \Gamma 1 R_w \Gamma 1^T$$
$$\Phi_c = \Phi - \Gamma K \tag{16}$$

When an estimator is required, the control covariance is given by

$$E\left\{ uu^T \right\} = K R_{\hat{x}} K^T \tag{17}$$

and $R_{\hat{x}}$ is the solution to

$$R_{\hat{x}} = \Phi_c R_{\hat{x}} \Phi_c^T + (P_m - P) \tag{18}$$

The square roots of the diagonal terms in Eqs. (15) and (17) are referred to as the "control RMS response" and those of R_x as the "state RMS response."

DOPTICON solves for K and L by the eigenvector decomposition of the associated Hamiltonians. The eigensystems of the open-loop and closed-loop systems are provided as well as the RMS state and control responses. Programs are provided for the computation of the zeros of the resulting designs and the evaluation of impulse and step-transient responses.

II. LANGUAGE

All four packages have been written in APL [5-7].

III. TYPICAL APPLICATION

The programs described herein can be used in the design of control systems for electromechanical servomechanisms, process control systems, and the design of aircraft autopilots to name a few. In a typical use of these packages, the user is prompted for input data either in the form of matrices or data elements. The design then proceeds by a call to the proper functions depending on the design in mind. Once the design has been completed, the programs determine the overall system matrices. The user can then look at the plots of the transient responses and compute the zeros of the system. Since the programs are interactive, iterations on the preliminary design can be carried out quickly and efficiently.

IV. LIMITATIONS

These programs have been written for systems up to order 10 to 20. DOPTICON assumes that the plant does not have any pure delays and the cost on the control is non-singular. If this is not the case, the problem can be formulated and solved as the solution to a generalized eigenvalue problem [8].

V. AVAILABILITY

A user's manual containing the description, source code, and an illustrative example is available at reproduction cost from the authors*. The documents are as follows:

(1) DIGICON: Interactive Design of Digital Controls
(2) CONCON: Interactive Design of Continuous Controls
(3) DOPTICON: An APL Workspace for the Interactive Design of Digital Optimal Controls
(4) OPTICON: Interactive Design of Continuous Optimal Controls.

The programs have been tested on the IBM 370/3033 under VS APL at Stanford University and have performed well for various examples.

VI. EXAMPLE

We present a simple example of the use of DIGICON for a double-integrator plant. The data are entered using the function INPUT as follows.

```
          INPUT
TO INPUT MATRICES, TYPE 1
TO INPUT DATA ELEMENTS, TYPE 2
⎕:
          2
HOW MANY STATES?
⎕:
          2
HOW MANY CONTROLS?
⎕:
          1
HOW MANY OUTPUTS?
⎕:
          1
ROWS OF F
ROW 1 = ?
⎕:
          0 1
ROW 2 = ?
⎕:
          0 0
 0 1
 0 0
 COLUMNS OF G
ELEMENTS OF ¯COLUMN 1 = ?
⎕:
```

* Please write to:

Office of Technology Licensing
Encina Hall Room 105
Stanford University
Stanford, CA 94305

```
          0 1
 0
 1
ROWS OF H
ROW 1 = ?
⎕:
          1 0
 1 0
ROWS OF J
ROW 1 = ?
⎕:
          0
```

The discrete model parameters are found using the function SAMPLEØ with the sampling period T = 0.5 sec.

```
          SAMPLEO
TO USE F AND G, TYPE 1
TO INPUT NEW SYSTEM, TYPE 2
⎕:
          1
SAMPLE PERIOD =
⎕:
          .5
SYSTEM DELAY LAMBDA =
⎕:
          0
0
PHI
¯1    0.5
 0    1
GAMMA
¯0.125
 0.5
HD
¯1 0
JD
¯0
```

This system has a zero at -1. Using the function CONLAW, the controller poles are placed inside the unit circle at radius .6 and an angle of 25° (corresponding to a $\zeta = 0.7$).

```
          CONLAW
TO USE PHI,GAMMA, TYPE 1
TO INPUT OTHER MATRICES,TYPE 2
⎕:
          1
TYPE   2 ROOTS AS RADIUS,ANGLE(DEGREES)
ROOT   1
⎕:
          .6 25
THE CONTROL GAIN IS K =
 1.089722622 1.552430656
CLOSED LOOP MATRIX IS PHIC =
 ¯0.8637846722  0.3059461681
 ¯0.5448613111  0.2237846722
```

The function REDUEST is called to design a reduced order estimator with its pole placed at the origin. If we cancel the estimator pole, the controller matrices are computed as follows.

```
      REDUEST
TO USE PHI,GAMMA,HD,JD,TYPE 1
TO INPUT NEW DISCRETE MATRICES,TYPE 2
⎕:
      1
TYPE  1 ROOTS AS RADIUS,ANGLE(DEGREES)
ROOT  1
⎕:
      0 0
TO ASSIGN ZEROES, TYPE   1
TO USE ERROR CONTROL, TYPE  2
TO CANCEL ESTIMATOR POLES, TYPE  3
⎕:
      3
SYSTEM MATRIX AO =
¯0.3881076639
INPUT MATRIX BO =
 4.732811482
OUTPUT MATRIX CO =
 1
DIRECT MATRIX DO =
¯4.194583933
REFERENCE INPUT MATRIX MO =
¯0.4229297012
REFERENCE DIRECT MATRIX NO =
 1.089722622
```

The function SYSMAT is called to obtain the overall system matrices. The step response of the closed-loop system is obtained using the function STEP. The system has a velocity error coefficient K_v = .7019.

```
      SYSMAT
OVERALL SYSTEM PHI MATRIX,    SPHI =
 0.4756770083  0.5           ¯0.125
¯2.097291967   1              0.5
 4.732811482   0             ¯0.3881076639
OVERALL SYSTEM INPUT MATRIX SGAMMA =
 0.1362153278
 0.5448613111
¯0.4229297012
OVERALL SYSTEM OUTPUT MATRIX SH =
 1 0 0
OVERALL SYSTEM DIRECT MATRIX SJ =
 0

      STEP
TO USE PHIC , GAMMA,HD,JD,  TYPE 1
TO USE SPHI,SGAMMA,SHD,SJD,TYPE 2
TO INPUT DISCRETE SYSTEM, TYPE 3
⎕:
      2
```

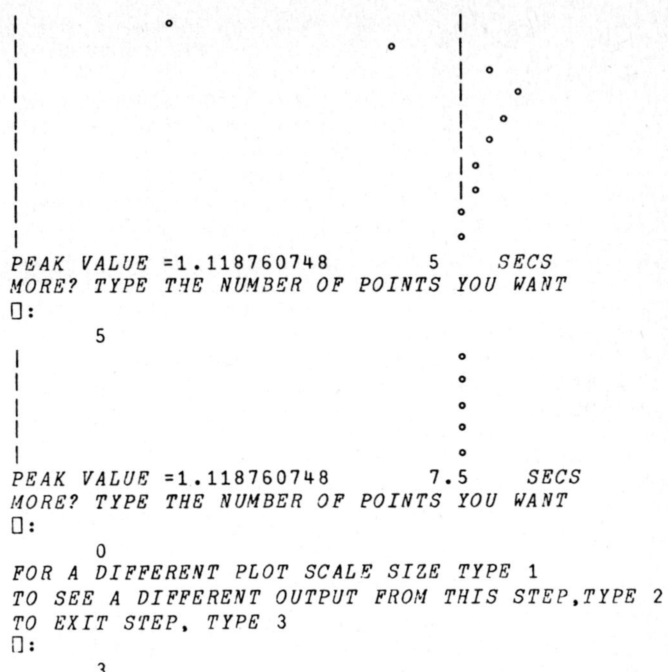

```
PEAK VALUE =1.118760748          5     SECS
MORE? TYPE THE NUMBER OF POINTS YOU WANT
⎕:
      5
PEAK VALUE =1.118760748          7.5     SECS
MORE? TYPE THE NUMBER OF POINTS YOU WANT
⎕:
      0
FOR A DIFFERENT PLOT SCALE SIZE TYPE 1
TO SEE A DIFFERENT OUTPUT FROM THIS STEP,TYPE 2
TO EXIT STEP, TYPE 3
⎕:
      3
```

If we use error control instead (the second choice in REDUEST) and after a call to SYSMAT, the step response of the system would be shown as below.

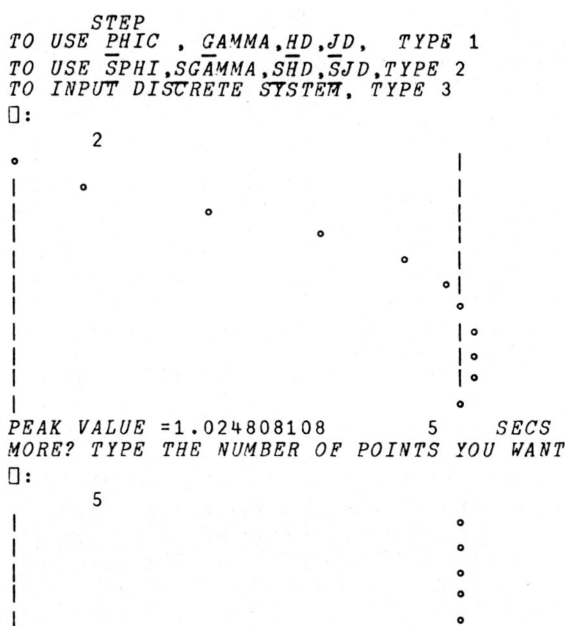

```
        STEP
TO USE PHIC , GAMMA,HD,JD,   TYPE 1
TO USE SPHI,SGAMMA,SHD,SJD,TYPE 2
TO INPUT DISCRETE SYSTEM, TYPE 3
⎕:
      2
PEAK VALUE =1.024808108          5     SECS
MORE? TYPE THE NUMBER OF POINTS YOU WANT
⎕:
      5
```

```
PEAK VALUE =1.024808108          7.5      SECS
MORE? TYPE THE NUMBER OF POINTS YOU WANT
⎕:
        0
FOR A DIFFERENT PLOT SCALE SIZE TYPE 1
TO SEE A DIFFERENT OUTPUT FROM THIS STEP,TYPE 2
TO EXIT STEP, TYPE 3
⎕:
        3
```

Notice that the system has a much higher overshoot than before. We can increase K_v by zero assignment (the first choice in REDUEST). For example, if we place a zero at 0.587544 and after a call to SYSMAT, the closed-loop step response is as follows.

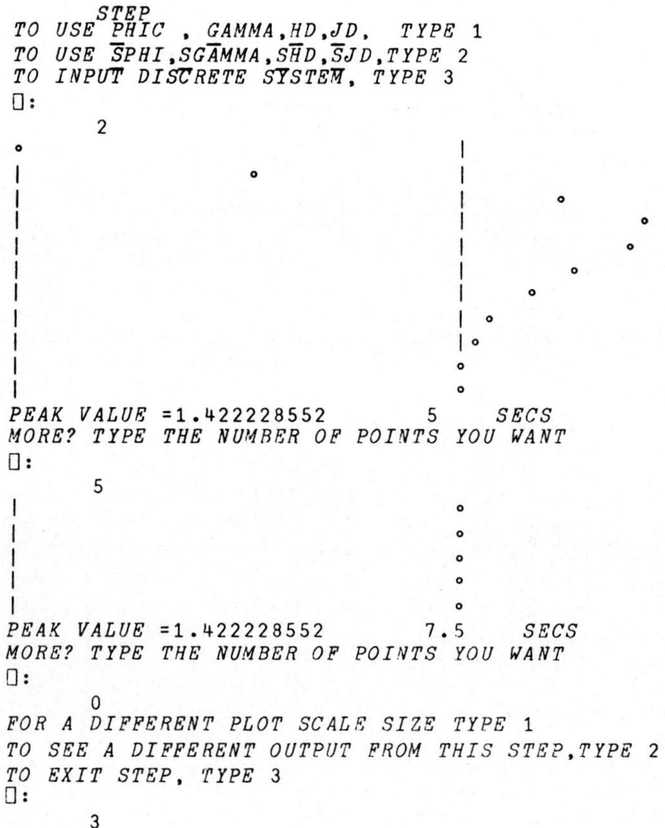

```
        STEP
TO USE  PHIC  , GAMMA,HD,JD,   TYPE 1
TO USE SPHI,SGAMMA,SHD,SJD,TYPE 2
TO INPUT DISCRETE SYSTEM, TYPE 3
⎕:
        2
PEAK VALUE =1.422228552          5      SECS
MORE? TYPE THE NUMBER OF POINTS YOU WANT
⎕:
        5
PEAK VALUE =1.422228552          7.5      SECS
MORE? TYPE THE NUMBER OF POINTS YOU WANT
⎕:
        0
FOR A DIFFERENT PLOT SCALE SIZE TYPE 1
TO SEE A DIFFERENT OUTPUT FROM THIS STEP,TYPE 2
TO EXIT STEP, TYPE 3
⎕:
        3
```

The velocity error coefficient K_v = 1.4038 and is double that of the first case. As a check, the closed-loop zeros and poles of the system are computed using the functions ZEROS and EIGEN, respectively.

```
      ZEROS
TO INPUT MATRICES, TYPE 1
TO INPUT DATA ELEMENTS, TYPE 2
⎕:
      1
SYSTEM MATRIX =
⎕:
      SPHI
  0.4756770083  0.5            0.125
 ¯2.097291967   1              0.5
  4.732811482   0             ¯0.3881076639
INPUT MATRIX =
⎕:
      SGAMMA
  0.3302542035
  1.321016814
 ¯2.577704506
OUTPUT MATRIX =
⎕:
      SH
 1 0 0
OUTPUT DIRECT MATRIX =
⎕:
      SJ
 0
TO SOLVE FOR INVARIANT ZEROS, TYPE 1
TO SOLVE FOR INPUT DECOUPLING ZEROS, TYPE 2
TO SOLVE FOR OUTPUT DECOUPLING ZEROS, TYPE 3
⎕:
      1
THE ZEROS ARE :
  0.587544   0        (real, imaginary)
 ¯1          0


     EIGEN SPHI
6.081535733E¯16  0.000000000E0
5.437846722E¯1   2.535709570E¯1     (real, imaginary)
5.437846722E¯1  ¯2.535709570E¯1
```

VII. ACKNOWLEDGEMENTS

We would like to thank Prof. David MacNeil for providing us with the APL functions for computing eigenvalues and eigenvectors, and Dr. Gürcan Aral for contributing the transient response program. This work was supported, in part, by NASA under Grant NGL 05-020-007, Project 6303.

REFERENCES

1. Franklin, G.F. and J.D. Powell, Digital Control of Dynamic Systems, Addison-Wesley, 1980.

2. Bryson, A.E. and Y.C. Ho, Applied Optimal Control, Blaisdell, Mass., 1969.

3. Bryson, A.E. and W.E. Hall, "Optimal Control and Filter Synthesis by Eigenvector Decomposition," Report SUDAAR No. 436, Dept. of Aeronautics and Astronautics, Stanford University, December 1971.

4. Kwakernaak, H. and R. Sivan, Linear Optimal Control Systems, J. Wiley, New York, 1972.

5. Iverson, K.E., A Programming Language, New York, Wiley, 1962.

6. Gilman, L. and A.J. Rose, APL: An Interactive Approach, Second Edition, New York, Wiley, 1976.

7. Grey, L.D., A Course in APL with Applications, Second Edition, Addison-Wesley, 1976.

8. Emami-Naeini, A. and G.F. Franklin, "Deadbeat Control and Tracking of Discrete-Time Systems," IEEE Trans. Auto. Control, Vol. AC-27, No. 1, pp.176-181, February 1982.

COMPUTER-AIDED CONTROL SYSTEMS ENGINEERING
M. Jamshidi and C.J. Herget (Editors)

LINEAR SYSTEMS ANALYSIS PROGRAM

Charles J. Herget
Diane M. Tilly

Lawrence Livermore National Laboratory
Livermore, CA 94550

The computer program LSAP (Linear Systems Analysis Program) is an interactive program with graphics capability that can be used for the analysis and design of linear control systems. Nearly all classical design tools are available, including manipulation of transfer functions and generation of root locus, time response, and frequency response plots. The program is capable of working with both continuous time systems and sampled data systems. For continuous time systems, the Laplace transform is used, and for sampled data systems, the Z-transform is used. The capability of converting from a Laplace transform to a Z-transform is provided.

I. Purpose

The program LSAP is primarily intended for the analysis and design of feedback control systems. Using the interactive graphics capability, the user can quickly plot a root locus, frequency response, or time response of either a continuous time system or a sampled data system. The system configuration or parameters can be easily changed, allowing the user to design compensation networks and perform sensitivity analyses in a very convenient manner.

This program is intended primarily for the analysis of feedback control systems as shown in Figures 1 and 2 for continuous time and sampled data control systems, respectively. The program is flexible enough to allow the user to configure more general systems than those shown in Figures 1 and 2 and the program is not limited to single-input/single-output systems.

A complete list of the available procedures is given in Table I.

II. Language

The main program and most subroutines are written in PASCAL. Some of the subroutines are written in FORTRAN.

* Work performed under the auspices of the U.S. Department of Energy by the Lawrence Livermore National Laboratory under contract number W-7405-ENG-48, and supported in part by the DOE Office of Basic Energy Sciences, Engineering Research Program.

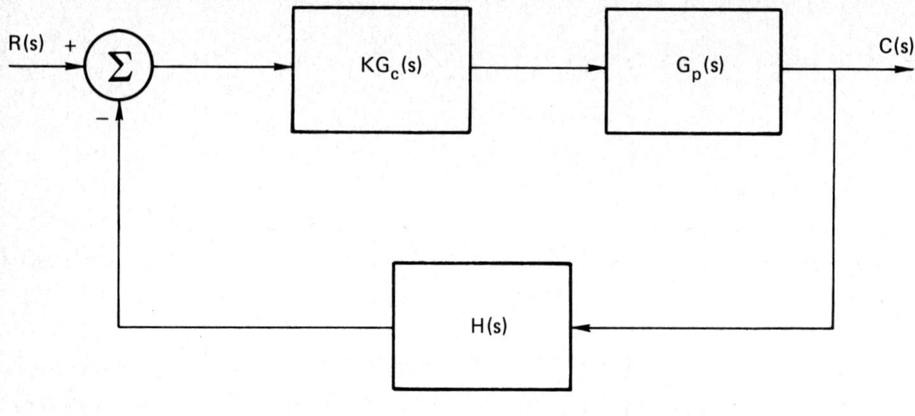

R(s) : Input
C(s) : Output
K : Gain
$G_c(s)$: Compensation
$G_p(s)$: Plant
H(s) : Feedback

Figure 1. Continuous time feedback system.

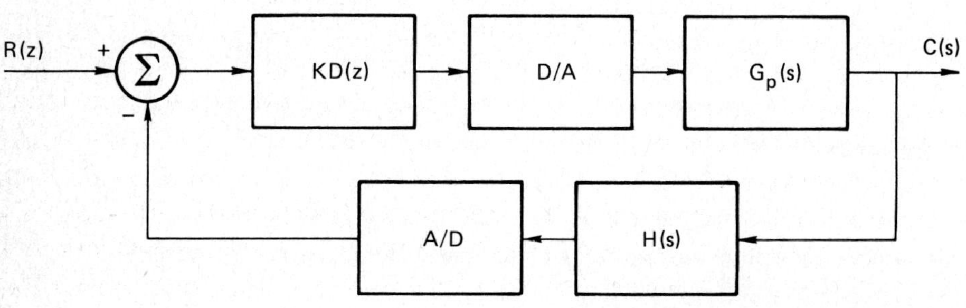

R(z) : Input
C(s) : Output
D(z) : Digital compensation
H(s) : Feedback
$G_p(s)$: Plant
A/D : Analog-to-digital
D/A : Digital-to-analog
K : Gain

Figure 2. Sampled-data feedback system.

III. Typical Application

Let us consider a typical use of the program. Assume we are to design the compensation network for the feedback control system shown in Figure 1. We assume the plant and feedback transfer functions are known and specifications on the closed loop performance, e.g., rise time, percent overshoot, and settling time for a step input, are given.

We would begin by executing the procedure Define. The program would prompt us to enter the name of the plant transfer function. In this case, we might enter GP. The program will continue to prompt us to enter information until the function is completely specified. For example, the plane, "s" or "z", must be specified, whether the function is to be entered in factored, unfactored, the order of the numerator and the order of the denominator, the coefficients of the numerator and denominator polynomials if the system is in unfactored form, and information concerning the poles and zeros if it is in factored form. Next, we would enter the feedback transfer function using the procedure Define. Let us assume the name H is selected.

We could then execute the procedure Compute to multiply GP times H to obtain the uncompensated open loop transfer function; let us call it GH. A frequency response and root locus of GH could then be obtained and some insight gained on how to select the compensation network. Suppose that an initial compensation is chosen and it is called GC. We could use the procedure Compute to multiply GC times GH to obtain the compensated open loop transfer function; let us call it GCH. We could again plot a frequency response and root locus of GCH. We could use the procedure Find Gain to determine the required value of K in Figure 1 from the root locus plot.

Next, we would execute the procedure Closed Loop to obtain the closed loop transfer function, $C(s)/R(s)$; let us call it T. We could then use the procedure Time Response to find the response of the closed loop system to a step input and see if the performance specifications were met. If they were not, we could go back and modify GC as needed and repeat the whole procedure as often as required.

Finally, we may wish to determine the sensitivity of the system to variations in parameters in the plant and feedback transfer functions. We could use the procedure Plot to put several time responses on one figure for various values of the system parameters. An example is included in another section.

IV. Limitations

The program is suitable only for systems which may be described by a collection of rational transfer functions which are either Laplace transforms or Z-transforms. The arrays for polynomials are presently dimensioned to permit up to thirtieth order polynomials.

V. Availability

A programmer's manual and a user's manual are available from the National Technical Information Service, the U. S. Department of Commerce, 5285 Port Royal Road, Springfield, VA 22161. The documents are as follows.

> D. M. Tilly, "Linear Systems Analysis Program, Programmer's Manual, Version 2.2 (VAX/VMS)" Lawrence Livermore National Laboratory, UCID-30183 Rev 1, April 20, 1984.
>
> C. J. Herget and D. M. Tilly, "Linear Systems Analysis Program User's Manual, Version 2.2 (VAX/VMS)" Lawrence Livermore National Laboratory, UCID-30184 Rev 1, April 20, 1984.

The source code is available from the National Energy Software Center, Argonne National Laboratory, 9700 South Cass Avenue, Argonne, IL 60439.

The program presently runs on a DEC VAX using the VMS operating system. The graphics is available on the Tektronix 4010 series terminals, the Tektronix 4025 and 4105, DEC VT-100 with Retrographics, and HP2648A. Drivers for additional terminals can be added.

VI. Example

As an example of a typical application of the program, let us consider the design of a feedback control system as shown in Fig. 1 where

$$G(s) = \frac{(s+10+10j)(s+10-10j)}{s(s+12)^3} \quad , \quad (j=\sqrt{-1}) \tag{6.1}$$

$$H(s) = \frac{s+24}{s^2+10s+24} \quad , \tag{6.2}$$

and

$$G_c(s) = 1 \quad . \tag{6.3}$$

In this simple example, we would like to investigate the step response of the closed loop system for several values of gain, K.

We would begin by defining the transfer function $G(s)$ using the procedure *DEFINE* in factored form. Next we would define $H(s)$ using the procedure *DEFINE* in unfactored from. We could then use the procedure *COMPUTE* to define $GH(s) = G(s) * H(s)$. We could then obtain the frequency response of $GH(s)$ using the procedure *FREQ R* as shown in Fig. 3. We could also obtain the root locus using the procedure *R LOCUS* as shown in Fig. 4. An expanded view of the root locus around the origin can be found using the procedure *ZOOM* as shown in Fig. 5.

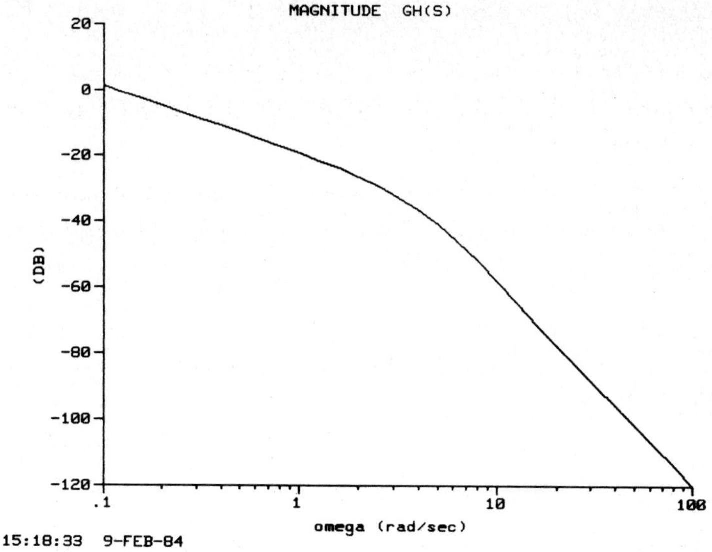

Figure 3(a). (a) Magnitude plot of $GH(s)$ generated by Frequency Response.

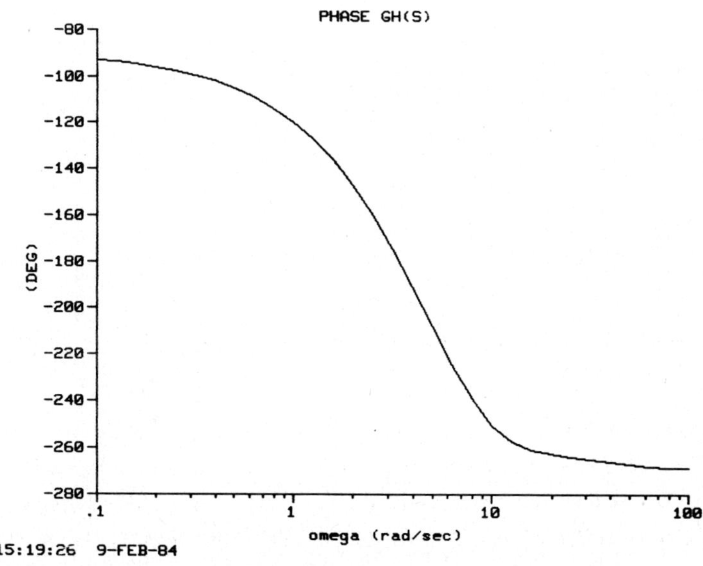

Figure 3(b). Phase plot of $GH(s)$ generated by Frequency Response.

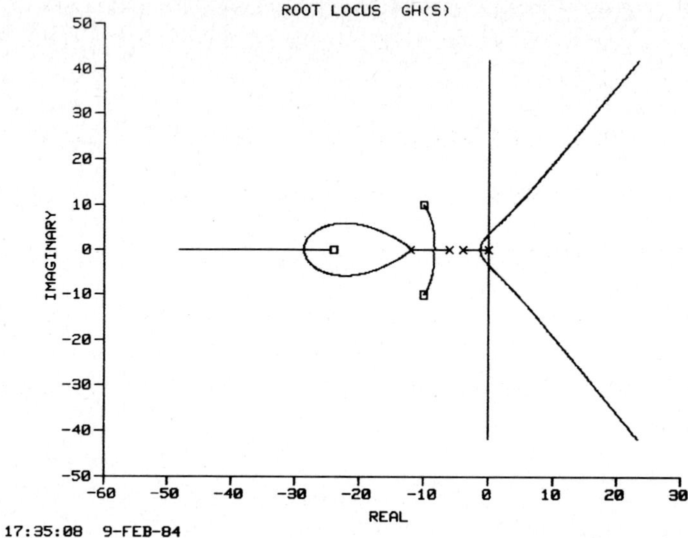

Figure 4. Root locus of an s-domain transfer function.

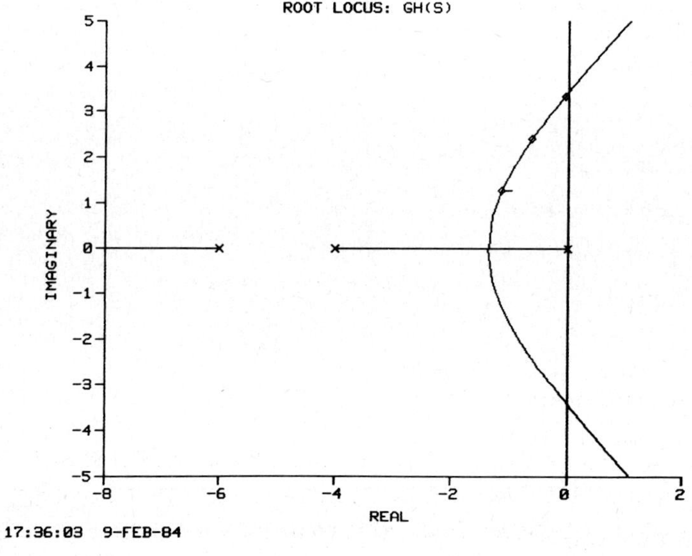

Figure 5. Find gain on the root locus.

By placing a cursor at various points on the root locus, the gains at these points can be found using the procedure *FIND G*. Values of gain were found at the points indicated by $\diamond$ on Fig. 5. The values are listed below.

THE GAIN AT THE POINT (−6.074798E−02, 3.332891E+00) IS 4.638094E+01

THE GAIN AT THE POINT (−6.261143E−01, 2.394239E+00) IS 2.145881E+01

THE GAIN AT THE POINT (−1.136168E+00, 1.274827E+00) IS 8.623815E+00

Next, the step response for selected values of gain can be plotted using the procedure *PLOT*. In Fig. 6, the step response for $K = 8.624$ and $K = 21.46$ are shown. Other time responses can be found by using as input any Laplace transform which can be represented as a rational transfer function.

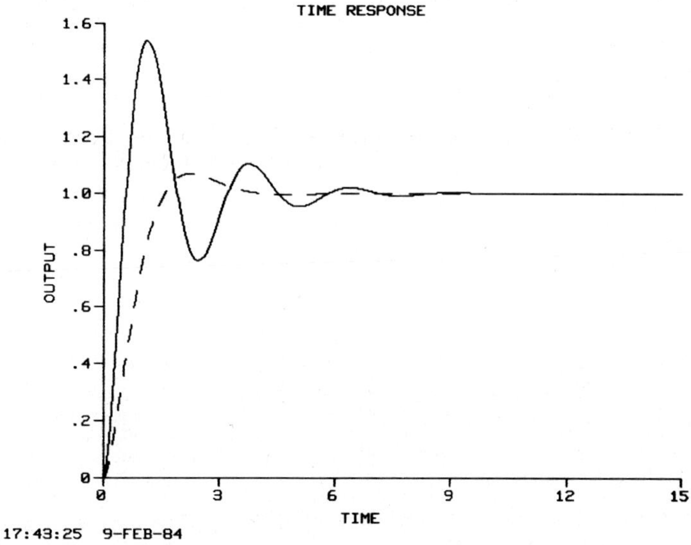

Figure 6. Plot of two time responses.

VII. Acknowledgments

This program is based on one developed by D. J. Duven at Iowa State University. That program was written in PL/I for batch use on IBM machines. The first interactive version was completely rewritten in PASCAL and FORTRAN by T. P. Weis with help from D. J. Balaban, H. R. Brand, S. M. Lanning, and R. K. Yamauchi. The control system engineer responsible for the development was C. J. Herget. The present version was written by D. M. Tilly for use on a VAX.

TABLE I. Brief description of the LSAP procedures.

1. Closed Loop:	Computes the closed loop transfer function for a system of the form shown in Figure 1 given K, G_c, G_p, and H. Any of these may be set equal to unity, and either Laplace or Z-transforms may be given.
2. Color:	Sets color for plot axes, markers, and multiple plot lines.
3. Compute:	Performs the algebraic operations of $+$, $-$, $\times$, and $\div$ on transfer functions.
4. Convert:	Converts a Laplace transform to a Z-transform.
5. Define:	Used to define rational fractions in either factored, unfactored, or 'Hertz' form, in either the s-domain (Laplace transforms) or the z-domain (Z-transforms).
6. Files:	Saves transfer functions in an external file for later use. Also reads in functions from a previously saved file.
7. Find Gain:	Used after plotting a root locus. By placing the cursor on the root locus, the gain (K) required to place the closed loop poles at that location will be given.
8. Frequency Response:	Determines the magnitude in dB and the phase in degrees of a transfer function in the s-domain (or the z-domain) as a function of $j\omega$ for ω over a specified range.
9. Halt:	Terminates execution of the program.
10. Help:	Causes all available procedures to be typed at the terminal.
11. List:	Gives a listing of all names of transfer functions currently defined and permits deletion of transfer functions.
12. Modify:	Used to make changes to previously defined transfer functions of any type.
13. Output:	Gives a listing at the terminal of any currently defined transfer function.
14. Partial Fraction Expansion:	Gives the partial fraction expansion of any currently defined transfer function.
15. Plot:	Permits the plotting of several time responses on one figure at the graphics terminal.
16. Root Locus:	Gives a root locus plot of any currently defined transfer function, either Laplace transform or Z-transform.
17. Time Response:	Plots the time response of any currently defined transfer function. An input may be used if desired. The input may be any previously defined transfer function, or the user may select from the standard inputs of step, ramp, and acceleration.
18. Zoom:	May be used after plotting a root locus to expand a portion for better viewing of details in a given area.

COMPUTER-AIDED CONTROL SYSTEMS ENGINEERING
M. Jamshidi and C.J. Herget (Editors)

COMPUTER-AIDED DESIGN OF SYSTEMS AND NETWORKS - Packages and Languages

M. Jamshidi, R. Morel*, T.C. Yenn, and J. Schotik
Laboratory for Computer-Aided Design of Systems and Networks
Department of Electrical and Computer Engineering
University of New Mexico
Albuquerque, NM 87131, U.S.A.

ABSTRACT

The past four years has been an especially active period for computer-aided design (CAD) of control systems in the U.S. The use of the computer as a design tool in integrated electronic circuits has become very common in both the industry and academic institutions. However, in the area of control system theory the past four years has been very critical. Today, there is hardly a university which does not have access to CAD software for control systems. Toward this goal, the Laboratory for Computer-Aided Design of Systems/Networks (LCADSN) in Electrical and Computer Engineering Department of The University of New Mexico has been active in developing and implementing various CAD software environments for design and analysis of linear control systems and networks. In this paper a brief report is being presented for 5 such CAD software packages and languages initiated at the University of New Mexico's LCADSN.

1. INTRODUCTION

Computer-aided design (CAD) of control systems design is emerging as an indispensible tool for the control system engineer. The CAD capability of control engineers would free the control engineer from routine and tiring tasks, but perhaps more importantly would make powerful numerical and complex algorithms available to him to handle challenging design and analysis problems. From the management point of view, CAD can be a very cost-effective tool for control system engineering activities. A good and reliable CAD package or language for control systems would require a great deal of cooperative work among control engineers, numerical analysts and computer engineers. Moreover, a good software package would need to be tested extensively by several users through a number of circumstances.

In this article, software for the design of control systems which are menu driven is referred to as a package. The activities can be divided into two basic categories: Packages and Languages. In this paper, a set of software programs which are interrelated together through a "menu" driver program in an interactive mode is called a "package". On the other hand, if the user has an access to individual elements of a collection of software with the capability of remembering the past transactions and saving all the future dealings with respect to a given system of network design analysis session is loosely called a "language". The packages and languages which are briefly described here are:

---

* Now with Los Alamos National Laboratory, Los Alamos, NM.

(i) **FREDOM** - A CAD package for linear classical control systems

(ii) **TIMDOM** - A CAD package for linear modern control systems

(iii) **LSSPAK** - A CAD package for large-scale linear control systems

(iv) **CONTROL_LAB** - A CAD language for linear multivariable control systems

(v) **NETWORK_LAB** - A CAD language for networks

2. CAD PACKAGES

The software for design of control system which are menu driven is referred to as a "package" in this article. In this section we will describe four CAD packages which are either developed or under development at the Laboratory.

2.1 FREDOM

FREDOM is a **FRE**quency-**DOM**ain CAD package for SISO (single-input single-output) described by a pair of transfer functions described by a feed forward transfer function G(s), described by [1]

$$G(s) = \frac{Af(s)}{Bf(s)} = \frac{s^m + a_{m-1}s^{m-1} + \dots + a_1 s + a_0}{s^n + b_{n-1}s^{n-1} + \dots + b_1 s + b_0} \tag{2-1}$$

and a feedback transfer function,

$$H(s) = \frac{Cf(s)}{Df(s)} = \frac{s^p + c_{p-1}s^{p-1} + \dots + c_1 s + c_0}{s^q + d_{q-1}s^{q-1} + \dots + d_1 s + d_0} \tag{2-2}$$

or they may be represented in the z-transfor domain for discrete-time systems as shown by:

$$G(z) = \frac{Af(z)}{Bf(z)} = \frac{z^m + a_{m-1}z^{m-1} + \dots + a_1 z + a_0}{z^n + b_{n-1}z^{n-1} + \dots + b_1 z + b_0} \tag{2-3}$$

$$H(z) = \frac{Cf(z)}{Df(z)} = \frac{z^p + c_{p-1}z^{p-1} + \dots + c_1 z + c_0}{z^q + d_{q-1}z^{q-1} + \dots + d_1 z + d_0} \tag{2-4}$$

where $m \leq n$ and $p \leq q$. There is no loss in generality by assuming,

$$a_m = b_n = c_p = d_q = 1 \tag{2-5}$$

since any of these coefficients can be normalized to one.

In sequel, a brief description of the capabilities of **FREDOM** will be presented.

Analysis: The analysis of linear control systems in **FREDOM** can be performed with several techniques. Figure 1 shows a tree structure of **FREDOM-TIMDOM/45** --the version on the HP9845 computer. The structure of **FREDOM** and **TIMDOM** on other computers such as the IBM PC/XT, HP9816, SUN Workstation or DEC 11/780 may differ to some extent. However, the essential elements of all versions of the packages are somewhat the same.

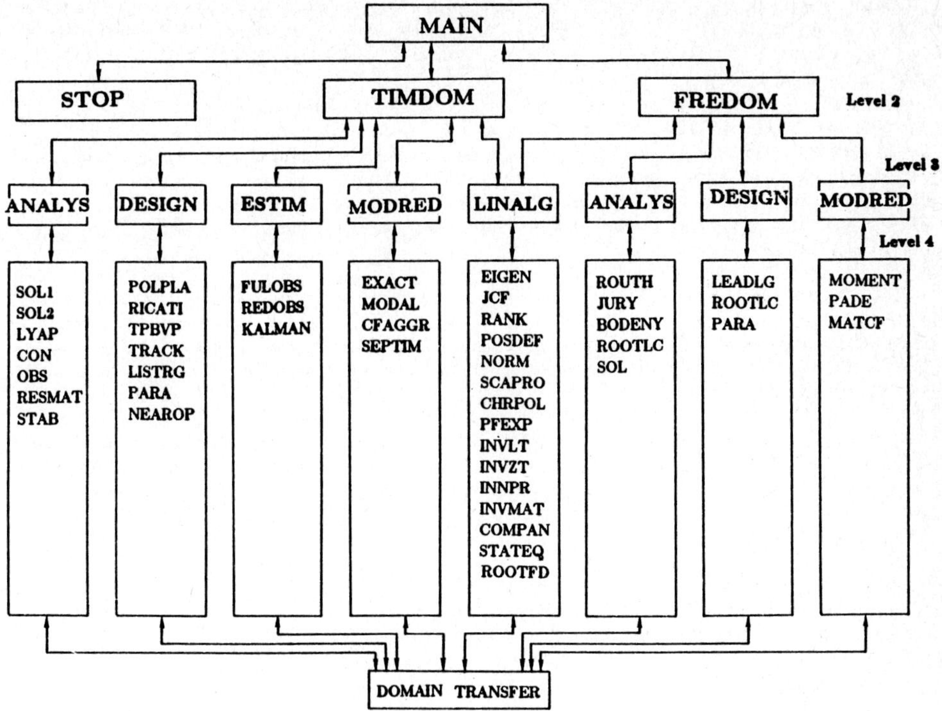

Figure 1. A tree structure for **FREDOM-TIMDOM/45**

When "ANALYS" is selected in **FREDOM/45** a menu with the following options:

COMMAND - *Description*
ROUTH - *Routh-Hurwitz stability criterion*
JURY - *Jury-Blanchard stability*
BODENY - *Bode-Nyquist plots*
ROOTLC - *Root-Locus plot*
SOL - *Time domain solution of transfer function*
EXIT - *Leave frequency analysis*

The first command, "ROUTH" performs a Routh-Hurwitz stability criterion for a SISO system. The second menu item is "JURY" which will construct the Jury-Blanchard table and, if necessary, the Raible table, for discrete-time system stability analysis. The characteristic equation for a discrete-time system is $1 + G(z)H(z) = 0$ where $G(z)$ and $H(z)$ are the z-transform representations of the forward and feedback transfer functions a SISO system. The right-half plane of the s-plane maps to the region outside the unit circle in the z-plane. Therefore, for a system in the z-transform domain to be stable all of its poles must be inside the unit circle. If they lie on the unit circle an oscillatory system is obtained as with poles on the $j\omega$-axis in the s-plane.

The next menu item "BODNEY" provides a means of obtaining Bode diagrams and the Nyquist diagram. Bode diagrams plot the log-magnitude of the open loop transfer function versus frequency and the phase versus frequency. The phase and gain margins can be obtained from the plots to determine stability characteristics. The Nyquist diagram is really a polar plot in which the concern is with the -1 point on the real axis. The Nyquist criterion states that a closed-loop system with p open loop poles in the right half plane will be stable if the Nyquist diagram of $G(j\omega)H(j\omega)$ encircles the point (-1, j0) exactly p times in the counter clockwise direction. Here $G(j\omega)$ and $H(j\omega)$ are the transfer functions of the system with $s = j\omega$.

The menu item "ROOTLC" produces a root-locus plot, another graphical method to check the system stability and help in a design problem. The system's open-loop gain is often susceptible to drifts and variations. It is important to be able to have an understanding of the locations of the roots of the system's characteristic equation as the open loop gain varies. The characteristic equation is represented by

$$1 + K\,G(s)\,H(s) = 0 \qquad (2\text{-}6)$$

where K is the variable dc gain parameter and G(s) and H(s) are the transfer functions of the system. The program will produce a plot of the root-locus from the given transfer function.

The final analysis technique is "SOL" which provides the complete response of the system. This is done by converting G(s) and H(s) into a state equation and appropriate integration and plotting routines are used to simulate the system. The above menue items will appear in other versions of **FREDOM**, e.g. **FREDOM/PC/XT-/16, -/WS, etc.**

Design: The design of systems in the frequency domain involve the classical approaches to design of SISO systems. When "DESIGN" in the frequency domain is selected the following menu will appear:

Design was chosen. Your options here are:

COMMAND - Description
LEADLG - Lead, Lag or Lead-Lag Compensation
ROOTLC - Root-locus Plots
PARAOP - Parameter Optimization
EXIT - Leave frequency design

The classical design methods provided by the package are cascade compensation ("LEADLG"), design via root-locus ("ROOTLC") and functional minimization via parameter optimization ("PARAOP"). Each of these capabilities will now be discussed in more detail.

The first menu item "LEADLG" is cascade compensation via Lead, Lag or a Lead-Lag compensating network. The next item is "ROOTLC" for root-locus design. The root-locus is a particularly useful design machanism when specifications such as percent overshoot, settling-time and natural frequency are the system parameters of interest. The root-locus can be used in conjunction with the cascade compensators. The one assumption made about systems in root-locus design is that the system has a pair of dominant poles. By this we mean that an n-th order system may have two of its closed-loop poles closer to the $j\omega$-axis as compared to the other n-z poles and zeros and therefore, may be effectively treated as a second order system.

The last menu item in frequency domain design is "PARAOP" which provides methods for functional minimization via parameter optimization. Any system that is to be controlled can be mathematically represented is which there is a dependence on a set of parameters. The value of these parameters determine the system characteristics in terms of stability and system response. The designer would like to obtain an optimum controller in order to obtain the "best" response possible. The choice of the optimum controller is based on a performance index. In many SISO linear control systems it is desired to regulate the output in accordance with some reference input. In this program, the system's performance index, say integral square error

$$ISE = \int_0^\infty e^2(t)dt \qquad (2\text{-}7)$$

is minimized by manipulating plant parameters such as gains, time constants, moments, etc.

Model Reduction: The methods of model reduction in the frequency domain are restricted to SISO systems except for the matrix continued fraction method. When the user selects "MODRED" in the frequency domain the following menu will appear when model reduction was chosen. Your options here are:

COMMAND - Description
PADE - Pade Approximation
MOMENT - Moment Matching
MATCF - Matrix Continued Fraction
EXIT - Leave frequency domain model reduction

The problem of model reduction described as follows. Consider a full-order medel's closed-loop transfer function given by,

$$G(s) = \frac{s^{n-1} + a_{n-2}s^{n-2} + \ldots + a_1 s + a_0}{s^n + b_{n-1}s^{n-1} + \ldots + b_1 s + b_0} \qquad (2\text{-}8)$$

and then seek a kth order reduced model, described by,

$$M(s) = \frac{s^{k-1} + c_{k-2}s^{k-2} + \ldots + c_0}{s^k + d_{k-1}s^{k-1} + \ldots + d_0} \qquad (2\text{-}9)$$

where the unknowns are $c_0, \ldots, c_{k-2}$ and $d_0, \ldots, d_{k-1}$, i.e. 2k-1 values.

The first menu item "PADE" refers to the method of Pade aproximation. A closely related scheme, "MOMENT", referring to moment matching scheme constitutes the second item on the menu. The idea in both cases is basically to try to find a reduced-order model such that the first few moments (or Pade approximants) of both models become equivalent. The last menu item is the matrix continued fraction method, "MATCF". This method will handle MIMO (multiple-input multiple-output) systems. The one restriction on this method is that the number of outputs must equal the number of inputs, i.e., $m = r$. Figure 2 shows a tree structure of **FREDOM/PC**.

FREDOM is available in various forms which are explained by the following table:

Name	*Language*	*Computer*
FREDOM/45	*Extended BASIC*	*HP9845*
FREDOM/16	*Extended BASIC*	*HP9816*
FREDOM/PC	*PC-BASIC*	*IBM/PC/XT and Compatibles*
FREDOM/WS	*FORTRAN/77*	*SUN Workstation (UNIX 4.2)*
FREDOM/VAX	*FORTRAN/77*	*DEC 11/780 VAX*
FREDOM/FPC	*MS-FORTRAN/77*	*IBM/PC/XT*

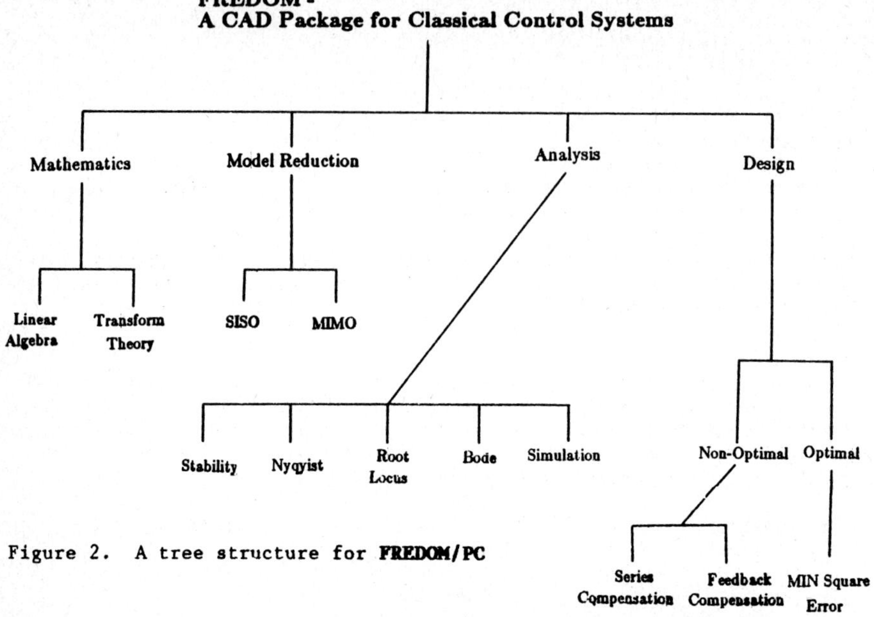

Figure 2. A tree structure for **FREDOM/PC**

2.2 TIMDOM

TIMDOM is a **TIMe-DOMain** CAD package for MIMO systems described by a quadruple of matrices (**A**,**B**,**C**,**D**) representing a linear system in state-space form:

$$\dot{\mathbf{x}} = \mathbf{A}\mathbf{x} + \mathbf{B}\mathbf{u}, \ \mathbf{x}(0) = \mathbf{x}_0 \qquad (2\text{-}10)$$

$$\mathbf{y} = \mathbf{C}\mathbf{x} + \mathbf{D}\mathbf{u} \qquad (2\text{-}11)$$

where **A** is m x n system matrix, **B** is n x m input matrix, **C** is r x m output matrix, **D** is r x m input-output matrix, and are n x 1 state, m x 1 control and r x 1 output reactors, respectively and $\mathbf{x}_0$ is the initial state.

In sequel, a brief description of the capabilities of **TIMDOM** will be presented.

Analysis: The analysis of linear control systems in **TIMDOM** can also be performed with several techniques. When "ANALYS" command of **TIMDOM** is provoked the following options would appear:

COMMAND - Description
SOL1 - Analytical solution of state transition matrix
SOL2 - Numerical solution of the state equation
LYAP - Lyapurov equation solution
CON - Controllability check
OBS - Observability check
RESMAT - Resolvent-matrix in Q(s)/P(s) form
STAB - Stability of the origin for continuous time and discrete-time systems
EXIT - Leave time domain analysis

The above commands are fairly self explanatory.

Design: Upon selection of "DESIGN" in the time domain the user is presented with the following menu:

Design was chosen. Your options here are:

COMMAND - Description
POLPLA - Pole placement
RICATI - Riccati equation solution
TPBVP - Two-point boundary-value problem
TRACK - Tracking problem
LISTRG - Linear state regulator problem
PARA - Parameter optimization
NEAROP - Near-optimum design
EXIT - Leave time domain design

The first item "POPLA" is concerned with the placement of poles for a closed-loop system, which provides a state feedback controller for a linear time-invariant SISO system. The remaining 5 commands are self-explanatory. The command "NEAROP" refers to the case where a controller (say state feedback) is designed for a reduced-order (aggregated) model

$$\dot{\mathbf{z}} = \mathbf{Fz} + \mathbf{Gu} \qquad (2\text{-}12)$$

say $\mathbf{u} = -\mathbf{Kz}$ and apply it to the full-order model

$$\dot{\mathbf{x}} = \mathbf{Ax} + \mathbf{Bu} \qquad (2\text{-}13)$$

while minimizing a quadratic cost function,

$$J = 1/2 \int_0^{\infty} (\mathbf{x}^T\mathbf{Qx} + \mathbf{u}^T\mathbf{Ru})dt \qquad (2\text{-}14)$$

The near-optimum controller is

$$\mathbf{u} = -\mathbf{KCx} \qquad (2\text{-}15)$$

where $\mathbf{C}$ is the aggregation matrix relating $\mathbf{x}$ and $\mathbf{z}$, i.e. $\mathbf{z} = \mathbf{Cx}$.

Model Reduction: The model reduction schemes of **TIMDOM** fall along the lines of large-scale systems order reduction schemes [2] "aggregation" and "perturbation". Without getting into the mathematical rigor of these schemes, one may define an "aggregated" model Eq. (2-12) of full-order model, say Eq. (2-13), as being one which has a "coarser" set of state variables, e.g. an average of the first three states being represented by one aggregated variable --$z_1 = (x_1 + x_2 + x_3)/3$. Perturbation, on the other hand, refers to the case when a system has variables which vary with different speeds, i.e. distinct gaps exist in system's eigenvalues to the extent that some variables, denoted as "fast", are varying very fast as compared to others called "slow". When provoking the "MODRED" command, the following sub-commands would typically appear in **TIMDOM**:

COMMAND - Description
EXACT - Exact aggregation
MODAL - Modal aggregation
CFAGGR - Continued fraction aggregation
SEPTIM - Separation of time scales
EXIT - Leave time domain model reduction

Here the first three commands are typical time-domain aggregation schemes, while the last one, "SEPTIM" corresponds to a perturbation method, whereby a system is checked whether its variables' TIMe scales can be SEParated into a "slow" and a "fast" subclass.

Estimation/Filtering: **TIMDOM** provides two techniques for state estimation and one for filtering. In a typical use of estimation/filtering submenu, called "ESTIM", the following options would appear:

COMMAND - Description
FULOBS - Full order observer
REDOBS - Reduced order observer
KALMAN - Kalman filtering
EXIT - Leave time domain estimation/filtering

The first item is essentially the design of a state estimator of the Luenberger-type for a linear time-invariant SISO system. The second one is the reduced-order version of it whereby only (n-r), when n is the number of states and r is the number of outputs, of the system's state variables are estimated. The last item is the design of a Kalman filter with a zero-mean white noise driven linear time-invariant discrete-time system. Figure 3 shows a tree structure of **TIMDOM/PC**.

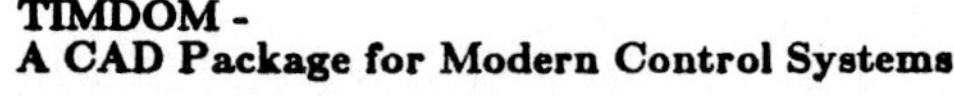

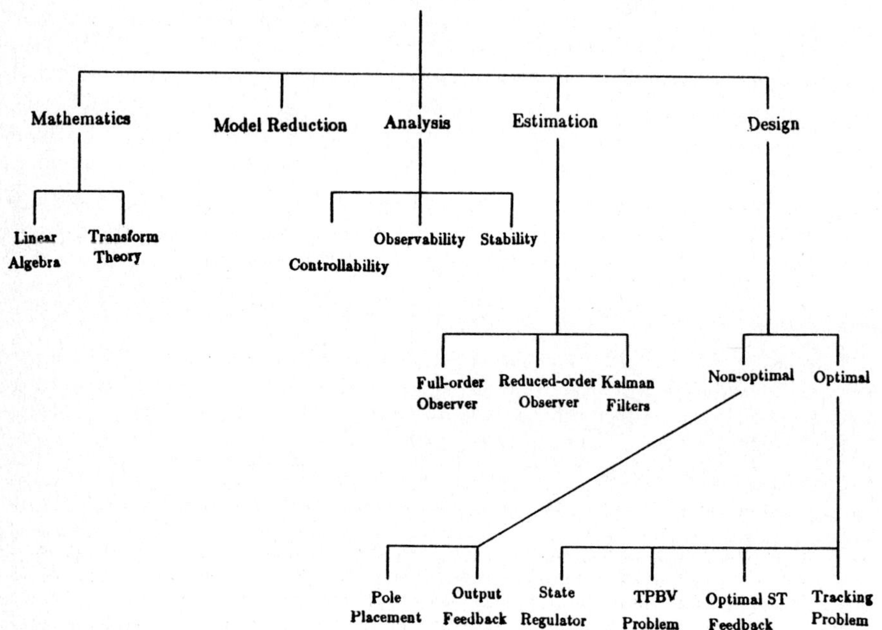

Figure 3. A tree structure for **TIMDOM/PC**

TIMDOM is available in various forms which are explained by the following table:

Name	*Language*	*Computer*
TIMDOM/45	*Extended BASIC*	*HP9845*
TIMDOM/16	*Extended BASIC*	*HP9816*
TIMDOM/PC	*PC-BASIC*	*IBM/PC/XT and Compatibles*
TIMDOM/WS	*FORTRAN/77*	*SUN Workstation (UNIX 4.2)*
TIMDOM/VAS	*FORTRAN/77*	*DEC 11/280 VAX*
TIMDOM/FPC	*MS-FORTRAN/77*	*IBM PC/XT and Compatibles*

2.3 LSSPAK

LSSPAK is a CAD package for modeling and control of large-scale linear systems. The name **LSSPAK** stems from **L**arge-**S**cale **S**ystems **PA**c**K**age. It consists of four main submenus: Linear Algebra (LINALG), Model reduction (MODRED), Analysis (ANALYS), and Design (DESIGN).

Consider a linear time-invariant system described by

$$\dot{\mathbf{x}} = \mathbf{A}\mathbf{x} + \mathbf{B}\mathbf{u}, \quad \mathbf{x}(t_o) = \mathbf{x}_o \tag{2-16}$$

where x is n x 1 state vector, u is m x 1 control vector, A and B are matrices of appropriate dimensions. It is assumed that the system is decomposable into N subsystems of the form,

$$\dot{\mathbf{x}}_i = \mathbf{A}_i\mathbf{x}_i + \mathbf{B}_i\mathbf{u}_i + \sum_{\substack{i=1 \\ j\neq i}}^{N} \mathbf{G}_{ij}\mathbf{x}_i, \quad \mathbf{x}_i(t_o) = \mathbf{x}_{oi} \tag{2-17}$$

for $i = 1, \ldots, N$

where $\mathbf{A}_i$, $\mathbf{B}_i$, and $\mathbf{G}_{ij}$ are n_i x n_i, n_i x m_i, and n_i x n_j dimensional matrices. The summation term in (2-17) represents the interaction term between the ith subsystem and the remaining (N-1) subsystems. The decomposed form (2-17) stems from the fact that one possible definition of a large-scale system which states that "a system is large in scale if it can be decomposed into a number of subsystems".

Another reformulation of a large-scale system (2-16) is the so-called "decentralized" form given by

$$\dot{\mathbf{x}} = \mathbf{A}\mathbf{x} + \sum_{i=1}^{N} \mathbf{B}_i\mathbf{u}_i, \quad \mathbf{x}(t_o) = \mathbf{x}_o \tag{2-18}$$

in which the system's control is being shared among N local controllers which act together in a decentralized fashion to control (stabilize, etc.) the large-scale systems. Due to the lack of space, we cannot get into the technical wealth of large-scale systems. The interested reader can consult a book by the first author [2]. Below is a brief description of **LSSPAK**'s structure.

Linear Algebra (LINALG) The linear algebra programs supported by **LSSPAK** is essentially the same as **TIMDOM** or **FREDOM** with exceptions of: (i) "Cheby" --a Chebyschev polynomial curve fitting program and (ii) "Genrank" -- a program to calculate the generic rank of a structured matrix for use in checking the structural controllability and observability of a large-scale linear interconnected system.

Model Reduction (MODRED) Under model reduction, **LSSPAK** supports both frequency-domain techniques of **FREDOM** and time-domain methods of **TIMDOM**. In addition it provides:

PADMOD - Pade-Modal method
PADROD - Pade-Routh method
CHAIN - Chained aggregation
BALANC - Balanced realization

Analysis (ANALYS) In the analysis sub-menu of **LSSPAK**, the following options are offered:

COMMAND - Description
STAB - Stability of a large-scale system via the Lyapurove method
STRCON - Structural controllability
STROBS - Structural observability
SIMUL - Simulation of a large-scale system

Design (DESIGN) The design of a large-scale system can take on two basic forms: "hierarchical" control which is associated with the decomposed form (2-16) of a large-scale system and "decentralized" control which is associated with the decentralized form (2-17). Here the following options are offered:

COMMAND - Description
GOALCR - Goal coordination algorithm of hierarchical control
INTPRO - Interaction prediction algorithm of hierarchical control
DYNCOM - Dynamic Compensation of decentralized control
ROBDEC - Robust decentralized controller design
DECSTM - Decentralized stabilization via the multi-level method

Figure 4 shows a tree structure of **LSSPAK/PC**.

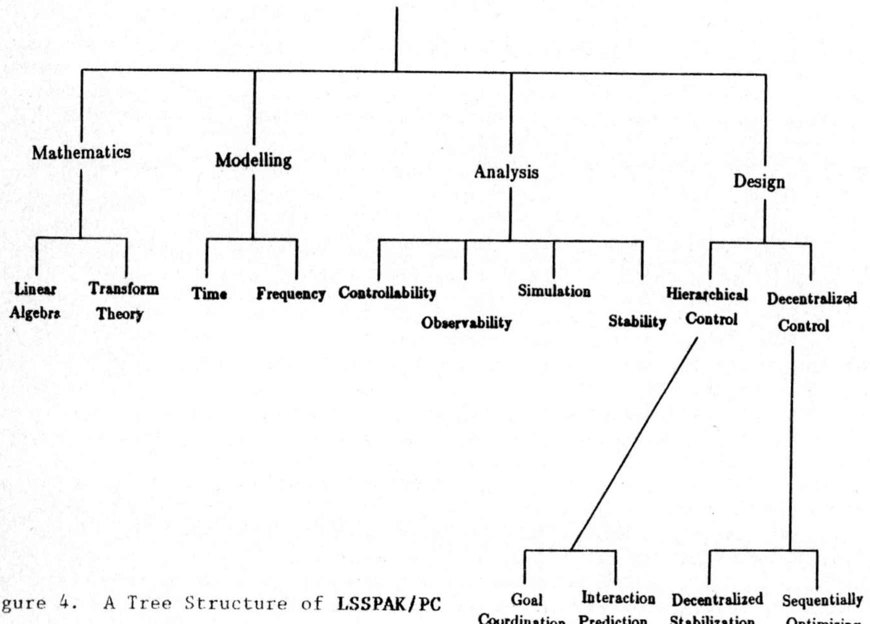

Figure 4. A Tree Structure of **LSSPAK/PC**

LSSPAK is available in various forms which are explained by the following table:

Name	*Language*	*Computer*
LSSPAK/16	*Extended BASIC*	*HP9816*
LSSPAK/PC	*PC-BASIC*	*IBM/PC/XT and compatibles*
LSSPAK/FPC	*MS-FORTRAN*	*IBM/PC/XT and compatibles*

3. CAD LANGUAGES

In this section two CAD languages are briefly introduced. One is explained in some detail and the other is very briefly described. The two languages are:

CONTROL. lab -- a CAD language for linear control systems; and

NETWORK. lab -- a CAD language for electrical networks.

3.1 CONTROL. lab

In this section a CAD language for linear multivariable systems will be introduced to the control system community. **CONTROL. lab** is designed to handle analysis, control, estimation, filtering and design of linear multivariable control systems with applications in many science and engineering fields. Using the capabilities of **MATLAB** [3] a great number of functions have been added to make **CONTROL. lab** a very powerful, user-friendly and computationally efficient language.

STRUCTURE

CONTROL. lab is designed to handle both linear continuous-time and discrete-time systems defined by a set of four matrices (A,B,C,D), i.e.

State: $\dot{\mathbf{x}} = \mathbf{Ax} + \mathbf{Bu}$ or $\mathbf{x(k+1)} = \mathbf{Ax(k)} + \mathbf{Bu(k)}$ (3-1)

Output: $\mathbf{y} = \mathbf{Cx} + \mathbf{Du}$ or $\mathbf{y(k)} = \mathbf{Cx(k)} + \mathbf{Du(k)}$ (3-2)

Once the system is defined, its description and all the subsequent calculations are stored and are on standby for future use.

Figure 5 shows a pictorial representation of the structure of **CONTROL. lab.**

Among the list of functions currently supported by **CONTROL. lab**, "LINALG" is basically **MATLAB** [2] which includes over seventy seven functions, six of which are:

NORM - matrix norm
RANK - matrix rank
INV - matrix inverse
POLY - matrix characteristic polynomial
DET - matrix determinant

as shown in Figure 5. Other sections of **CONTROL. lab** shown would provide the following:

(i) **Analysis** (ANALYS):

LYAP - Solution of a continuous-time matrix Lyapunov equation.
COMP - Companion form transformation.

RESM - Analytical solution of the resolvent matrix.
ROSN - Rosenbrock's functional minimization,
etc.

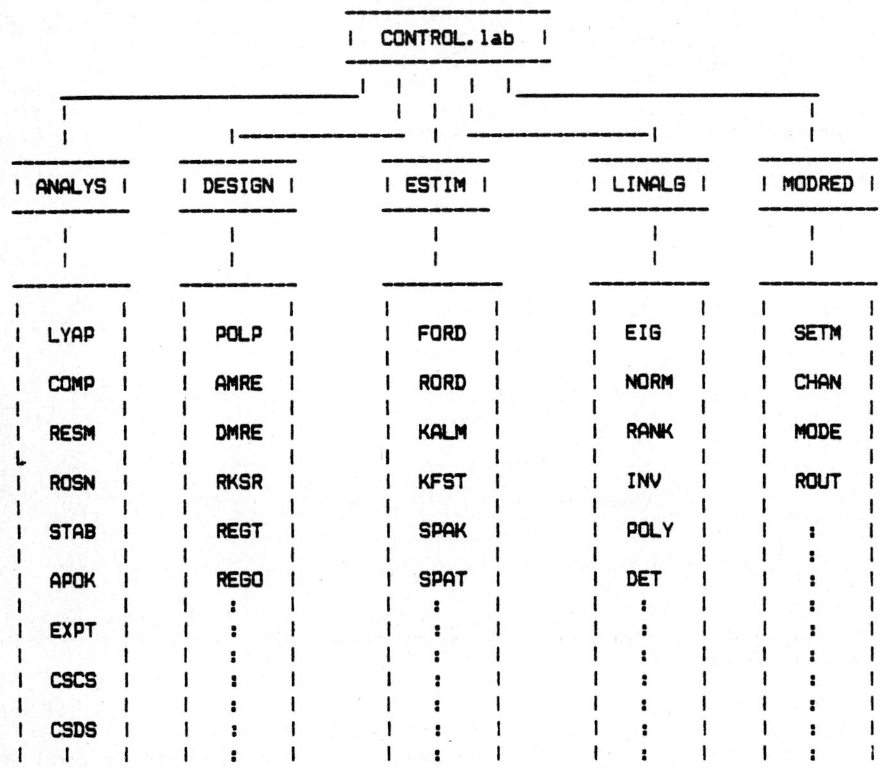

Figure 5. A pictorial structure of **CONTROL. lab**

(ii) **Design** (DESIGN):

POLP - Pole placement.
AMRE - Algebraic matrix Riccati equation.
RKSR - Solution of algebraic matrix Riccati equation by Schur transformation.
REGT - State regulator problem.
REGO - Output regulator problem,
etc.

(iii) **Estimation** (ESTIM):

FORD - Full-order observer.
RORD - Reduced-order observer.
KALM - Kalman filter design.
KFST - Discrete-time standard Kalman filter with zero mean Gaussian noise.
SPAK - Discrete-time decentralized Kalman filter via partitioned two-subsystems with zero mean Gaussian noise,
etc.

(iv) **Model Reduction** (MODRED):

SETM - Separation of singularly perturbed time scales.
CHAN - Chained aggregation.
MODE - Modal aggregation.
ROUT - Aggregation via Routh approximation,
etc.

In yet another schematic form, Figure 6 shows a diagram showing the relation of **MATLAB**, [3] **LINPAK** [4], **EISPAK** [5] and **PLOTV** [6], a graphical package developed at the University of New Mexico which runs with Digital's GIGI terminal.

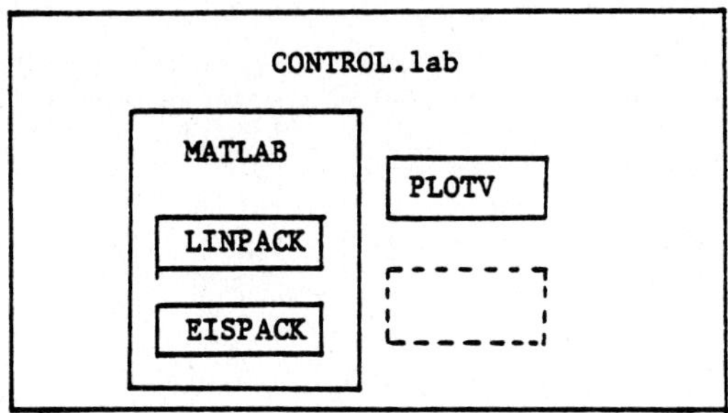

Figure 6. A schematic of **CONTROL. lab** and its relations with other packages.

The dashed box in the figure represents the future incorporation of frequency-domain packages such as **FREDOM/FPC** computers.

EXAMPLES:

Below are three examples of use of **CONTROL. lab** to solve: (1) a Lyapunov equation, (2) an algebraic matrix Riccati equation and (3) the state regulator problem:

Example 1: LYAP

Consider the Lyapunov equation

$$\mathbf{A'P + PA + Q = 0} \tag{3-3}$$

Then, in **CONTROL. lab** LYAP(A,Q) provides a solution for it:

$$\langle\rangle A = \begin{matrix} -10 & 0 & 0 \\ 0 & -0.5 & 0 \\ 0 & 0 & -8 \end{matrix} \tag{3-4}$$

$$\langle\rangle Q = \begin{matrix} 3 & 2 & 0.6 \\ 2 & 5 & 0.2 \\ 0.6 & 0.2 & 10 \end{matrix} \tag{3-5}$$

$$\langle\rangle P = \text{LYAP (A,Q)} \tag{3-6}$$

$$\langle\rangle P = \begin{matrix} 0.1500 & 0.1905 & 0.0333 \\ 0.1905 & 5.0000 & 0.0235 \\ 0.0333 & 0.0235 & 0.6250 \end{matrix} \tag{3-7}$$

LYAP is written based on an interactive scheme described by an algorithm in reference [2].

Example 2: RKSR

Consider the continuous-time algebraic Riccati equation:

$$A'K + KA - KBR^{-1}B'K + Q = 0 \tag{3-8}$$

then in **CONTROL. lab**, RKSR(A,B,Q,R) will give the results:

```
              0     1    0
   <>A =      0     0    1                 (3-9)
           -0.5   0.5 -0.5

              0     1
   <>B =      1     1                      (3-10)
              1     0

              1     0    0
   <>Q =      0     1    0                 (3-11)
              0     0    1

              1     0
   <>R =      0     1                      (3-12)

                              0.9858 0.1099 -0.2108
<>K = RKSR (A,B,Q,R) =        0.1099 0.6899  0.1845 (3-13)
                             -0.2108 0.1845  0.6582

<>KT = (A)'*K + K*A - K*B*INV(R)*(B')*K + Q

                        0.0     -0.1527  0.0173
     = 1.0d - 15*      -0.0416 -0.0278 -0.0139   (3-14)
                       -0.0312 -0.0833 -0.0555
```

RKSR has been written based on the Sehur transformation method due to Laub [7].

Example 3: REGT

Below is a test run of REGT of **CONTROL. lab** in solving a linear state regulator problem:

Given a linear time-invariant system with a quadratic cost function,

$$\dot{x} = Ax + Bu, \quad x(t_0) = x_0 \tag{3-15}$$

$$J = 1/2 \int_0^{\infty} (x'Qx + u'Ru)dt \tag{3-16}$$

where **A**, **B**, **Q**, and **R** are nxn, nxm, nxn positive semi-definite and mxm positive definite constant matrices. For this example run we have:

```
              0    1    0
   <>A =      0    0    1
            -.5   .5  -.5
```

```
         0   1
<>B  =   1   1                 1
         1   0         <>X  =  1
                               1
<>C  =   1   1   1
                       <>KF =  1   0   0   1   0   1
         1   0   0
<>Q  =   0   1   0     <>IC =  0   2   .1   .00001
         0   0   1
                       <>REGT (A,B,C,Q,R,X,KF,IC)
<>R  =   1   0
         0   1
```

Figures 7-9 show a graphical representation of the state regulator solution for $0.0 \leq t \leq 2.0$ seconds using the capabilities of **PLOTV** within the **CONTROL. lab.** In the input data of **REGT** the vector IC is defined as IC = (t_o, t_f, dt, Err), where t_o, t_f and dt are initial time, final time and step size of integration and Err is an error test constant.

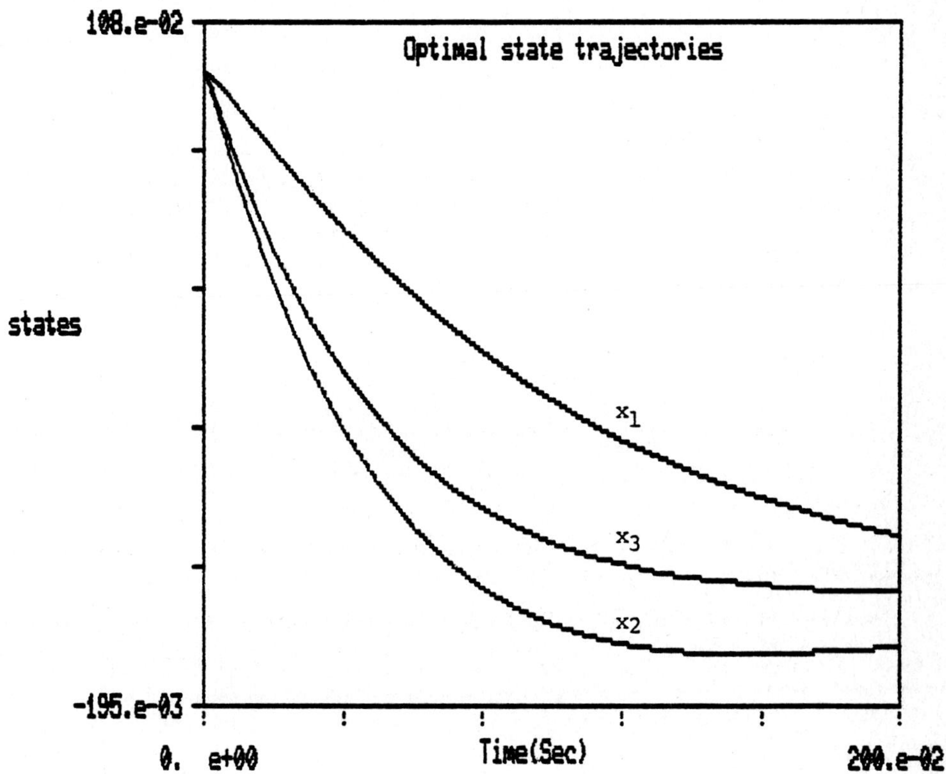

Figure 7. Graphical presentation of the optimal states in the sample us of REGT

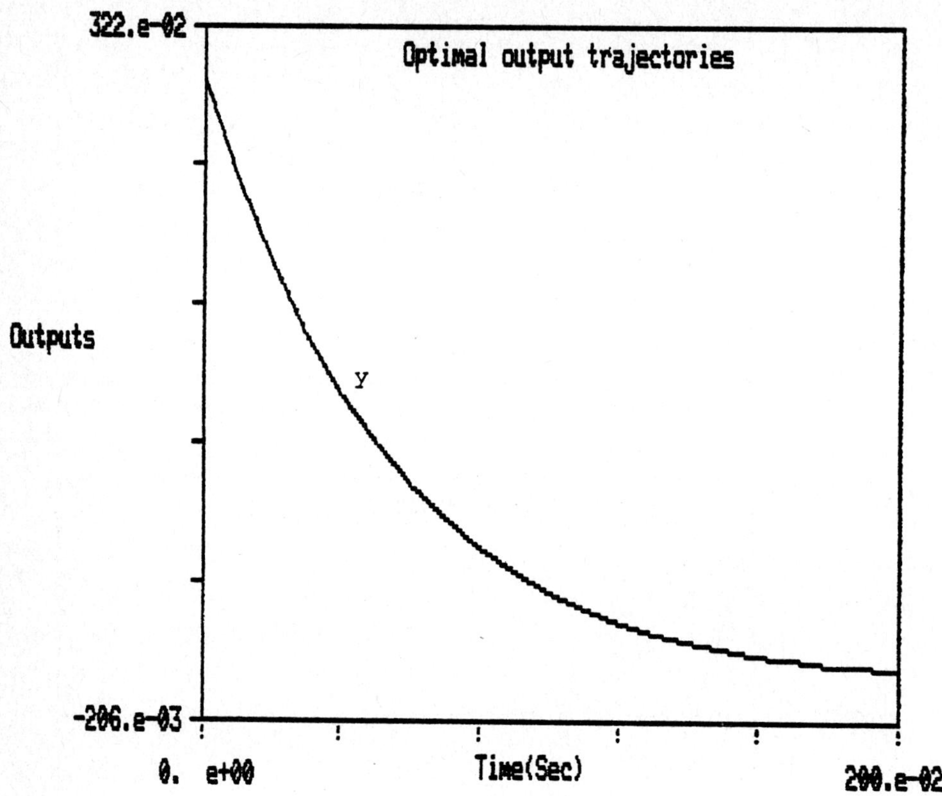

Figure 8. Graphical presentation of the optimal output in the sample use of REGT

3.2 NETWORK. lab

In this section another CAD language for the analysis of electrical networks is described. **NETWORK. lab** is a computer language which adds many capabilities to **MATLAB**. **NETWORK. lab** is designed to accomodate AC and DC, node admittance, transient, state space, sensitivity, and nonlinear analysis of networks. In addition, filter approximation and synthesis are possible in **NETWORK. lab**.

STRUCTURE:

NETWORK. lab is designed to analyze linear and nonlinear networks given as input which is similar to that of **CORNAP**. From this point on, in this particular analysis, the functions of **NETWORK. lab** operate on the initial data or on the results of previously applied functions. A pictorial representation of **NETWORK. lab** is shown in Figure 10.

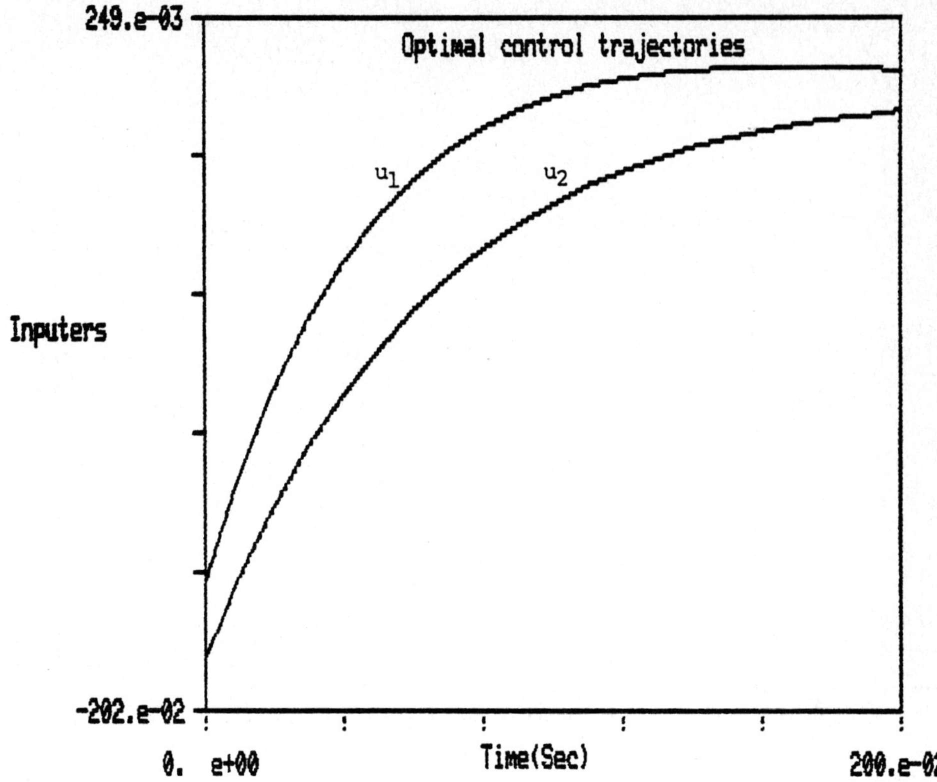

Figure 9. Graphical presentation of the optimal controls in the sample use of REGT

As seen in Figure 10, the sections of **NETWORK. lab** are Filter design and synthesis, mathematics, analysis, sensitivity, and device modeling. A more detailed description of the functions contained within these sections is given below:

(i) **Filter Design and Synthesis**

BUTR: Design and synthesize a Butterworth filter
CHEB: Design and synthesize a Chebychev filter
ELPT: Design and synthesize an Elliptic filter

(ii) **Analysis - Linear**

AC: Performs AC analysis
DC: Performs DC analysis
BODE: Plots the phase and magnitude frequency response
TRAN: Performs transient analysis

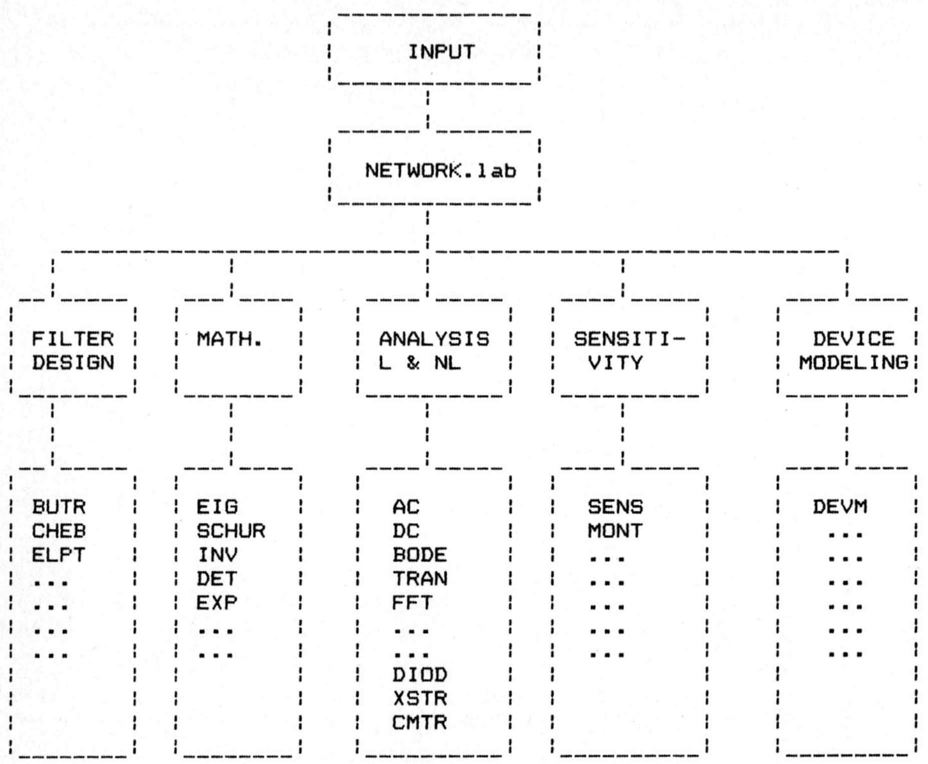

Figure 10. Structure of **NETWORK. lab**

(iii) **Analysis - Nonlinear**

DIOD: Allows the analysis of circuits with 1 or more diodes
XSTR: Allows the analysis of circuits with 1 or more large signal model transistors
CMTR: Allows the analysis of circuits with 1 or more charge controlled transistors

(iv) **Sensitivity**

SENS: Performs a statistical (Gaussian) analysis
MDNT: Performs a Monte Carlo analysis

(v) **Device Modeling**

DEVM: This function allows the construction of non standard circuit elements

4. FUTURE EFFORTS

One of the areas where some efforts have been made is the design of a CAD package for nonlinear control systems. To this end, the Laboratory is working on a nonlinear control package, called **NONLIN-ctr.** The primary object of this package is to be able to design and analyze nonlinear control systems with

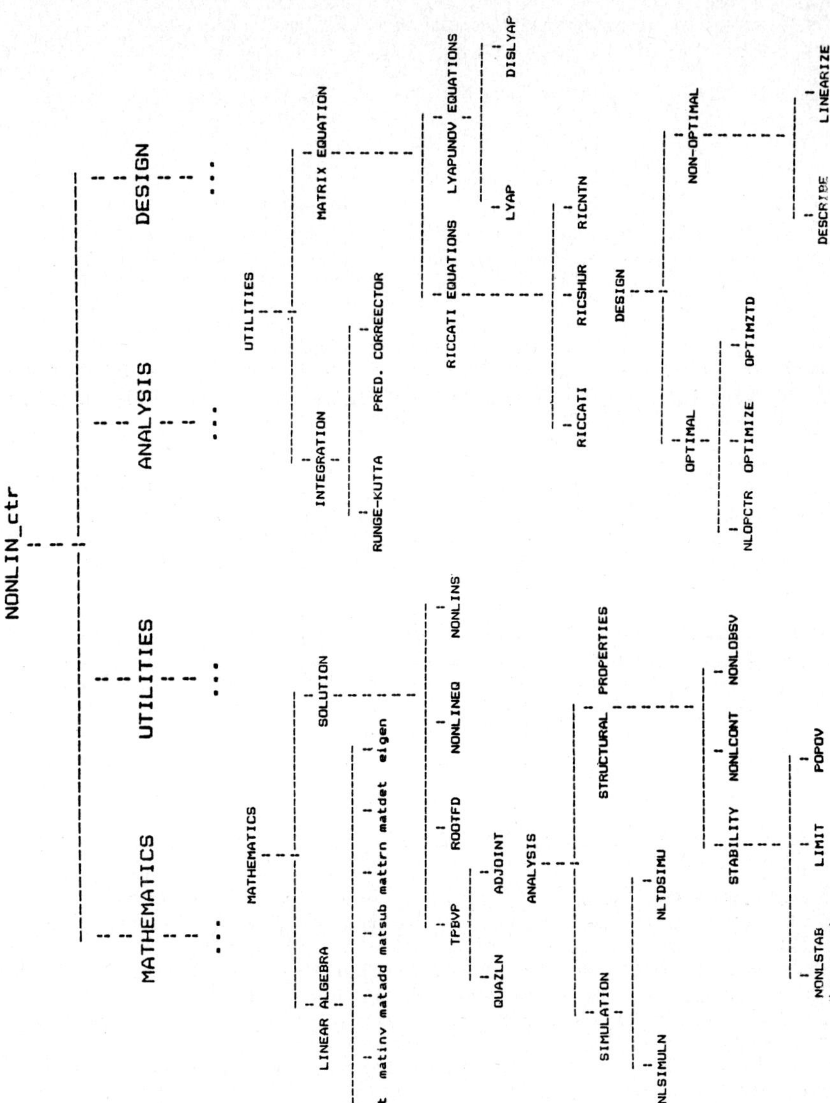

Figure 11. A perspective tree structure for **NONLIN-Ctr.**

continuous or discontinuous nonlinearities and delays in state or control. The initial structure of this package was put together in a recent report [8] by the first author. The preliminary structure of NONLIN-ctr. is shown in Figure 11. At the present time about 25% of this package has been written and efforts are underway to complete it. Some of the key programs of **NONLIN-ctr.** as placed in its structure in Figure 11 are briefly defined below:

Mathematics

EIGEN	a general-purpose routine to compute eigenvalues/eigenvectors of any n x n matrix.
ROOTFD	a program to evaluate all the roots of a general polynomial.
NONLINEQ	a program to solve a nonlinear algebraic equation via, e.g. Newton-Raphson method.
NONLINSY	a program to solve a system of simultaneous nonlinear equations i.e. vector form of Newton-Raphson's method.
ADJOINT	a program to solve a system of simultaneous nonlinear differential equations with 2-point boundary conditions via the adjoint method.
QUAZLN	a program to solve a system of simultaneous nonlinear differential equations with 2-point boundary conditions via the quazalinearization method.

Utilities

RICNTN	a routine to solve the algebraic matrix Riccati equation via the Newton's method.
DISLYAP	a routine to solve the discrete-time Lyapunov equation via an iterative method.
LYAP	a routine to solve the continuous-time Lyapunov equation via an iterative method.

Analysis

NLSIMULN	a general-purpose routine to simulate a nonlinear system.
NLTDSIMU	a general-purpose program to simulate a nonlinear time-delay system.
NONLSTAB	a program to check the stability of a nonlinear system around an equivalent point.
NONLCONT	a program to check for the controllability of a nonlinear system.
NONLOBSV	a program to check the observability of a nonlinear system.
POPOV	a program to check the stability of a nonlinear system via the Popov's method.
LIMIT	a program to check the stability of a nonlinear system using the limit cycle.

Design

DESCRIBE	a general-purpose routine to design a nonlinear system via the describing function method.
LINEARIZE	a genral-purpose program to nearly optimize a nonlinear system via a linear controller.
OPTIMIZE	a general-purpose program to optimize a nonlinear system via a TPBV approach.
OPTIMZTD	a general-purpose program to optimize a nonlinear system with time delay via sensitivity methods.

5. CONCLUSIONS

In this paper a brief description of three computer-aided design (CAD) packages and two CAD languages for control systems and electrical networks are presented. The three packages, **FREDOM**, **TIMDOM**, and **LSSPAK** are written for various computers including the IBM Personal Computer and its compatibles. Except for the graphics segments most of these packages' programs should run on the IBM compatibles. Further insights in the algorithm used and source listing of most of **FREDOM-TIMDOM/45** can be found in reference [9]. The two CAD languages -- **CONTROL. lab** and **NETWORK. lab** are designed for larger computers such as DEC 11/750/780 VAX systems under UNIX 4.2 operating system. A version of these languages are also being implemented on the SUN Microsystem workstations which would take advantage of split screens and graphics shells. For future efforts, not only the present software would be continuously improved but also new packages such as **NONLIN-ctr.** -- a CAD package for nonlinear control systems are being developed.

REFERENCES

[1] R.S. Morel, "FREDOM-TIMDOM/45 -- A Descriptive Guide", Tech. Report LCAD - 84-02 Laboratory for CAD Systems/Networks, Electrical & Computer Engineering Department, University of New Mexico, Albuquerque, NM, April 1984.

[2] M. Jamshidi, Large-Scale Systems: Modeling and Control, Elsevier North-Holland, New York, 1983.

[3] C.B. Moler, "MATLAB Users' Guide", Department of Computer Science, University of New Mexico, Albuquerque, NM, June 1981.

[4] J.J. Dongarra, C.B. Moler, J.R. Bunch and G.W. Stewart, "LINPAK Users' Guide", SIAM, Philadelphia, PA, 1979.

[5] EISPAK, "Eigensystem Workshop", Argone National Laboratory, Argone, IL, 1973.

[6] D.M. Etter, J.K. McDowell and L.J. White, "A Multi-Purpose Plot Package for GIGI Color Terminals", Technical Report No. ECE 2-83 NSF-0271-1, Department of Electrical and Computer Engineering, the University of New Mexico, Albuquerque, NM, 1981.

[7] A.J. Laub, "A Schur method for solving algebraic Riccati equations," IEEE Trans. Auto. Contr., Vol. AC-24, pp. 913-921, 1979.

[8] M. Jamshidi, "NONLIN-ctr. ... a Computer-Aided Design Package for Nonlinear Control Systems", Internal Progress Report, Electronics Department, GM Research Laboratories, Warren, MI, August 1984.

[9] M. Jamshidi and M. Malek-Zavarei, "LINEAR CONTROL SYSTEMS --a Computer-Aided Approach", Oxford, England: Pergamon Press, Ltd., 1985.

COMPUTER-AIDED CONTROL SYSTEMS ENGINEERING
M. Jamshidi and C.J. Herget (Editors)

LCAP2 - LINEAR CONTROLS ANALYSIS PROGRAM

Eugene A. Lee

The Aerospace Corporation
El Segundo, California

The computer program LCAP2 (Linear Controls Analysis Program) provides the analyst with the capability to numerically perform classical linear control analysis techniques such as transfer function manipulation, transfer function evaluation, frequency response, root locus, time response and sampled-data transforms. It is able to deal with continuous and sampled-data systems, including multiloop multirate digital systems, using s, z and w transforms. This program has been used extensively for years at the Aerospace Corporation in the analysis and design of satellite and launch vehicle control systems.

PURPOSE

The program LCAP2 was designed to provide the control system analyst with most of the classical analysis tools needed for analyzing complex continuous and sampled-data systems by transform techniques. Primary considerations in the development of this program were ease of use and computational accuracy. A set of transfer function and polynomial operators has been defined in a fashion similar to the instruction set of a computer. Transfer function and polynomial arrays are defined to be referenced with indices so that they may be easily addressed by the operators. The combination of this set of LCAP2 operators and the form of the data structure provides a very flexible and easy to use program.

The LCAP2 program is the successor to LCAP[1] which is a batch program utilizing card inputs. This original version did not have the flexibility to allow the user to easily develop code to automate, for example, a complete stability analysis task beginning with the input of raw data to the generation of the stability plots. This is a very desirable feature in an industrial environment. The batch version of LCAP2 provides this flexibility since each LCAP2 operator is coded as a FORTRAN subroutine. An interactive version of LCAP2 is also available.

There are over seventy LCAP2 operators. A partial list of these operators is given in Table 1.

LANGUAGE

The program is written entirely in FORTRAN with the exception of one subroutine which is written in assembly language. The batch and interactive versions of this program run on the CDC 176 and 835 computers, respectively, and are described in [2] - [4]. Since publication of these reports, a VAX version of Interactive LCAP2 has been developed jointly by Mark Tischler of Ames Research Center, Moffett Field, California, and this author.

DESCRIPTION AND CAPABILITY

The data structure of the program includes (1) s, z and w plane transfer functions designated as $SPTF_i$, i=1,2,..., $ZPTF_i$, i=1,2,... and $WPTF_i$, i=1,2,..., respectively, and (2) polynomials designated as $POLY_i$, i=1,2,... . Operations on these transfer functions or polynomials are specified by references to their indices. For example, to add $SPTF_1$ to $SPTF_2$ and store the results in $SPTF_3$, the LCAP2 operation would be SPADD(3,1,2). In the batch version this operation is specified by the FORTRAN statement CALL SPADD(3,1,2). In the interactive version this operation can be SPADD(3,1,2) or, if the arguments were left out, the program will prompt the user for them.

The transfer functions are represented as ratios of polynomials. The user can load data into a transfer function using either the coefficient or the root form representation. An example is the operator SPLDC(i) which is an s plane load of coefficient data into $SPTF_i$. The program can save both the coefficient and the root form representation for each transfer function so that arithmetic operations can be implemented with greater accuracy. For example, to multiply two transfer functions the product is obtained by collecting the roots of the multiplicand and the multiplier rather than by multiplying the coefficients. Another example is addition where common roots of the denominators, if there are any, are first factored out before the addition and rationalization is carried out.

A typical use of these operators for a simple system would be to reduce a block diagram to a single open or closed loop transfer function using the add, subtract, multiply and divide operators. Then one of the operators used to implement the classical controls analysis techniques can be applied. For example, if $SPTF_i$ is the open loop transfer, then the operator SFREQ(i) can be used to compute the frequency response so that the system can be evaluated.

For complex continuous systems, the above block diagram reduction method may not be practical to apply. Cramer's method for transfer function evaluation could then be used. For a set of Laplace transformed differential equations represented in matrix form by,

$$\underline{M}(s) * \underline{X}(s) = \underline{B}(s) * u(s) \tag{3.1}$$

where $\underline{M}(s)$ is a matrix of polynomial elements, $\underline{X}(s)$ is a vector of system variables and $\underline{B}(s)$ is the vector relating the input u(s) to the set of equations, Cramer's method for computing the transfer function between u(s) and the jth element of $\underline{X}(s)$ is given by

$$\frac{x_j(s)}{u(s)} = \frac{\det \underline{M}_j(s)}{\det \underline{M}(s)} \tag{3.2}$$

where $\underline{M}_j(s)$ is equal to $\underline{M}(s)$ with column j replaced by the vector $\underline{B}(s)$ and $\underline{M}_0(s)$ is defined to be equal to $\underline{M}(s)$. To apply this method the user first enters the data for $\underline{M}(s)$ and $\underline{B}(s)$. Then the operator DTERM(i,j) is used with $i=i_2$ and j=0. This will compute the determinant of $\underline{M}_0(s)$ and store the resultant polynomial into $POLY_{i_2}$. The operator DTERM is called again with $i=i_1$ and $j \neq 0$. This will compute the determinant of $\underline{M}_j(s)$ and store the resultant polynomial into $POLY_{i_1}$. After both the denominator and numerator polynomials have been computed, they can be copied into an s-plane transfer function $SPTF_i$ by using the CPYPS(i,j,k) operator with $j=i_1$ and $k=i_2$.

For sampled-data systems computation of the z and w plane transfer functions from the s plane description are provided by the operators SZXFM(i,j) and SWXFM(i,j), respectively. Both of these transform operators allows the inclusion of time delay and the zero-order hold. For transformation between the z and w plane, the bilinear transformation operators ZWXFM(i,j) and WZXFM(i,j) are available. Analysis of small order systems can be performed in either the z or w plane. The analyst may prefer to perform the analysis in the w plane since this would allow the use of the Bode design techniques. However, if the order of the system is high, the analysis must be performed in the w plane since z plane transfer function coefficients cannot be as accurately represented by the computer.

Several operators are available for use in analyzing multirate sampled-data systems. For the analysis of the fast to slow rate sampler where the ratio of the sampling rates is an integer n, Sklansky's [5] frequency decomposition method can be implemented by the operators ZMRFQ(i,n), WMRFQ(i,n), ZMRXFM(i,j) or WMRXFM(i,j). This decomposition method for the sampler

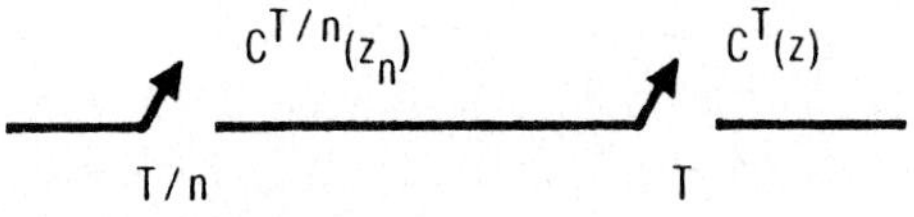

$C^{T/n}(z_n)$ $C^T(z)$

T/n T

Figure 1. Slow to Fast Sampler

expresses the transform of the slower output transform $C^T(z)$ in terms of the faster input transform $C^{T/n}(z_n)$ as;

$$C^T(z) = \frac{1}{n}\sum_{k=0}^{n-1} C^{T/n}\left(z_n e^{j\frac{2\pi k}{n}}\right) \tag{3.3}$$

If $C^{T/n}(z_n)$ is stored in $ZPTF_i$, the ZMRFQ(i,n) operator will compute the frequency response of $C^T(z)$ by numerically evaluating (3.3). The ZMRXFM(i,j) operator will compute the rational representation of $C^T(z)$ in (3.3) given $C^{T/n}(z_n)$. Both of these two types of operators has its advantages and disadvantages. The frequency response form will always yield the correct results but does not provide information on the zeros of $C^T(z)$. The rational form computes $C^T(z)$ using a root finding technique but is subject to inaccuracies for higher order transfer functions. An example using these operators will be given later.

The program includes many features to reduce the amount of data which the user must enter. For frequency responses, automatic frequency selection is available which will choose all the frequency points required to produce a continuous smooth plot without skipping over lightly damped modes. Gain and 180 degree crossover frequencies are also automatically found so that the printout of the response will include the gain and phase margins. For root locus plots, gains are automatically incremented linearly or geometrically with provision for the user to specify some of the values. Automatic scaling for all plot variables is available.

The interactive version of the program provides extensive prompting so that a user's manual is not required. For the experienced user though, the amount of prompting may be excessive. Provisions are made to reduce the amount of prompting by allowing the user to enter some of the parameters directly without the use of prompts.

LIMITATIONS

Data structures for the transfer functions and polynomials require that the order of the polynomials be less than fifty. An arbitrarily large number of transfer functions and polynomials are available to the user since disk storage is utilized when the number becomes too large. The dimension of the matrix data used for transfer function evaluation must not be greater than 30x30 and the polynomial elements must be fourth order or less. Since the determinant of this matrix will be stored in one of the polynomials, the order of the resultant determinant polynomial must be less than fifty.

Although the data structure used for transfer functions limits the use of rational functions to less than fiftieth order, the program has a limited capability to accommodate both higher order transfer functions and nonrational functions through the use of user supplied FORTRAN functions.

AVAILABILITY

The source code and documentation [2] - [4] of this program are available to agencies supporting United States Department of Defense projects and studies. A nominal fee will be charged for reproduction and handling. Requests are to be addressed to: Administrator, Information Processing Division, The Aerospace Corporation, P. O. Box 92957, Los Angeles, California 90009.

The FORTRAN programs have been compiled on the CDC (Control Data Corporation) FORTRAN EXTENDED 4 Compiler. The batch version runs on the CDC 176 using the SCOPE 2.1.5 Operating System. The interactive version runs on the CDC 835 using the CDC INTERCOM Version 5.0 System. Batch jobs typically require 140K-240K words. No overlays are currently used. For the interactive version though, the segment loader is used since only 120K words are available to the user on the CDC 835.

The program will require some modification if it is to be operated on another system. Primary effort will be to replace the non ANSI FORTRAN 4 ENCODE and DECODE statements. A future CDC FORTRAN 5 version of this program will replace these statements with ANSI standard internal write and read statements. A FORTRAN version of the single subroutine written in assembly language is available, although it executes much more slowly.

When the user's manual of the VAX-11 version of Interactive LCAP2 is completed, both the manual and source code will be available on the same basis as the CDC version.

EXAMPLES

Example 1: The FORTRAN code to compute the open loop frequency response of the following sampled data system is to be determined.

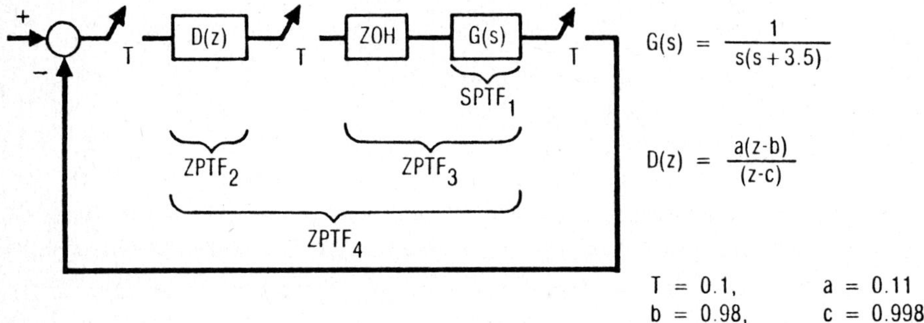

Figure 2. Single Rate Sampled-Data System

The block diagram is labeled with LCAP2 s and z plane transfer functions to be used in the analysis. The FORTRAN code to compute the open loop transfer function, along with the description in quotes, is:

```
POLYN(1)=0              "deg. of num."
POLYN(2)=1.             "coeff. of s**0"
POLYD(1)=2              "deg. of denom."
POLYD(2)=0.             "coeff. of s**0"
POLYD(3)=3.5            "coeff. of s**1"
POLYD(4)=1.             "coeff. of s**2"
CALL SPLDC(1)           "load G(s) into SPTF1"

A=.11
B=.98
C=.998
POLYN(1)=1              "deg. of num."
POLYN(2)=-A*B           "coeff. of z**0"
POLYN(3)=A              "coeff. of z**1"
POLYD(1)=1              "deg. of denom."
POLYD(2)=-C             "coeff. of z**0"
POLYD(3)=1.             "coeff. of z**1"
CALL ZPLDC(2)           "load D(z) into ZPTF2"

SAMPT=.1                "sampling period, T"
ZOH=1                   "≠0 to include ZOH for SZXFM"
DELAY=0                 "time delay for SZXFM"
CALL SZXFM(3,1)         "compute z transform of
                         SPTF1 and store into ZPTF3"

CALL ZPMPY(4,2,3)       "ZPTF4=ZPTF2*ZPTF3, open
                         loop transfer function
                         computed"
```

```
NOMEG=4                  "number of omega values"
OMEGA(1)=.01             "OMEGA(1)=first value for response"
OMEGA(2)=.1              "value used for response"
OMEGA(3)=1.              "value used for response"
OMEGA(4)=10.             "OMEGA(NOMEG)=last value of
                          response"
FBODE=1                  "≠0 for Bode plot"
FNICO=1                  "≠0 for Nichols plot"
RAD=1                    "≠0 for rad/sec, otherwise HZ"
CALL ZFREQ(4)            "compute frequency response of
                          ZPTF4"
```

The range of the frequency values to be used in computing the z plane frequency response is specified in the s plane. The program will perform the conversion to the z plane when evaluating the response. Four values of OMEGA were used for this example to specify the frequency points. Since the program will dynamically choose its own frequency points for computing the response to insure that a continuous smooth plot is produced, it most likely will not select the frequencies of 0.1 and 1.0. The inclusion of frequency points between OMEGA(1) and OMEGA(NOMEG) by the user allows the explicit specification of some of the values to be used in computing the response.

Since user input for the batch version of LCAP2 is FORTRAN code, the user can easily perform tasks such as parametric or sensitivity studies with the use of FORTRAN DO loops, or implementation of algorithms to select design values.

Examples 2: The use of the LCAP2 operators ZMRFQ(i,n), ZVCNG(i,j,n), and ZMRXFM(i,j) will be applied to the analysis of the system in Figure 3.

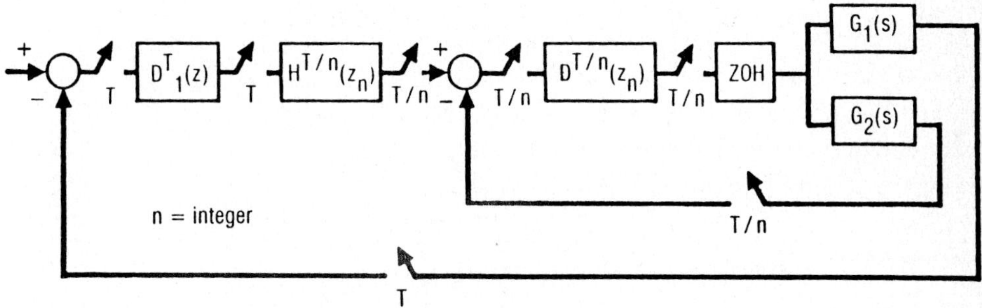

Figure 3. Multirate Multiloop Sampled-Data System

The open loop frequency response of the outer loop with the inner loop closed will be determined. The first part of the analysis is to simplify the block diagram. For the forward loop at the slower sampling rate, the transform

$$D_1^T(z) * H^{T/n}(z_n) \tag{6.1}$$

can be simplified if $D_1^T(z)$ is first expressed in terms of the faster z_n variable so that transfer function multiplication can be applied. The operator ZVCNG(i,j,n) provides this capability. If $D_1^T(z)$ was loaded in $ZPTF_1$ the operator ZVCNG(2,1,n) will transform $D^T_1(z)$ to the faster rate by replacing z with z_n^n and storing it into $ZPTF_2$. If $H^{T/n}(z_n)$ is loaded into $ZPTF_3$, the operator ZPMPY(4,2,3) will compute the desired transform and store the results into $ZPTF_4$. The faster inner loop can be simplified to a single transfer function by first computing the z transform of $G_2(s)$ with the zero order hold and then computing the closed loop transfer function using block diagram reduction. If this closed loop transfer function is stored in $ZPTF_8$, the block diagram can be simplified as

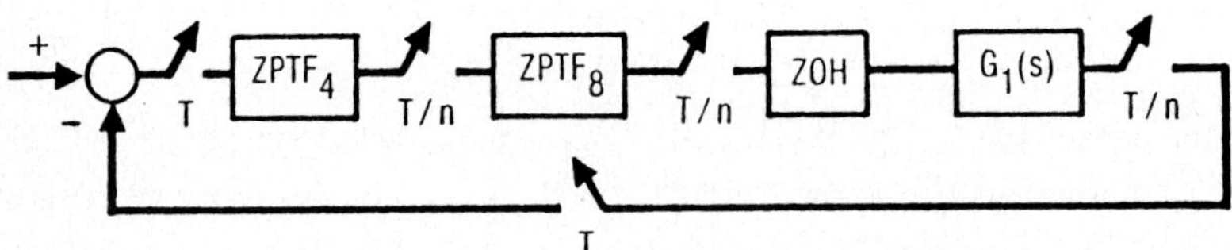

Figure 4. Simplified Block Diagram

The inclusion of the fictitious T/n sampler after $G_1(s)$ does not affect the sampling process. It does, however, simplify the analysis. If the z transform of $G_1(s)$ with the zero order hold is computed and stored, the product of this transfer function with $ZPTF_4$ and $ZPTF_8$ can be computed. If this resultant product is stored in $ZPTF_{10}$ and is also defined as $G_{10}^{T/n}(z_n)$, the desired open loop transfer function is given by

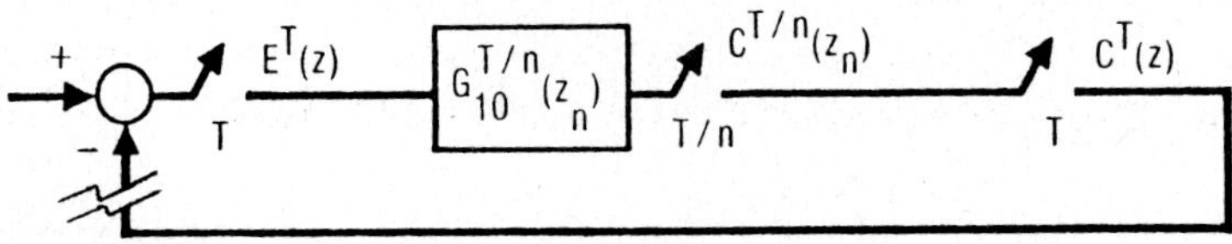

Figure 5. Open Loop Transfer Function

By the frequency decomposition method the output at the slower rate, $C^T(z)$, is given by

$$C^T(z) = \frac{1}{n}\sum_{k=0}^{n-1} C^{T/n}\left(z_n e^{j\frac{2\pi k}{n}}\right) \tag{6.2}$$

The output at the faster rate, $C^{T/n}(z_n)$, is given by

$$C^{T/n}(z_n) = G_{10}^{T/n}(z_n) * E^T(z) \tag{6.3}$$

If $E^T(z)$ is expressed in terms of the faster z_n variable as

$$E^{T/n}(z_n^n) = E^T(z) \tag{6.4}$$

substitution of Eq. (6.3) in Eq. (6.2) yields

$$C^T(z) = \frac{1}{n}\sum_{k=0}^{n-1} G_{10}^{T/n}\left(z_n e^{j\frac{2\pi k}{n}}\right) * E^T(z) \tag{6.5}$$

The open loop transfer function is thus given by

$$\frac{C^T(z)}{E^T(z)} = \frac{1}{n}\sum_{k=0}^{n-1} G_{10}^{T/n}\left(z_n e^{j\frac{2\pi k}{n}}\right) \tag{6.6}$$

Since $G_{10}^{T/n}(z_n)$ is stored in $ZPTF_{10}$, the operator ZMRFQ(10,n) can be used to evaluate the open loop frequency response. An alternate method for obtaining the open loop frequency response is to first compute the rational form of $C^T(z)/E^T(z)$ using the operator ZMRXFM(i,j) with i=11 and j=10. This will transform $ZPTF_{10}$ at the faster sampling rate to $ZPTF_{11}$ at the slower sampling rate. The single rate z plane frequency response operator ZFREQ(11) can then be used to compute the desired open loop frequency response. As discussed previously, the ZMRXFM operator is not very accurate for higher order transfer functions.

TABLE 1. TYPICAL LCAP2 OPERATORS

	OPERATOR	DESCRIPTION
*	SPADD(i,j,k)	S PLANE TRANSFER FUNCTION ADD $SPTF_i = SPTF_j + SPTF_k$
*	SPSUB(i,j,k)	S PLANE TRANSFER FUNCTION SUBTRACT $SPTF_i = SPTF_j - SPTF_k$
*	SPMPY(i,j,k)	S PLANE TRANSFER FUNCTION MULTIPLY $SPTF_i = SPTF_j * SPTF_k$
*	SPDIV(i,j,k)	S PLANE TRANSFER FUNCTION DIVIDE $SPTF_i = SPTF_j / SPTF_k$
*	SPLDC(i)	S PLANE TRANSFER FUNCTION LOAD, COEFFICIENT FORM $SPTF_i$ = POLYN / POLYD
*	SPLDR(i)	S PLANE TRANSFER FUNCTION LOAD, ROOT FORM $SPTF_i$ = ROOTN / ROOTD
*	SFREQ(i)	S PLANE FREQUENCY RESPONSE OF $SPTF_i$
*	SLOCI(i)	S PLANE ROOT LOCUS OF $SPTF_i$
	STIME(i)	INVERSE LAPLACE TRANSFORM AND TIME RESPONSE OF $SPTF_i$
*	ZMRFQ(i,n)	Z PLANE MULTIRATE FREQUENCY RESPONSE (SKLANSKY'S FREQUENCY DECOMPOSITION METHOD) OF $ZPTF_i$
*	ZMRXFM(i,j)	Z PLANE MULTIRATE TRANSFORM BY FREQUENCY DECOMPOSITION $ZPTF_i$ = MULTIRATE TRANSFORM OF $ZPTF_j$
*	SZXFM(i,j)	S TO Z PLANE TRANSFORM $ZPTF_i$ = Z TRANSFORM OF $SPTF_j$
	ZVCNG(i,j,n)	Z TO Z**n TRANSFORM $SPTF_i$ = Z TO Z**n TRANSFORM OF $ZPTF_j$
*	ZWXFM(i,j)	Z TO W PLANE BILINEAR TRANSFORMATION $WPTF_i$ = BILINEAR TRANSFORM OF $ZPTF_j$
	DTERM(i,j)	DETERMINANT EVALUATION FOR TRANSFER FUNCTION EVALUATION BY CRAMER'S METHOD $POLY_i$ = DETERMINANT OF $\underline{M}_j(s)$
*	CPYPS(i,j,k)	COPY POLYNOMIALS INTO S PLANE TRANSFER FUNCTION SPTF(i) = POLY(j) / POLY(k)

* Equivalent Operators Available In The z and w Plane

REFERENCES

[1] Lee, E. A., "Linear Controls Analysis Program (LCAP) User's Guide," The Aerospace Corporation, TOR-0077(2443-23)-1, 5 October 1976.

[2] Lee, E. A., "LCAP2 - Linear Controls Analysis Program, Volume I: Batch LCAP2 User's Guide," The Aerospace Corporation, TR-0084(9975)-1, Vol. I, 15 November 1983.

[3] Lee, E. A., "LCAP2 - Linear Controls Analysis Program, Volume II: Interactive LCAP2 User's Guide," The Aerospace Corporation, TR-0084(9975)-1, Vol. II, 15 November 1983.

[4] Lee, E.A., "LCAP2 - Linear Controls Analysis Program, Volume III: Source Code Description," The Aerospace Corporation, TR-0084(9975)-1, Vol. III, 15 November 1983.

[5] Sklansky, J., "Network Compensation of Error-Sampled Feedback Systems," Ph.D. Dissertation, Dept. of Elect., Eng., Columbia University, New York, New York, 1955.

COMPUTER-AIDED CONTROL SYSTEMS ENGINEERING
M. Jamshidi and C.J. Herget (Editors)

ON THE DEVELOPMENT OF ELECTRICAL ENGINEERING ANALYSIS AND DESIGN SOFTWARE FOR AN ENGINEERING WORKSTATION

Gordon K. F. Lee
Electrical Engineering Department
Colorado State University
Fort Collins, CO 80523

Howard Elliott
Department of Electrical and Computer Engineering
University of Massachusetts
Amherst, MA 01002

A desktop computer workstation provides the user with a valuable tool for interactive analysis and design. As an alternative to the large main-frame computers, microprocessor-based desktops provide a single-user interactive environment which enhances the engineering learning process. Many systems include graphics capabilities including a CRT, light pen or pen plotter. The software support for these desktop computer systems usually is implemented within a prompt-driven or turn-key mode so that the user can work in an interactive friendly environment.

Several analysis and design packages have been developed at Colorado State University which may be incorporated into an engineering workstation environment. These modules provide the design engineer with software capabilities for digital and analog control design, linear systems analysis and digital filter design. This paper shows how some of these modules may be embedded into an interactive engineering workstation environment. Typical graphical outputs and system menu selections are provided to illustrate the interactive friendliness of these software packages.

INTRODUCTION

Computer-aided engineering (CAE), and in particular, computer-aided design (CAD), is already impacting the engineering work environment. However, rapid advances in microelectronics, leading to faster and more economical computing power, seem to indicate that we have only seen the tip of the iceberg in terms of the potential impact which computers can give on the daily work routine of the engineer.

The introduction of computer tools into the working environment may be hindered, however, for a number of reasons. Many institutions rely on central time-share computing facilities to service the user's needs. If an engineer wants to use such a facility, he must overcome a number of obstacles. First, he must familiarize himself with the available hardware and may have to learn the details of using the computer's operating system. Second, he must either invest his own time to develop special software for his needs, or he must search the country for software which is both compatible with his needs and with the hardware which is available at the computing center. Since CAD software usually relies heavily on computer graphics, there is rarely a perfect hardware match and some time must be devoted to conversion and familiarization with the software if this latter route is undertaken; hence, it takes a very dedicated user to undertake such an endeavor.

In order to address these issues of computer-aided analysis and design, the Department of Electrical Engineering at Colorado State University and Hewlett-Packard Desktop Computer Division at Ft. Collins have established a long-term commitment towards developing interactive engineering software for the HP9835, HP9845, HP9826, and HP9836 computers. The goal of this endeavor is to provide the engineer with design and analysis modules within a friendly interactive setting. Graphics, light pen entry and key functions are used to enhance the interactions between user and computer.

Several analysis and design packages have been developed which may be incorporated into an engineering workstation environment. These modules provide the design engineer with software capabilities for analog and digital control analysis and design. This paper discusses the capabilities of these packages and investigates how these modules may be integrated into an engineering workstation. The section entitled "Software Available" provides a brief summary of the software available for linear single input, single output analysis, multivariable systems analysis and digital control design-hybrid simulation. Typical graphical outputs and system menu selection are provided to illustrate the interactive friendliness of the software. In the section on "Development of a Computer-Aided Workstation" the philosophy of the engineering workstation is discussed and possible methods of incorporating the control software packages into the workstation environment are provided.

SOFTWARE AVAILABLE

The heart of a computer-aided engineering workstation is the interactive software modules available to the user. In this section, several interactive software packages are discussed, which are implemented on the Hewlett-Packard 9845 and 9836S desktop computers. These control systems analysis and design tools provide a user-friendly environment. Graphics, prompt-driven structures, and key functions allow one to use his expertise within an interactive mode.

Linear Systems Analysis Package

The Linear Systems Analysis Software contains a set of modules which can be used for analysis and design of single-input, single-output continuous, linear, time-invariant systems. These modules are coordinated by a main supervisory module which also provides data entry and graphical or tabular output.

Upon start-up, the user observes a menu of the software capabilities. This is shown in Figure 1. By pressing an appropriate key function, operations can be performed for data entry or editing, time domain analysis or frequency response generation for the system being investigated. The output can be displayed graphically or in tabular form.

The system to be analyzed is assumed to be represented by a single transfer function or by a set of interconnected transfer functions. The structure of these interconnections must be parallel, cascade or feedback in nature. Furthermore, the numerator and denominator transfer function polynomials must be of degree 19 or less, for this software.

In entering a simple transfer function or a block diagram containing parallel, cascade or feedback interconnections, a binary tree structure is employed. Hence the user transforms a block diagram into a binary tree structure made up of nodes and connecting branches. Each node corresponds to either a single transfer function or a cascade, parallel or feedback interconnection of subsystems.

Upon entering an interconnected transfer function structure, an overall transfer function may be generated. This overall transfer function can be analyzed through time domain (unit step or impulse response) or frequency domain (Bode frequency plots, Nyquist Diagram or Root Locus) techniques.

Graphical outputs may be displayed on a multitude of devices including a CRT, plotter, or thermal printer. Further automatic or self-scaling may be selected to enhance the plots. The lettering feature allows the user to document each graph.

Re-plotting and multi-plotting provides the user with graphics capabilities for analyzing or evaluating several system designs at the same time. Further the digitization feature allows the user to obtain finer detail (specific data points) on the graph through cursor locations.

As an example of the graphics capabilities of the Linear Systems Software package, consider a second-order system given by

$$T(s) = \frac{w_n^2}{s^2+2\delta w_n s+w_n^2}$$

where $w_n = 4$ and δ is chosen as 1, 0.5 and 0.25. It is desired to investigate the effects of the damping coefficient on time and frequency responses.

Figures 2 through 6 illustrate this effect using the multiplot option for the step and impulse responses, Bode plots, Nyquist diagram and root locus for a variable feedforward gain in cascade with T(s) and with a negative feedback structure.

Multivariable Systems Analysis

One may analyze linear, time-invariant multivariable systems using the Multivariable Systems Analysis Software. Upon start-up, the user observes the main menu as illustrated in Figure 7. There are three multivariable system representations from which to select: state-space, transfer matrix or ordinary differential equation form.

Upon pressing the state-space representation key, the state-space menu is shown as in Figure 8. One can build, edit or display the matrices {A,B,C,D} corresponding to the system representation:

$$\dot{\underline{x}}(t) = A\underline{x}(t) + B\underline{u}(t)$$

$$\underline{y}(t) = C\underline{x}(t) + D\underline{u}(t)$$

One feature of the entry or edit keys is the information summary at the upper right corner of the display. In addition to displaying the name and dimension of the matrix being operated on, the position of the element is displayed graphically. The matrix is represented by an array of dots, and the element being operated on by an asterisk.

Once the system matrices have been entered several analysis tools can be invoked to investigate the system. For example, the eigenvalues can be calculated for stability analysis. By pressing the controllability key function, controllability of the system is tested, the controllability index associated with each input is displayed and the controllable canonical form is computed. Similar information is provided for the observability operation.

Time and frequency responses for any input-output pair can be generated. Graphical capabilities similar to the Linear Systems Analysis Sotware are also provided. Furthermore the transfer matrix representation can also be generated from the state-space description.

If one presses the transfer matrix operation key, one observes the menu as shown in Figure 9. Elements of the transfer matrix can be entered or edited in a similar fashion as that of the state-space realization. Furthermore, stability analysis, controllability and observability tests and associated state-space representations, and time and frequency responses can be generated similar to that of the state-space operation.

In the ordinary-differential equation representation the menu shown on the CRT is as illustrated in Figure 10. The component polynomial, $p_{ij}(s)$ and $q_{ij}(s)$ of the polynomial matrices P(s) and Q(s) for the O.D.E. representation:

$$P(s)\underline{Y}(s) = Q(s)\ \underline{U}(s)$$

can be entered or edited in a similar fashion as the state-space operations. When P(s) is row proper and the maximum polynomial degree in each row of Q(s) is less than or equal to the maximum degree in the corresponding row of P(s), the O.D.E. representation is completely analogous to the state-space representation in observable form. Hence to analyze this model, one can first generate this canonical form by pressing the state-space subsystem (operations) using the transfer key.

In addition to the analysis software provided for the three system representations, the Multivariable Systems Software Package provides the user with a set of utility operations. The utility menu is shown in Figure 11. Observing this menu, one can perform basic matrix operations such as addition, multiplication or inversion, find the eigenvalues of a square matrix, find the solution of linear equations, solve a Lyapunov matrix equation, find the steady-state matrix Ricatti equation solution or design state variable feedback gain matrices for pole placement.

Digital Control and Hybrid Simulation

With the advent of low-cost, high speed, flexible microprocessors and the development of sophisticated discrete algorithms, much interest has focused on the design of microprocessor-based digital controllers. The approach in digital control is to substitute a continuous control strategy (in which the input and output values are known continuously) by an equivalent digital computer and interfaces (in which the input and output values are known only at sampled instances of time).

The Digital Control and Hybrid Simulation Software Package provides a single-user, friendly environment for the digital control designer. The interactive, single-user atmosphere assists the control engineer in designing, simulating, and evaluating a digital control system with a minimal amount of effort. The program assumes a linear single-input, single-output plant is to be controlled. Some of the features of the pack include five digital design algorithms, analog to digital conversions of both state-space and transfer function models, transfer function to state-space conversions, analog, digital, and hybrid system simulations and analog and digital frequency response plots.

The objective of the Digital Control and Hybrid Simulation software package is to integrate digital analysis, design and simulation algorithms into one interactive computer-aided environment. This software allows the design engineer to work in both the analog and digital mode; simulation can be done to investigate the use of digital hardware (with finite word length constraints) in controlling analog systems.

The software package is organized in a function key driven, menu oriented configuration. Upon start-up, a menu is displayed as shown in Figure 12. The invocation of each function key transfers the system to one of several operations.

Several function keys are employed for entering the system parameters. A block diagram of the system model is shown in Figure 13. Each block could represent an analog or digital system in state-space or transfer function form. Furthermore, the user could enter one or more of these blocks; the remaining blocks may be computed through the design procedures discussed below.

System entry operations include data entry, data display, editing data, storing data on a mass storage media and loading data from mass storage. Through prompts, the user is asked the appropriate data values when required.

In digital control design, it is desired to control a process such that performance measures on the output are met. Usually this measure uses the error signal between the output and a reference input.

As shown in Figure 13, controllers may be located in the feedforward, feedback or error path of the overall system. This versatility allows one to implement and test many types of control designs.

Digital control design methodologies can be divided into two philosophies: analog invariance methods and direct digital design methods. In the analog invariance approach, one designs an analog controller that satisfies system specifications, then approximates the analog controller by a digital one using some matching criterion.

Although these techniques may work well in many applications, performance of these controllers usually heavily depends on the sampling period selection.

This software package also contains controller design methods based upon the direct digital approach. All algorithms, therefore require the digital representation of the plant to be calculated for the design.

The five design algorithms included in this pack are: step deadbeat, pole-placement (tracking-error feedback), minimum delay deadbeat, weighted input design, and model reference.

Simulation is done in a closed-loop fashion in a analog, digital or hybrid mode. The effect of finite word-length analog to digital converters may also be analyzed in a hybrid system. In order to provide maximum flexibility, the user selects the sampling period, analog integration step size, number of bits and voltage bounds on the A/D's and D/A's, initial and final time values, and the type of input for each simulation.

The simulation outputs may be either graphical or tabular. The graphical outpus allow the user to plot: output signal, superimpose two output simulations, input signal, superimpose two input signals, superimpose the input and output signals, and superimpose reference and output signal. The tabular output provides a listing of the input and the output signals for each simulation.

As seen with the various design methods certain algorithms require specific system representations to complete the design. Thus, analog to digital conversions, state-space to transfer function and transfer function to state-space algorithms have also been implemented.

The analog to digital transfer function transformation techniques incorporated in the package are the Impulse Invariant, Step Invariant, and Bilinear Z transformation.

The frequency analysis algorithms in this pack provide the user with a variety of options. Both analog and digital magnitude and phase analysis are available for the controllers, plant or closed-loop system.

The output may be either graphical or tabular. The graphical output provides the user with: a) Analog plots, b) Digital plots, and c) superposition of analog and digital plots.

The plot may be log or linearly scaled with the limits either user-selected or computed automatically. The tabular output provides the user with a listing of the magnitude and phase of the analysis desired.

As an example to illustrate the interactive design and analysis features of this software package, consider the design of a digital controller for a direct current motor for a tape device.

The sampling period is chosen as one second. The time response of the overall hybrid system should be as close to deadbeat as possible and only one controller in the error feedback path is desired. It is assumed that 8-bit A/D and D/A devices will be used. A 10-volt saturation limitation is imposed for this problem.

A linear model of the magnetic tape drive motor was obtained by experimentally measuring the response at various frequencies. Using the physical properties of the motor dynamics, various transfer functions were constructed and compared to the experimental response curves. From the Bode frequency plots (see Figure 14), a fourth-order model was found sufficient to describe the tape motor system. The transfer function of the plant is:

$$T(s) = \frac{1.66 \times 10^6(s^2 + 281s + 880{,}000)}{(s+1.3)(s+2705)(s^2+520s+1.19\times10^6)}$$

Using this model of the motor system, it is desired to design several digital controllers which satisfy system specifications. For example, using the Digital Control and Hybrid Simulation Software Package, one can choose the step deadbeat and minimum delay deadbeat methods for design. Figures 15-17 show the input and output responses of the analog plant and digital controller.

As can be seen from these plots, the step deadbeat controller gives the best results. In order to investigate the effects of finite wordlength in implementation, the hybrid system can be simulated with the A/D and D/A devices included in the overall system. The simulation results are shown in Figure 18 and a comparison with the infinite precision case is shown in Figure 19. One observes that the finite wordlength constraints result in a different steady-state value than that of the ideal case.

Development of a Computer-Aided Workstation

In this section, the philosophy of an engineering workstation is discussed. Using this approach, the analysis and design software discussed in the previous section can be integrated into an interactive computer aided system which can enforce the design process. The computer-aided signal processing and control experimentation laboratory at Colorado State University is discussed as an example of implementing such a workstation within the academic environment.

In designing a working environment for a control facility, installation requirements based upon user comfort and hardware limitations need to be evaluated. In particular, these requirements may take the form of:

a) an individual private workstation area in which the user can use the computer-aided facilities without distraction and with proper space to lay out schematic drawings or initial sketches;

b) a close proximity to common input/output facilities such as printers, plotters, and disk devices;

c) low costs for maintenance (air conditioning, ventilation, and lighting); and

d) adequate space for potential growth.

In order to satisfy these and other requirements for students at Colorado State University who will use the computer-aided laboratory, the workstation concept was implemented. Applying this philosophy, each user would have available a personal desktop computer with essential peripherals. In the case of housing the software for control systems analysis and design, the essential peripheral for the desktop is the graphic table used for selecting blocktypes for block diagram entry, time or frequency domain response operations, or the various design algorithms one can choose.

Common peripherals which the workstations share include printers, plotters, floppy and hard disc drives, and real-time processing devices.

Upon decision for a workstation philosophy with an enumeration of the equipment desired, the macro-design or overall layout of the computer-aided laboratory can be developed. Figure 20, illustrates the laboratory layout for the Department of Electrical Engineering at Colorado State University. Six workstations, with an HP9836S desktop computer and HP graphics tablet at each mode, comprises the central computer-aided area of operation. Although each workstation has its own memory units internal to the computer, a control coordination unit, the shared resource management system (SRM), is used to tie all of the stations together. Hence common software (such as VLSI layout or hybrid simulation packages requiring much computation) can be shared by workstations if desired. The SRM is housed on a Winchester disc system.

Input-output peripherals in close proximity to each workstation includes a graphics printer and an eight color-pen plotter. Along the walls of the laboratory include workspace for students, the analog computer, real-time processing devices, and more stand-alone systems. Manuals and other reference material are stored in a cabinet near the workstation configuration.

With this six-workstation structure, it was felt that the computer-aided software developed for control systems design could be available for all those who required it; furthermore, the software is presented in a friendly one-to-one interactive environment which enhances results rather than frustrates the design process for the user.

Concluding Remarks

The goal of computer-aided engineering is to provide the user with the opportunity to experiment and evaluate, analyze, and design systems within a friendly interactive environment. In the area of control systems design, a computer-aided workstation can provide this feature.

The essential components of the workstation are the interactive software packages in linear system analysis, multivariable systems and digital control/hybrid simulation. These modules can be incorporated into a user-friendly workstation structure which minimized the user's frustration level by providing a one-to-one tool for the design experience.

```
              HEWLETT-PACKARD LINEAR SYSTEMS ANALYSIS PACKAGE VOL.1

KEY  0: 'Menu'-----KEY DESCRIPTIONS
KEY  1: 'Trebld'---BUILDS A NEW TREE
KEY  2: 'Tabprnt'--PRINTS INFORMATION TABLE FOR A TREE
KEY  3: 'Tredit'---EDITS AN EXISTING TREE
KEY  4: 'Tresave'--SAVES TREE INFORMATION IN A FILE
KEY  5: 'Treload'--LOADS TREE INFORMATION FROM A FILE
KEY  6: 'Trngen'---GENERATES OVERALL TRANSFER FUNCTION FOR A TREE
KEY  7: 'Step'-----GENERATES STEP RESPONSE FOR AN OVERALL TRANSFER FUNCTION
KEY  8: 'Imp'------GENERATES IMPULSE RESPONSE FOR AN OVERALL TRANSFER FUNCTION
KEY  9: 'Bode'-----GENERATES BODE PLOT FOR AN OVERALL TRANSFER FUNCTION
KEY 10: 'Nyqu'-----GENERATES NYQUIST DIAGRAM FOR AN OVERALL TRANSFER FUNCTION
KEY 11: 'Rootlc'---GENERATES ROOT LOCUS FOR AN OVERALL TRANSFER FUNCTION
KEY 12: 'Letter'---ALLOWS MANUAL TITLING OR LABELING OF THE PREVIOUS PLOT
KEY 13: 'Replot'---RE-FORMS PREVIOUS OUTPUT
KEY 14: 'Mltiplot'-SUPERIMPOSES NEW OUTPUT ON MOST RECENT PLOT
KEY 15: 'Prtdev'---SELECTS CRT OR THERMAL PRINTER
KEY S0: 'Dmpgrph'--DUMPS PLOTS ON THE THERMAL PRINTER
NOTE:  S INDICATES SHIFT
```

Figure 1: Linear Systems Analysis Menu

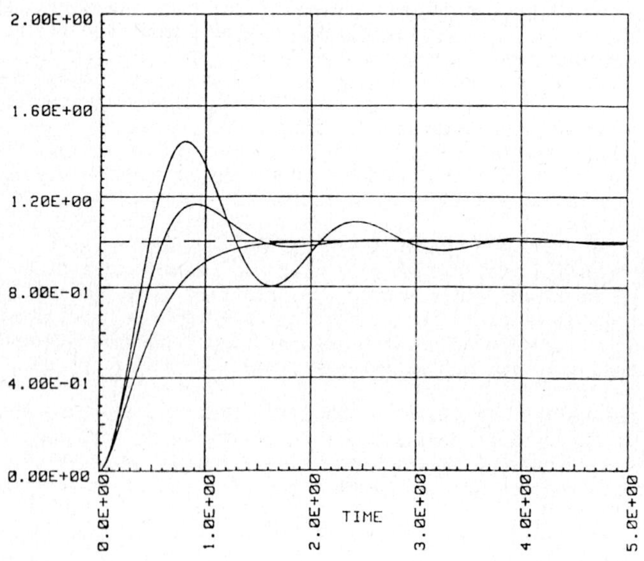

Figure 2: Effects of Damping Coefficients on Step Response

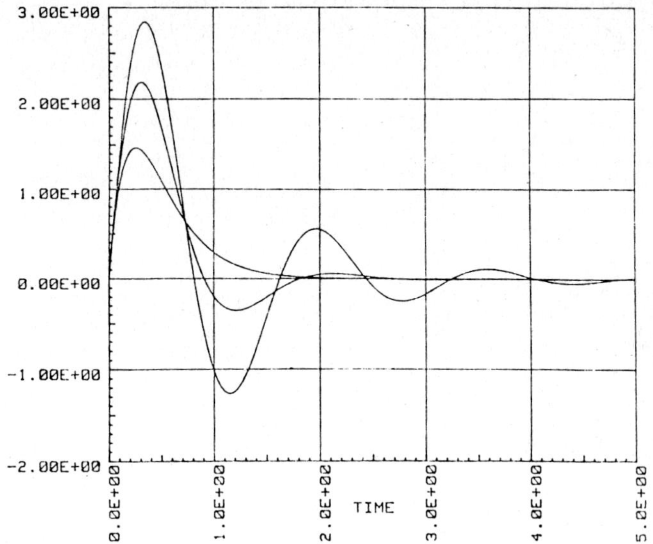

Figure 3: Effects of Damping Coefficients on Impulse Response

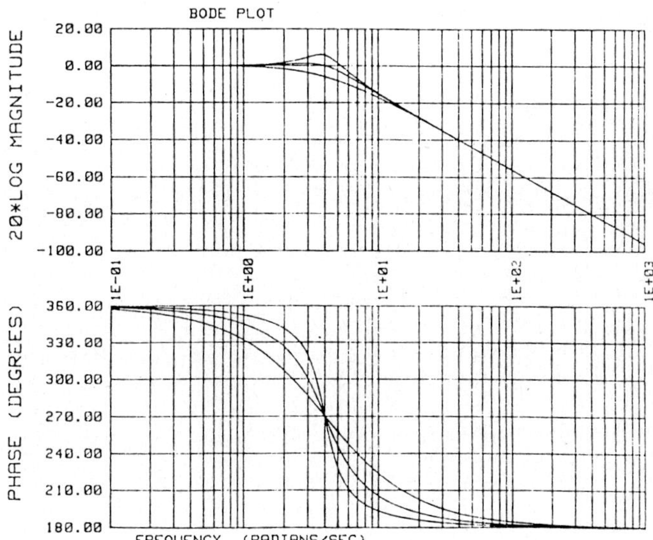

Figure 4: Effects of Damping Coefficients on Bode Plots

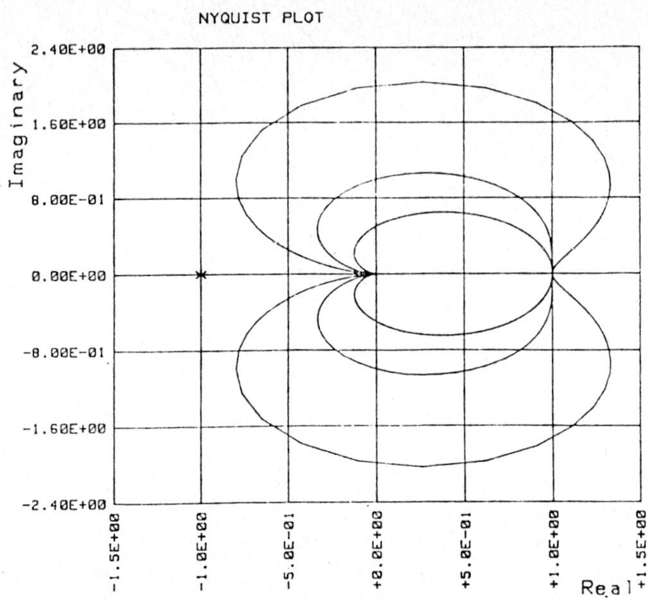

Figure 5: Effects of Damping Coefficients on Nyquist Plots

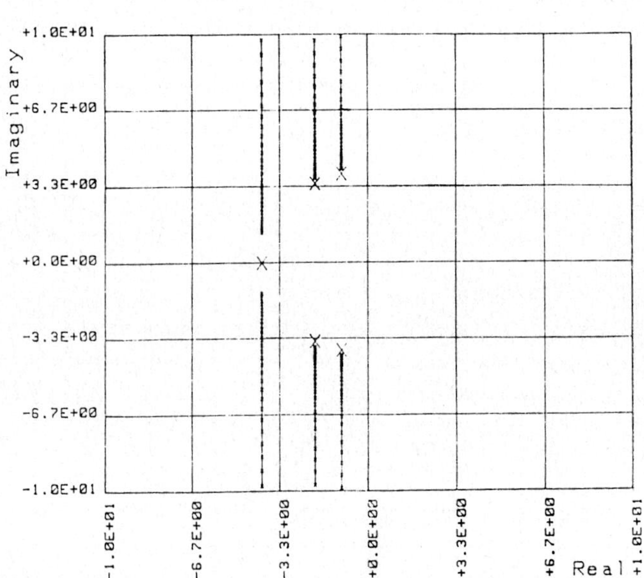

Figure 6: Effects of Damping Coefficients on Root Locus

```
                    Hewlett-Packard
           Multivariable Systems Analysis
                      MAIN MENU


KEY 0  :'MENU'........................Print key definition

KEY 1  :'STATE SPACE '................State Space Subsystem

KEY 2  :'TRANSFER MATRIX'.............Transfer Matrix Subsystem

KEY 3  :'O.D.E.'......................Ordinary Differential Equation Subsystem

KEY 4  :'UTILITIES'...................Matrix Utilities Subsystem

KEY 15:'EXIT'.........................Stop program
```

Figure 7: Multivariable System Analysis Main Menu

```
                    Multivariable Linear System Analysis
                              STATE SPACE MENU
KEY 0  :'MENU'.............Print key definition
KEY 1  :'BUILD'............Enter the matrices A,B,C and E
KEY 2  :'EDIT'.............Modify the matrices A,B,C and E
KEY 3  :'DISPLAY'..........Display the data and the results
KEY 4  :'LOAD'.............Load data previously stored
KEY 5  :'STORE'............Store data
KEY 6  :'STABILITY'........Compute the modes of the system
KEY 7  :'CONTROLLABILITY'...Check controllability and calculate canonical form
KEY 8  :'OBSERVABILITY'.....Check observability and calculate canonical form
KEY 9  :'STEP'.............Step response between specified inputs and outputs
KEY 10:'IMPULSE'...........Impulse response between specified inputs and outputs
KEY 11:'TRANSFER MATRIX'...Compute the transfer function matrix
KEY 12:'BODE'..............Frequency resp. between specified input and output
KEY 13:'REPLOT'............Replot one of the available outputs
KEY 14:'DUMP GRAPHICS'.....Dump on the internal printer the most recent plot
KEY 15:'EXIT'..............Go back to the main menu
KEY S0:'DEVICE'............Input-Output device selection
KEY S1:'LETTER'............Titling or labeling of plots
```

Figure 8: Multivariable System Analysis State Space Menu

```
                    Multivariable Linear System Analysis

                          TRANSFER MATRIX MENU
KEY 0  :'MENU'.............Print key definition
KEY 1  :'BUILD'............Enter the matrix T
KEY 2  :'EDIT'.............Modify the matrix T
KEY 3  :'DISPLAY'..........Display the data and the results
KEY 4  :'LOAD'.............Load data previously stored
KEY 5  :'STORE'............Store data
KEY 6  :'STABILITY'........Compute polynomial roots and state space modes
KEY 7  :'CONTROLLABILITY'...Compute a minimal controllable representation
KEY 8  :'OBSERVABILITY'.....Compute a minimal observable representation
KEY 9  :'STEP'.............Step response between specified inputs and outputs
KEY 10:'IMPULSE'...........Impulse response between specified inputs and outputs
KEY 12:'BODE'..............Frequency resp. between specified input and output
KEY 13:'REPLOT'............Replot one of the available outputs
KEY 14:'DUMP GRAPHICS'.....Dump on the internal printer the most recent plot
KEY 15:'EXIT'..............Go back to the main menu
KEY S0:'DEVICE'............Input-Output device selection
KEY S1:'LETTER'............Titling or labeling of plots
```

Figure 9: Multivariable System Analysis Transfer Matrix Menu

```
Multivariable Linear System Analysis

O.D.E. MENU

KEY 0  :'MENU'..............Print key definition
KEY 1  :'BUILD'.............Enter the matrices P and Q
KEY 2  :'EDIT'..............Modify the matrices P and Q
KEY 3  :'DISPLAY'...........Display the data and the results
KEY 4  :'LOAD'..............Load data previously stored
KEY 5  :'STORE'.............Store data
KEY 7  :'STATE SPACE'.......Compute an observable state space representation
KEY 8  :'XFER'..............Transfer to state space mode if a
                            state space representation has been computed
KEY 15:'EXIT'...............Go back to the main menu
KEY S0:'DEVICE'.............Input-Output device selection
```

Figure 10: Multivariable System Analysis ODE Menu

```
Multivariable Linear System Analysis
UTILITIES MENU

KEY 0  :'MENU'..............Print key definition
KEY 1  :'BUILD'.............Enter the matrices A,B,F,Q and R
KEY 2  :'EDIT'..............Modify the matrices A,B,F,Q and R
KEY 3  :'DISPLAY'...........Display the data and the results
KEY 4  :'LOAD'..............Load data previously stored
KEY 5  :'STORE'.............Store data
KEY 6  :'MATRIX ARITHMETIC'.Perform arithmetic operations on A,B,F,Q,R,X or Y
KEY 7  :'INV(A)'............Compute X=INVERSE(A)
KEY 8  :'INV(A)*B'..........Compute X such that A*X=B
KEY 9  :'EIGEN(A)'..........Compute the eigenvalues of A
KEY 10:'POLE PLACEMENT'.....Compute F such that A+B*F has specified eigenvalues
KEY 11:'EIGEN(A+B*F)'.......Compute the eigenvalues of A+B*F
KEY 12:'RICATTI'............Ricatti equation (given A,B,F(stable guess) and Q,R)
KEY 13:'LYAPUNOV'...........Lyapunov equation (given A and Q)
KEY 15:'EXIT'...............Go back to the main menu
KEY S0:'DEVICE'.............Input-Output device selection
```

Figure 11: Multivariable System Analysis Utilities Menu

```
DIGITAL CONTROL PACK MAIN MENU

KEY 0....KEY DEFINITION
KEY 1....ENTRY-ENTER TF OR ST-SP DATA
KEY 2....DISPLAY-DISPLAY ENTERED DATA
KEY 3....EDIT-EDIT TF OR ST-SP DATA
KEY 4....STORE-STORE DATA ON MASS STORAGE
KEY 5....LOAD-LOAD DATA FROM MASS STORAGE
KEY 6....ANALOG TO DIGITAL
KEY 7....STATE-SPACE/TRANSFER FUNCTION CONV.
KEY 8....DESIGN-DESIGN A DIGITAL CONTROLLER
KEY 9....SIMULATION-STEP,RAMP, OR SINEWAVE
KEY S7...POLE AND ZERO VALUES
KEY S8...FREQUENCY RESPONSE
KEY S9...SIMULATION RESULTS-PLOT OR TABULAR
S INDICATES SHIFT
```

Figure 12: Digital Control and Hybrid Simulation Menu

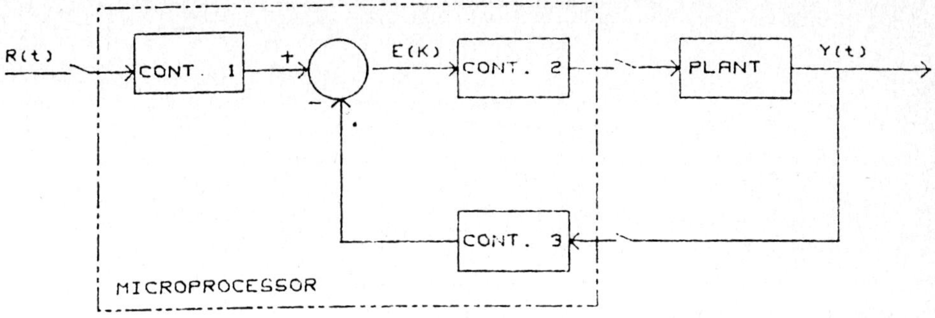

Figure 13: Structure of Control System

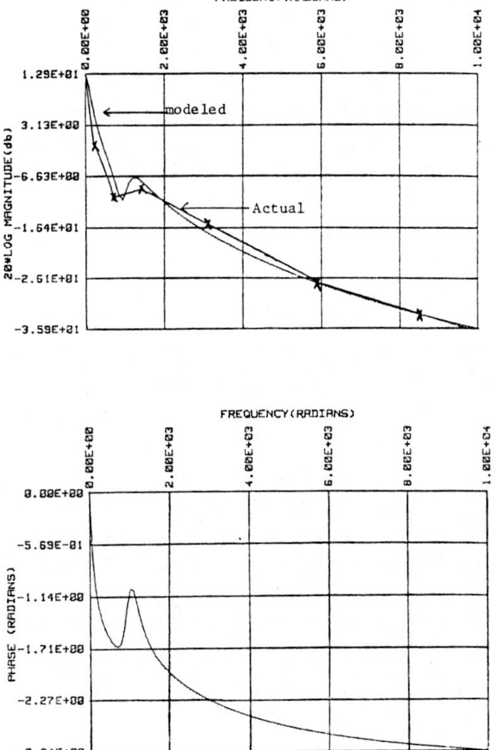

Figure 14: Frequency Response of Actual and Modeled Plant

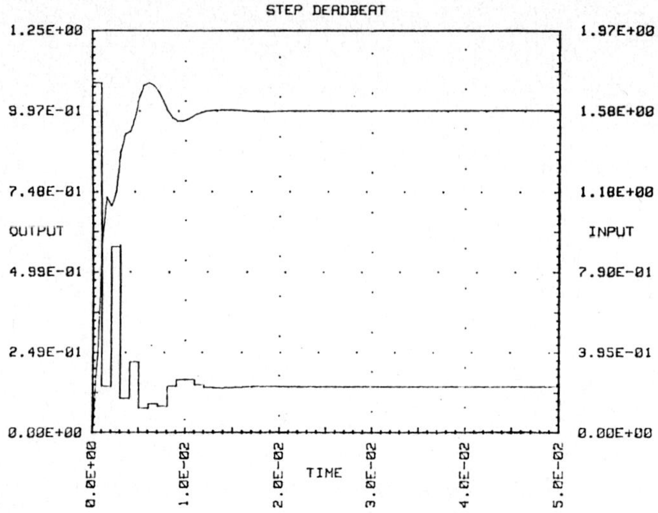

Figure 15: Step Response for Step Deadbeat Design

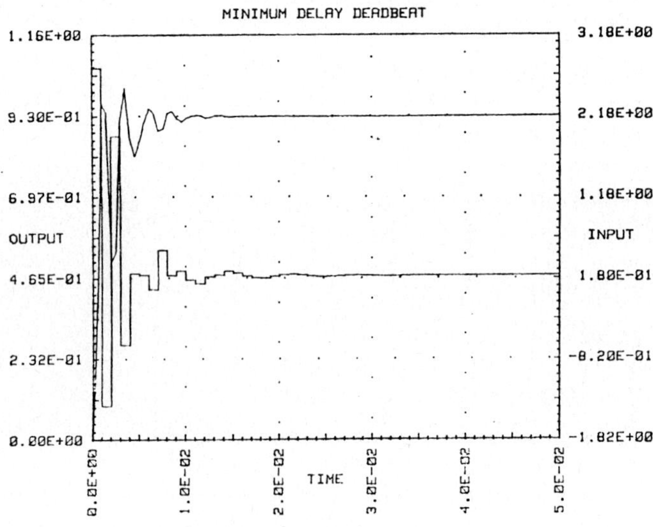

Figure 16: Step Response for Minimum Delay Deadbeat Design

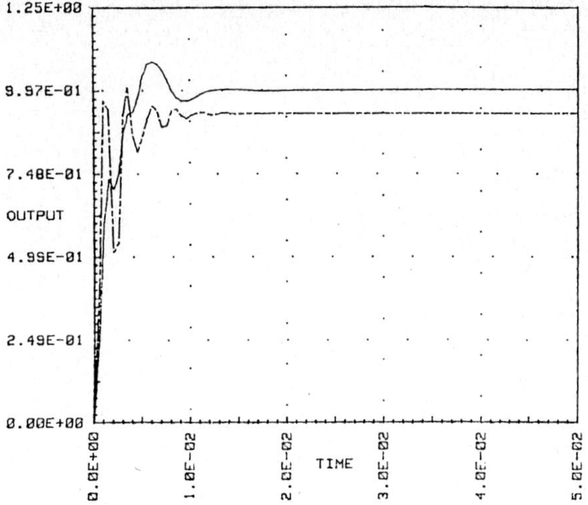

Figure 17: Superposition of Step Responses for the Minimum Delay and Step Deadbeat Designs

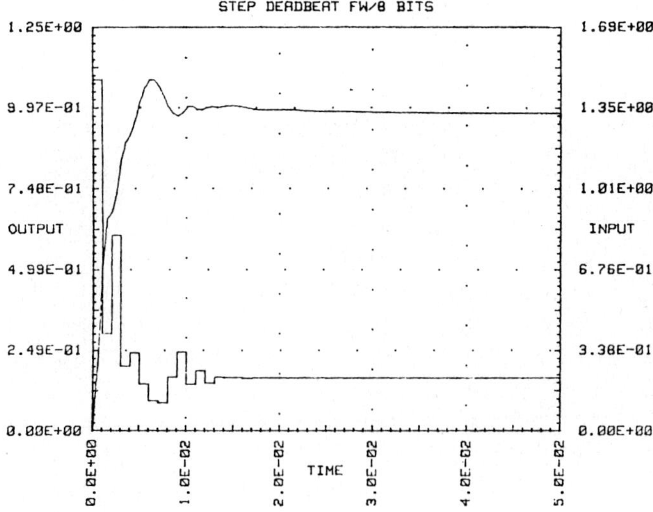

Figure 18: Step Deadbeat Simulation which Includes Finite Wordlength Effects

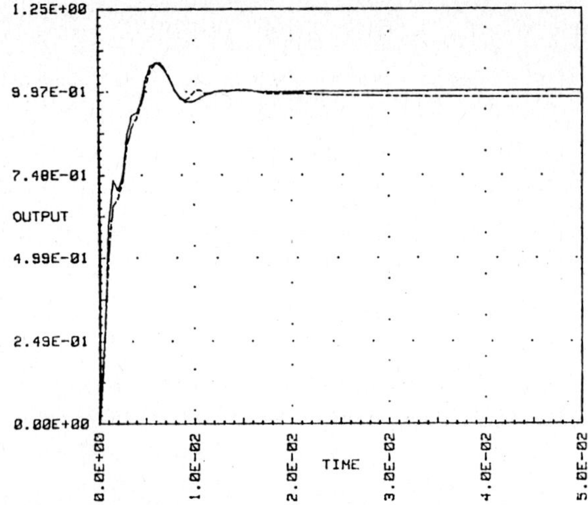

Figure 19: Superposition of Step Responses for the Step Deadbeat Design with and without Finite Wordlength A/D and D/A

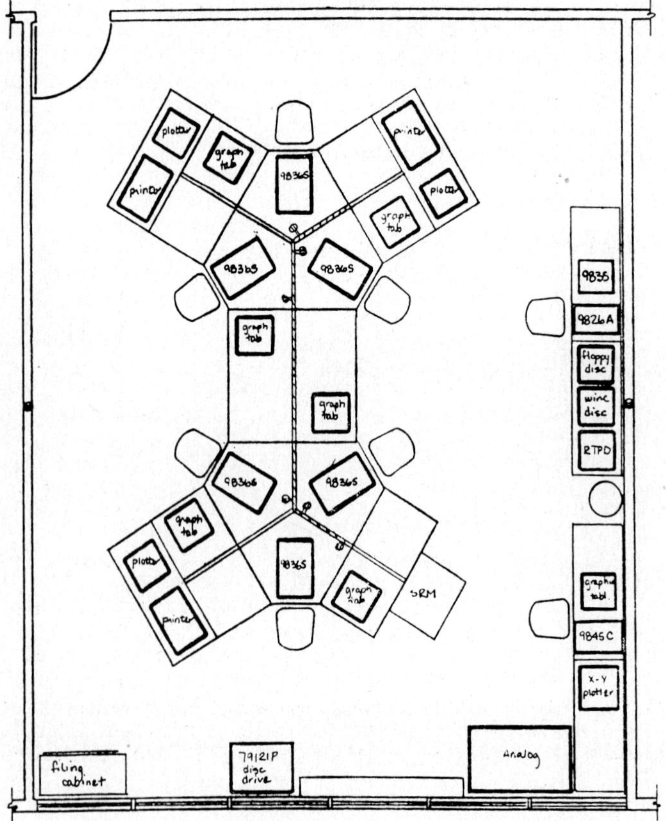

Figure 20: CA-SPACE Laboratory Layout

COMPUTER-AIDED CONTROL SYSTEMS ENGINEERING
M. Jamshidi and C.J. Herget (Editors)
Elsevier Science Publishers B.V. (North-Holland), 1985

CTRL-C AND MATRIX ENVIRONMENTS FOR THE COMPUTER-AIDED DESIGN OF CONTROL SYSTEMS

J.N. Little, A. Emami-Naeini, and S.N. Bangert
Systems Control Technology, Inc., 1801 Page Mill Rd., Palo Alto, CA 94303, USA

Abstract

A computer-aided control system design package, called CTRL-C, provides a matrix workbench for the analysis and design of multivariable systems. CTRL-C is an interactive environment with a comprehensive set of tools for analysis, identification, design, and evaluation.

A unified software system is possible for matrix analysis, engineering graphics, control system design, and digital signal processing. A common thread in these disciplines is the role of a single data object: the complex matrix. CTRL-C demonstrates that a matrix environment can lead to a powerful, natural, and extensible software system.

1.0 INTRODUCTION

A Workbench is a collection of tools (a "toolbox") and a suitable environment in which to perform a job. Several computer workbenches have been available for some time under operating systems like Unix. Professional writers have a writer's workbench. The writer's workbench is a collection of tools that include editors, spelling checkers, grammar critiquing and document preparation. Professional programmers have a programmer's workbench. The programmer's workbench provides editors, beautifiers, verifiers, timing analyzers, and source code control systems. Both of these workbenches exist in an environment (Unix) that provides excellent file handling and text manipulation capabilities. Inspired by these two examples, CTRL-C is intended to be a control designer's workbench.

A useful workbench for a control designer should not be limited to control design. His workbench should encompass other important, related fields. The fields considered by CTRL-C include: Matrix Analysis; Engineering Graphics; Control System Design and Analysis; and Digital Signal Processing.

Historically, separate, stand-alone programs have been employed within each of these disciplines. A unified approach to these disciplines is possible, however, based upon a simple observation: matrices are important objects in all four fields. That is, a single data type, a rectangular matrix with complex elements, can be used to represent the important objects in each of these fields. Scalars, when needed, are simply represented as 1-by-1 matrices, while 1-by-n and m-by-1 matrices represent row and column vectors.

The principle goals of this paper are: (1) to demonstrate the CTRL-C system and (2) to show how the use of a matrix environment can lead naturally to a unified, versatile software system. Section 2 of this paper describes the fundamental principles and concepts that were used in the development of the CTRL-C interactive environment. Sections 3-6 show the use of CTRL-C for each of: (1) matrix analysis, (2) engineering graphics, (3) control design and analysis, and (4) digital signal processing. Section 7 describes extensibility concepts in CTRL-C, and the use of CTRL-C as a programming language. Section 8 concludes with a description of some of the important numerical algorithms.

2.0 PRINCIPLES

In the 1970s, the first computer-aided control system design environments emerged. These early packages were most often menu-driven or of the question/ answer dialog variety. Recognizing the limitations of these primitive environments, more advanced command-driven environments have been developed. Unfortunately, most of these are specialized and limited, often utilizing complex and arbitrary data structures. The data structures are not generally transportable between programs, nor are they usually understood by the casual user. The result is that most environments are not extensible, they do what they are designed to do, and very little more.

To try and overcome these difficulties, related computer science fields have been examined. The result is a set of four principles upon which the CTRL-C interactive environment is based: easy matrix manipulation; uniform file handling; direct manipulation; and extensibility.

Master the Matrix

Since the matrix is an important data object, then an appropriate interactive environment should be one where matrices are treated naturally. After learning a few simple concepts for matrix manipulation, the user becomes able to work throughout four disciplines. It is not necessary to learn four separate environments. The idea that matrix manipulation environments are powerful is not new; the small but dedicated group of APL users have been saying so since the 1960s. For numerous reasons, however, APL has not been widely popular within the four disciplines. One reason is that APL does not use the standard ASCII character set. Another, more serious reason is that APL code is often too concise and subtle. For these reasons, APL has been referred to as a "write-only" language.

A matrix program called MATLAB offers an alternative to APL. CTRL-C is based on MATLAB, a program which was originally developed by Cleve Moler of the University of New Mexico [1]. MATLAB was written as a convenient "laboratory" for computations involving matrices. Applying some of the concepts of SPEAKEASY to APL resulted in an environment where the only primitive data object is the complex array. It is command driven; that is single line commands are accepted from the user, processed immediately, and the result displayed.

Uniform File Handling

In the CTRL-C environment, all variables are stored in a large stack. This stack resides in semiconductor memory (or managed virtual memory). It is necessary, however, to allow data communication between this stack and disk files. It is important for the utility of the system that file manipulation commands be powerful yet simple. To provide a uniform user interface, all commands that read or write disk files use the Unix-like notation [2] of left and right angle bracket symbols < and >, and the hyphen "-" for switches. Roughly translated, the brackets mean "get input from" and "send output to", respectively. Thus, file operations, which are cumbersome in a pure matrix environment, can be accomplished using commands that excel at file operation.

Direct Manipulation

A principle described as direct manipulation [3] has been used to characterize traits often associated with popular software. It has been observed that some systems evoke "glowing enthusiasm" from their users, while others result in "grudging acceptance or outright hostility". The good systems usually are easy to learn, inspire confidence in their use, install an eagerness to teach others, and develop a desire to explore. Examples of these types of systems include display editors (e.g. EMACS, EDT, VI, FSE, WORDSTAR), spreadsheet programs (VISCALC, 1-2-3), and certain operating systems or languages (UNIX, LOGO). Many of these systems are aptly described with the expression "what you see is what you get". For all of these systems, the user is able to apply intellect directly to the task; the tool seems to disappear.

A feeling of direct manipulation is found in CTRL-C. For systems involving matrices, "what you see is what you get", that is, matrix algebra is performed naturally.

Extensibility

Certain computer languages inspire a unique view of programming. In the traditional FORTRAN sense, programming consists of writing a main program, and then writing subroutines. In other languages, including LISP, LOGO, and FORTH, there is a subtle, but important difference in the approach to programming. The user thinks of programming as consisting of creating new "words" in the language. Once a new word is created, it is used the same way as a permanent word. This principle can be described as extensibility of the environment. In CTRL-C this is achieved through the Define Function capability.

With these concepts in mind, the next three sections demonstrate the CTRL-C system for (1) matrix analysis, (2) engineering graphics, 3) control system design and analysis, and (4) digital signal processing. The examples are intended to show how the use of a matrix environment can lead naturally to a useful and simple interaction with the computer.

3.0 MATRIX ANALYSIS

CTRL-C has a natural matrix environment. To enter a matrix, a simple list is used. The list is surrounded by brackets, '[' and ']', and uses the semicolon ';' to indicate the ends of the rows. For example, the input line
a = [1 5 9 13; 2 6 11 14; 3 7 11 16; 4 8 12 18] results in the output

```
A =
        1.    5.    9.   13.
        2.    6.   11.   14.
        3.    7.   11.   16.
        4.    8.   12.   18.
```

The matrix A will be saved for later use.

In CTRL-C, matrix algebra is easy -- it is accomplished the way it is normally written on the back of an envelope. For example, the matrix transpose is obtained as b = a' which results in

```
B =
        1.    2.    3.    4.
        5.    6.    7.    8.
        9.   11.   11.   12.
       13.   14.   16.   18.
```

Matrix multiplication is obtained by typing c = a * b which produces

```
C =
      276.  313.  345.  386.
      313.  357.  393.  440.
      345.  393.  435.  488.
      386.  440.  488.  548.
```

Simple matrix functions are easily obtained, for example, the determinant is found by typing det(a) which results in ANS = 4.0000.

A complete set of common matrix functions is available in CTRL-C. Largely inherited from MATLAB, they represent the basic tools for matrix analysis. Typical functions include:

```
eig(x)    - eigenvalues and eigenvectors
geig(a,b) - generalized eigenvalues
exp(x)    - matrix exponential
inv(x)    - inverse
svd(x)    - singular value decomposition
schur(x)  - schur decomposition
```

Polynomials can be represented in a matrix environment as row vectors containing the coefficients ordered by descending powers.

Polynomial multiplication may be accomplished using convolution. If A and B are polynomials, then Y = CONV(A,B) calculates the polynomial product. For example, typing

```
[> a = [1 2 1];   b = [1 2];
[> c = conv(a,b)                  yields the polynomial product

C =  1.    4.    5.    2.
```

Polynomial division, root finding, and other polynomial operations are similarly accomplished.

In summary, a matrix environment allows matrix algebra operations to be written directly, with no cumbersome syntax. Dimensioning of variables is accomplished automatically by the software. Polynomials can be represented in a matrix environment and polynomial arithmetic is performed readily.

4.0 ENGINEERING GRAPHICS

Graphics abilities are a requirement for a useful computer-aided engineering package. Rather than being an afterthought, as with many CAD packages, the graphics facilities in CTRL-C are useful in their own right as a standalone system. Data are graphed using the same natural syntax with which matrices are manipulated.

Engineering X-Y plots are created with separate commands for data plotting, titling and labeling. For example, a sine curve might be generated, plotted, and titled with

```
[> t = 0:.05:4*pi;
[> y = sin(t);
[> plot(t,y)
[> title('sine(t)')
```

The first statement generates a vector consisting of elements running from 0.0 to 4pi in increments of 0.05. The second statement creates a vector y containing the sine of each of the elements of t. The third statement plots y versus t and, together with the fourth statement, results in Figure 1.

Three-dimensional surface plots can be useful to "look at" large matrices. An intuitive understanding of the structure of a matrix can often be found that is not clear from just looking at numbers. For example, the state dynamics matrix of a 59th order aircraft model is too large to display conveniently on a CRT screen. The command p3d(a) produces the 3-dimensional plot of Figure 2, where the value of each element represents the height Z

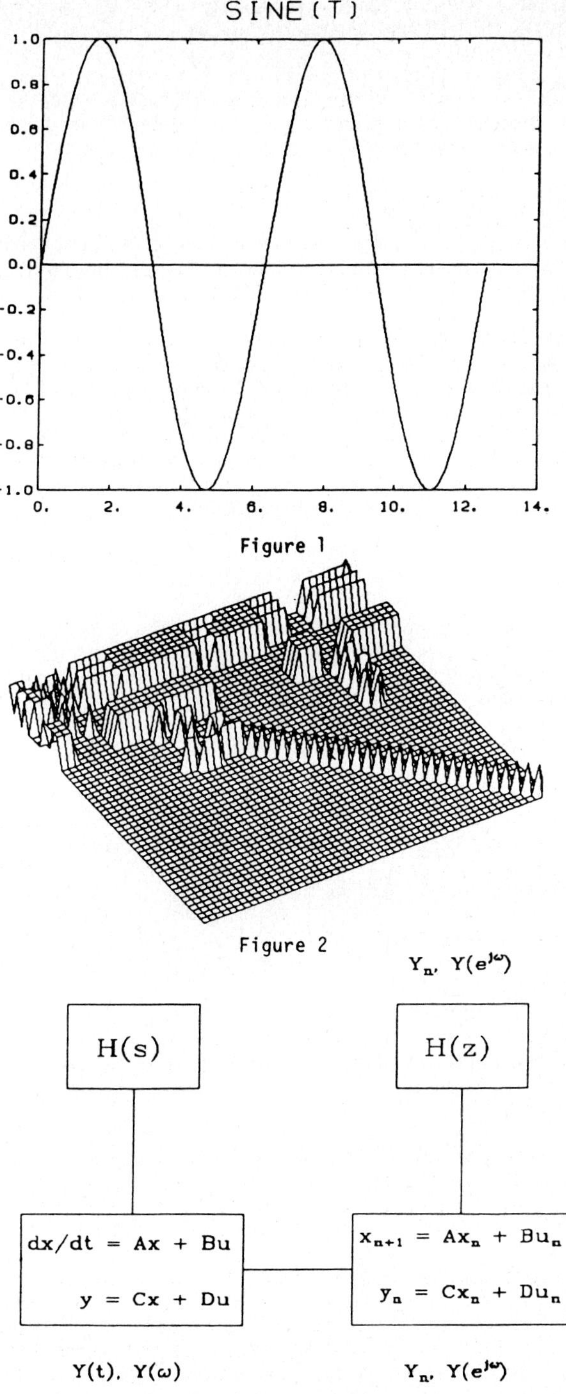

Figure 1

Figure 2

Figure 3

above the X-Y plane. This yields a perspective on the matrix structure not evident from looking at 3600 numbers.

In summary, some simple plots have been created. Other commands and options are available in CTRL-C for log-log plots, overplots, axis labeling and other basic graphics functions. It is demonstrated that engineering graphics can be a natural extension to a matrix environment.

5.0 CONTROL SYSTEMS

Matrix environments are particularly convenient for working with linear systems that can be represented in state-space form. Systems may be described in discrete-time or in continuous-time. Systems may also be described in polynomial notation as a Laplace transfer function for continuous time, or as a Z-transform transfer function for discrete time.

In CTRL-C, transformations between these representations are provided, as well as tools for the calculation of time and frequency domain measures. Other primitives implement various control design algorithms. Pictorially, the system representations in CTRL-C are shown in Figure 3. The remainder of this section consists of two simple examples, each selected to illustrate basic concepts of the use of matrix environments and CTRL-C for control design and analysis.

Example 1

The first example demonstrates the input of a system described by a Laplace transfer function, the conversion to state-space, the calculation of time and frequency responses, and finally the design of a simple controller. Consider the system described by a simple Laplace transfer function in Figure 4. To describe this system in CTRL-C, the numerator and denominator coefficients for the first block are entered:

```
[> num = [1 2]; [> den = [1 .4 1]
```

If it is desired to find the pole locations, ROOT is used. Typing:

```
[> dr = root(den)     results in

DR =

   -0.2000 + 0.9798i
   -0.2000 - 0.9798i
```

The natural frequency and damping factor are easily found:

```
[> Wn = abs(dr)

WN    =

   1.0000
   1.0000

[> Zeta = cos(imag(log(dr)))

ZETA  =

  0.2000
  0.2000
```

These commands show some of the power of complex arithmetic in a matrix environment. To cascade the second block of Figure 4, the denominator term is formed for the new block:

```
[> den2 = [1 1.96]
```

The series connection is achieved by polynomial multiplication (convolution) of the two denominators. Typing:

```
[> den  = conv(den,den2)     results in

DEN =

   1.0000    2.3600    1.7840    1.9600
```

This combined system can be transformed to state-space using the transfer function to state-space primitive. Typing

```
[> [a,b,c,d] = tf2ss(num,den)
```

results in the controller canonical form description:

```
D =     0

C =     0.    1.    2.

B =     1.
        0.
        0.

A =

      -2.3600    -1.7840    -1.9600
       1.0000     0.0000     0.0000
       0.0000     1.0000     0.0000
```

With the system in state-space, a variety of common time and frequency domain measures can be calculated. The first step in the calculation of a time response is to define the time base. In CTRL-C, this is done using the colon ":" operator. The command

```
[> t=0:.1:10;
```

creates a vector of points from 0.0 to 10.0 seconds in increments of 100 ms. Impulse and step responses are found by typing:

```
[> yi = impulse(a,b,c,1,t);
[> ys = step(a,b,c,d,1,t);
```

Matrices yi and ys now contain output time histories. The rows correspond to the rows of the output vector, while the columns correspond to the successive time points from 0.0 to 10.0 seconds. The two responses are graphed by typing

```
[> plot(t,yi,t,ys)        which results in Figure 5.
```

Frequency response measures are calculated in a manner similar to time response functions. First, a frequency vector is formed. The function LOGSPACE is provided to create a vector with points evenly spaced in frequency between two decades. Typing

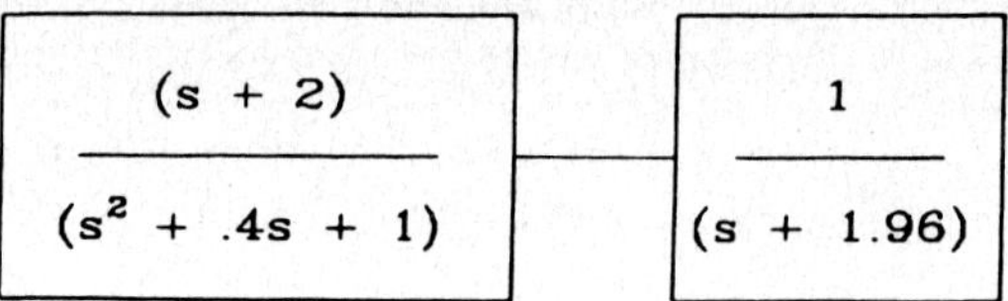

Figure 4

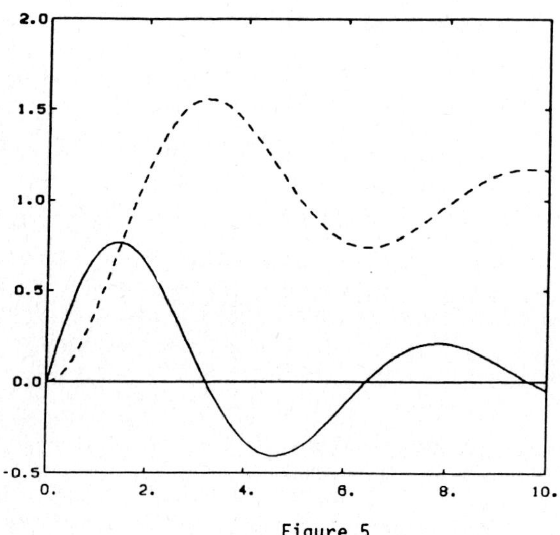

Figure 5

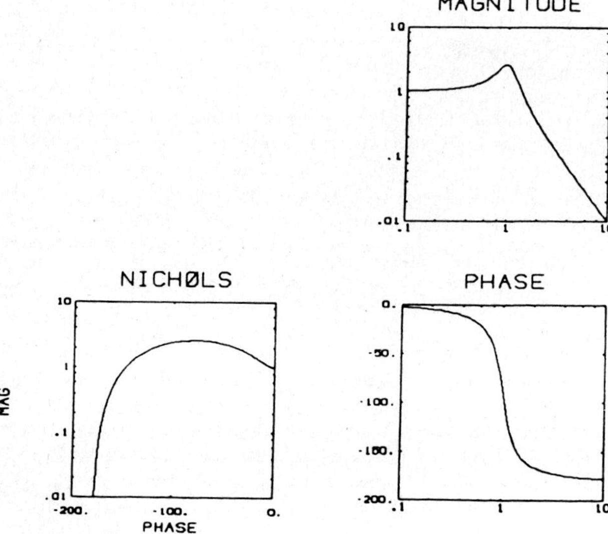

Figure 6

```
[> w=logspace(-1,1);
[> [mag,phas]=bode(a,b,c,d,1,w);
```

creates matrices mag and phas containing the magnitude and phase responses at the frequencies in vector w. This magnitude response is plotted on log-log scales and titled in the upper right corner of the screen with the commands

```
[> window('222')
[> plot(w,mag,'loglog')
[> title('magnitude')
```

Similar commands plot the phase and Nichols responses, resulting in Figure 6.

The pole placement formula of Ackerman allows arbitrary pole placement for single input systems. In CTRL-C the primitive PLACE is used on a vector P containing the desired pole locations, to calculate the gain vector K:

```
[> p = 3 * [-1; (-1 + i); (-1 - i)];
[> k = place(a,b,p)

K =

   6.6400    34.2160    52.0400
```

We can check the closed loop eigenvalues:

```
[> e = eig(a - b*k)

E =
    -3.0000 + 3.0000i
    -3.0000 - 3.0000i
    -3.0000 + 0.0000i
```

and indeed they are at the prescribed locations.

The reference feedforward matrix N is calculated to provide unity DC gain

```
[> n = 1/(d-(c-d*k)/(a-b*k)*b)

N =
    27.0000
```

The closed loop system matrices are built within CTRL-C:

```
[> Ap = a-b*k;   Bp = b*n;   Cp = c-d*k;   Dp = d*n;
```

and the closed loop impulse and step responses are found

```
[> yi = impulse(Ap,Bp,Cp,1,t);
[> ys = step(Ap,Bp,Cp,Dp,1,t);
[> plot(t,yi,t,ys)
```

which results in Figure 7.

Optimal Control Solutions

The standard linear quadratic regulator problem is solved with the CTRL-C function LQR. For the linear system described by:

$$\dot{x} = Ax + Bu$$

Typing [> k = lqr(a,b,q,r)
finds the gain matrix K such that the control law u = -Kx minimizes the quadratic cost function:

$$J = 1/2 \int [x'u'] \begin{bmatrix} Q & N \\ N' & R \end{bmatrix} \begin{bmatrix} x \\ u \end{bmatrix} dt$$

The LQR function is a good example of how optional arguments and calling sequences provide flexibility within CTRL-C. For example, the cross weighting term N often is not needed, so it is an optional input argument. It can be included as: K = LQR(A,B,Q,R,N). The Riccati solution matrix S is an optional output argument. It is obtained with [K,S] = LQR(A,B,Q,R).

A method is provided to switch between different algorithms. Typing LQR('qz') switches to the QZ algorithm. Typing LQR('qr') changes back to the default QR algorithm.

Primitives to solve the optimal estimator problem, the discrete time problem, and the implicit model following and output weighting formulations are also available.

In summary, two examples have been shown that introduce some of the CTRL-C control design and analysis primitives. The state-space representation of systems lends itself naturally to a matrix environment. CTRL-C provides primitives to convert to and from other representations, and to perform common analysis and design tasks. Many control design methodologies are possible using a combination of the matrix primitives and the control primitives. The matrix environment results in a very simple dialogue with the computer.

6.0 DIGITAL SIGNAL PROCESSING

Digital signal processing (DSP) is concerned with the representation and processing of signals that are represented by sequences of numbers. The purpose of processing a signal, in general, can be to identify some model or model parameters that characterize the signal. It can also be to enhance a signal or to remove undesirable components of the signal.

A matrix environment is ideal for the development and use of signal processing techniques. Vectors are used to represent arbitrary sampled-data signals. The natural mathematical interaction with vectors provided in a matrix environment makes it very convenient to process and manipulate sampled data sequences. Primitives for filtering, FFT analysis, identification, and other digital signal processing calculations become very conversational using the complex (Real + Imaginary) vector manipulation concepts.

Consider the implementation of a simple digital filter. The difference equation for a general causal linear time-invariant (LTI) digital filter is given by

```
y(n) = b(1)*x(n) + b(2)*x(n-1) + ... + b(nb)*x(n-nb+1)
                 - a(2)*y(n-1) - ... - a(na)*y(n-na+1)
```

where x is the input signal, y is the output signal, and the constants b(i), i=1,2,3,...,nb , a(i), i=1,2,3,...,na are the filter coefficients. In CTRL-C, if the numerator and denominator filter coefficients are contained in vectors B and A,

```
B = [0.1042   0.2083   0.1042];  A = [1.0000   -1.1430   0.5596];
```

then a data sequence x is filtered with a "tapped delay-line" filter by typing

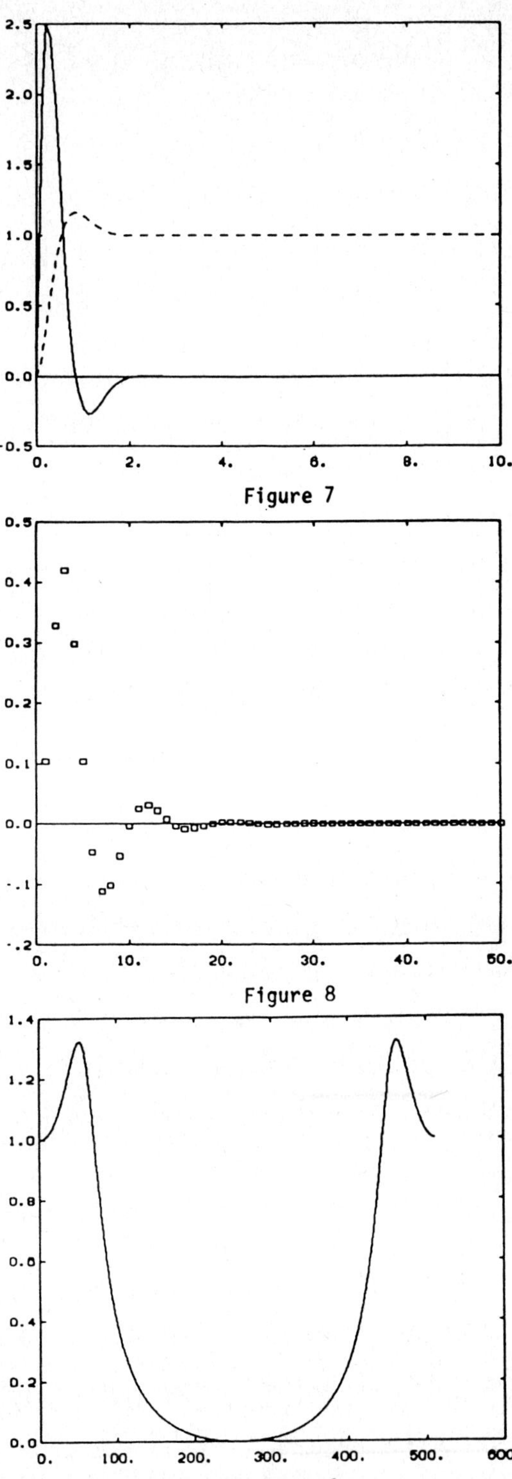

Figure 7

Figure 8

Figure 9

```
[> y = tdlf(x,a,b);
```

Suppose the impulse response of a digital filter is desired. An input vector representing a unit sample is created:

```
[> u = [1 0.*ones(1,511)];
```

which in this case is of length 512 points. The impulse response is found and graphed with

```
[> y = tdlf(u,a,b); plot(y(1:40),'point=3')
```
producing Figure 8.

The frequency response of the filter is easily found using a fast Fourier transform (FFT):

```
[> yy = fft(y); plot(abs(yy))
```
which results in Figure 9.

This simple filter example demonstrates the versatility of a matrix environment. In CTRL-C, other DSP primitives allow filter design, system identification, and power spectrum estimation.

7.0 EXTENSIBILITY

CTRL-C is most often used in a command driven mode; the user types single-line commands, CTRL-C processes them immediately, and the results are displayed. CTRL-C is also capable of executing sequences of commands that are grouped together to form a short "procedure".

In some other CAD packages, the word "Macro" is used to describe what is referred to as a Procedure or a User-Defined Function in CTRL-C.

There are three different types of procedures available in CTRL-C:

(1) DO Procedures; (2) User-Defined Functions; and (3) Text Macros

The first type operates globally on the workspace. It works by simply re-directing the input from the keyboard to a disk file. The second type allows the user define his own functions, complete with local and global variables, and argument passing. Once defined, the new functions are indistinguishable from the native CTRL-C primitives. The third is a simple facility for interpreting the text contained in a CTRL-C variable.

Together, these three procedure types form a powerful interpretive environment. Other syntax and commands form a complete programming language, similar in spirit to other popular interpretive languages. This notion of extensibility is one of the most powerful features of CTRL-C. Many applications can be developed directly in the CTRL-C language, without resorting to time-consuming "low-level languages" like Fortran.

8.0 ALGORITHMS

Careful attention has been paid to the selection of reliable algorithms. Subroutines from EISPACK and LINPACK provide state-of-the-art algorithms for matrix analysis, decompositions, and eigenvalue problems.

The numerically stable staircase algorithm [4] is used to compute controllable (observable) and uncontrollable (unobservable) modes as well as the Kronecker (controllability) indices for a system. The method also provides an orthogonal basis for the associated subspaces. The staircase algorithm can be used together with SVD to provide a reliable algorithm for minimal realization as well as complete canonical decomposition of a linear system [4].

The matrix pencil reduction algorithm [6] is used to compute the transmission and decoupling zeros of multivariable systems. This algorithm treats the most general case of the problem as it handles non-square and degenerate problems as well.

The Lyapunov and Riccati equations arise in many control and estimation problems. The most reliable and efficient algorithm to solve the Lyapunov equation is the method of Bartels and Stewart [7]. A modification of this method solves the unsymmetric Lyapunov equations in an efficient manner. For the Riccati equation, the extended generalized eigenvalue approach [4], [10] is used. The Schur vector approach [5] is also available. These algorithms are the latest techniques for solving Riccati equations; they provide a good balance between numerical reliability and efficiency.

Various pole-placement algorithms exist in the literature. CTRL-C contains an algorithm for robust eigenstructure assignment [11]. The poles of a multivariable system can be assigned while the eigenvectors are selected using various strategies. One possibility is to find a set of eigenvectors which are as close to orthogonal as possible.

For frequency response and various singular value measures, CTRL-C uses the efficient algorithm based on the Hessenberg form [9]. This algorithm is numerically stable and avoids problems encountered when methods based on the Jordan structure are used.

The "squaring-down" algorithm [5] is used to compute discrete equivalents of continuous systems. This is a reliable algorithm and compares very well to the numerous other techniques of discretizing continuous systems. Transfer function computations are based on the numerically stable algorithm in Ref. 14.

The identification algorithms in CTRL-C include the Levenberg-Marquardt modification of Gauss-Newton method for maximum likelihood identification. Trankle [13] has suggested a further modification of the algorithm which makes it extremely efficient.

9.0 CONCLUSIONS

A unified software system is possible for matrix analysis, control system design, digital signal processing, and engineering graphics. A common thread in these disciplines is the role of a single data object: the complex matrix. The use of a matrix environment produces a powerful, natural, and extensible software system.

The concept of direct manipulation should be a design goal for the development of user interfaces for computer-aided control system design (CACSD) packages. The user must be able to apply intellect directly to the task; the tool should seem to disappear.

The identification of the basic tools required is a crucial step in the design of a CACSD package. A well-designed CACSD system has a minimum set of reliable baseline primitives, plus a mechanism for extensibility. Extensibility is a method whereby new primitives are constructed out of existing ones. It must be possible to use the new primitives as if they were baseline primitives.

REFERENCES

1. Moler, C. and C. Van Loan, "Nineteen Dubious Ways to Compute the Exponential of a Matrix", SIAM REVIEW, 20, 4, 1978.

2. Thomas, R. and Yates, J., "A User Guide to the UNIX System," Osborne/ McGraw-Hill, Berkeley, CA, 1982.

3. Shneiderman, B., "Direct Manipulation: A Step Beyond Programming Languages," IEEE Computer Magazine, August 1983.

4. Emami-Naeini, A., "Application of the Generalized Eigenstructure Problem to Multivariable Systems and the Robust Servomechanism for a Plant which Contains an Implicit Internal Model," Ph.D. Dissertation, Dept. of Electrical Engineering, Stanford University, April 1981.

5. Franklin, G.F. and J.D. Powell, Digital Control of Dynamic Systems, Addison-Wesley, 1980.

6. Emami-Naeini, A. and P. Van Dooren, "Computation of Zeros of Linear Multivariable Systems," Automatica, Vol. 18, No. 4, pp. 415-430, July 1982.

7. Bartels, R.H. and G.W. Stewart, "A Solution of the Equation AX+XB=C," Commun. ACM, Vol. 15, pp. 820-826, 1972.

8. Laub, A.,J., "A Schur Method for Solving Algebraic Riccati Equations," Laboratory for Information and Decision Systems Report 859, MIT, October 1978.

9. Laub, A.J., "Efficient Multivariable Frequency Response Computations," IEEE Transactions on Automatic Control, Vol. AC-26, No. 2, April 1981.

10. Van Dooren, P., "A Generalized Eigenvalue Approach for Solving Riccati Equations," SIAM J. Sci. Stat. Comput., Vol. 2, pp. 121-135, 1981.

11. Kautsky, J. et al., "Numerical Methods for Roust Eigenstructure Assignment in Control System Design," in Proc. Workshop on Numerical Treatment of Inverse Problem..., Heidelberg, 1982.

12. Moler, C.B., MATLAB User's Guide, University of New Mexico, Computer Science Department, 1981.

13. Trankle, T.L., Vincent, J.H., Franklin, S.N., "Systems Identification of Nonlinear Aerodynamic Models," AGARDOgraph No. 256, Advances in the Techniques and Technology of the Application of Nonlinear Filters and Kalman Filters, 1982

14. Emami-Naeini, A. and P. Van Dooren, "On Computation of Transmission Zeros and Transfer Functions," in Proc. IEEE Conf. Dec. Contr., pp.51-55, 1982.

COMPUTER-AIDED CONTROL SYSTEMS ENGINEERING
M. Jamshidi and C.J. Herget (Editors)
Elsevier Science Publishers B.V. (North-Holland), 1985

CLADP: THE CAMBRIDGE LINEAR ANALYSIS AND DESIGN PROGRAMS

J.M. Maciejowski

A.G.J. MacFarlane

Cambridge University Engineering Department
Trumpington Street
Cambridge CB2 1PZ
England

ABSTRACT:

An Interactive software facility for designing multivariable control systems is described. The paper discusses the desirable characteristics of such a facility, the particular capabilities of CLADP and the numerical algorithms which lie behind them.

1. INTRODUCTION

The problem of creating a feedback controller for a plant described in terms of a given dynamical model has three aspects, conventionally called analysis, synthesis and design. In developing a synthesis technique the aim is to formulate a desired objective as a sharply-defined mathematical problem having a well-founded solution which is expressible in terms of a workable, efficient and robust computer algorithm. In principle then, one loads the synthesis problem description into the computer and the answer duly emerges. The disadvantages of a purely synthetic approach to design are obvious in an engineering context since the role of the designer, particularly the exercise of his intuitive judgement and skill, is severely reduced. An even greater drawback is that, at the beginning of his investigations, the designer simply may be unable to specify what he wants because he lacks information on what he will have to pay, in engineering terms, for the various aspects of desired final system performance. In developing a design technique, one seeks to give a practising and experienced design engineer a set of manipulative and interpretative tools which will enable him to build up, modify and assess a design put together on the basis of the physical reasoning within the guidelines laid down by his engineering experience. Thus design inevitably involves both analysis and synthesis and hence, in the development of design techniques, consideration of the way in which a designer interacts with the computer is vitally important. It is imperative to share the burden of work between computer and designer in such a way that each makes an appropriate contribution to the overall solution.

In developing the Cambridge Linear Analysis and Design Programs (CLADP) the aims have been to:

(i) allow the designer to fully deploy his intuition, skill and experience while still making an effective use of powerful theoretical tools; and

(ii) to harness the manipulative power of the computer to minimise the level of detail with which the designer has to contend.

The designer communicates with the computer through an _interface_. This allows him to _interpret_ what the computer has done and to _specify_ what he wishes it to do next. In general terms we will call anything which is presented to the designer by the computer, and which is relevant to the design process, an _indicator_. The designer must operate within an appropriate _conceptual framework_, and any powerful interactive design package must present the designer with the full set of indicators required to specify his needs and interpret his results in the context of his conceptual framework.

The computer is used for calculation, manipulation and optimisation. In any fully-developed interactive design package the "tuning" of controller parameters is best done by a systematic use of appropriate optimisation techniques. Generally speaking, in the design process the designer will be doing analysis and the computer will be doing synthesis. That is to say the computer will be used to solve a series of changing and restrictively- specified synthesis problems put to it by the designer as he works his way through a range of alternatives, between which he chooses on the grounds of engineering judgement, as he travels towards his final design.

Since the designer will usually want to think in the most physical way possible about the complex issues facing him, a high premium is placed on developing a conceptual framework for him to work in which makes the maximum use of his _spatial_ intuition, that is on one which is formulated as much as possible in geometric and topological terms. For this reason heavy emphasis is placed in CLADP on generalized frequency-response methods. Generalized Nyquist diagrams and multivariable root-locus diagrams are used as indicators of stability. These are derived from frequency-dependent characteristic decompositions of transfer function matrices. While such a decomposition gives accurate stability information, singular-value decompositions (of Nyquist or Bode arrays) are needed for an accurate assessment of performance and robustness. Bode plots of principal gains (derived from frequency-dependent singular value decompositions) are used as indicators for performance and in investigations of robustness. These indicators enable a natural extension of the classical gain/phase approach to feedback system design to be made to the multivariable case. For complex plants they can be used to derive a realistic closed-loop specification, which can then be achieved using appropriate parameter optimization techniques.

2. SUMMARY OF CLADP CAPABILITIES

Analysis Facilities

Some analysis is required both before and after attempting the design of a multivariable feedback system. Before starting the design, it is necessary to see whether the system is stable, and, if not, how many unstable poles it has. It is useful to check for any right-half plane zeros, which will impose an upper bound on the attainable closed-loop bandwidth. The Nyquist array, in Bode form, will disclose whether there are any sharp resonances which can be expected to give trouble, and whether the system transmission paths are strongly cross-coupled; it will also reveal features such as the cross-over frequency of each transmission path. CLADP provides facilities for performing this analysis easily, as well as the post-design analysis described below.

After proposing a feedback design, the most important property to be checked is closed-loop stability. This is most conveniently done by computing and displaying the characteristic loci of the compensated system, and applying the Generalised Nyquist theorem [1]. It is also possible to display a root-locus diagram, but this is usually much less useful for multivariable systems than it is for single-loop systems, since it gives virtually no guidance on how the design may be improved (for example, to attain greater stability margins). For the same reason, the direct evaluation of closed-loop characteristic roots (which is easily done in CLADP) is of limited use.

It is usually simple enough to achieve closed-loop stability. To achieve acceptable closed-loop performance as well is much more difficult. For single-loop systems the Nyquist locus, in any of its usual forms, gives reliable information about stability, performance, and robustness of the design in the face of large parameter variations. However, in the case of multivariable systems, the characteristic loci carry all this information only for the special case of so-called "normal" systems, i.e. those for which the return-ratio matrix $Q(s)$ satisfies

$$Q(s)Q^*(s) = Q^*(s)Q(s).$$

(Where * denotes "complex-conjugate transposed"). In general, the characteristic loci can give misleading information about performance and robustness, and it is necessary to compute and display other indicators in order to assess these.

CLADP allows the designer to display the "principal gains" of the open-loop return-ratio $Q(s)$ (i.e. the singular values of $Q(j\omega)$, evaluated over a range of frequencies) [2] [3], as well as the principal gains of the closed-loop transfer function $(I+Q)^{-1}Q$, and of the sensitivity function $(I+Q)^{-1}$. These displays allow aspects of performance, such as "velocity constant","output-disturbance rejection bandwidth", etc, to be accurately quantified. For example, Fig.1 shows the two principal gains, as well as the gains of the characteristic loci, of $(I+Q)^{-1}Q$ for a 2-input, 2-output system. By adopting an arbitrary (but quite usual) definition of bandwidth, namely the "-3dB frequency", we can define the "tracking bandwidth", ω_1, up to which tracking of reference signals can be guaranteed to be good, and the "noise-transmission bandwidth", ω_2, beyond which transmission of sensor noise can be guaranteed to be small. Note that the smallness of $\omega_2-\omega_1$ gives one measure of the efficiency of the design, which would be greatly overestimated by looking at the characteristic gains alone. Principal gains are also useful for assessing robustness [3] [4].

A complementary means of assessing performance and robustness is to display the Nyquist arrays of $(I+Q)^{-1}Q$ and of $(I+Q)^{-1}$ [5]. In Bode magnitude form these give performance assessments which are particularly useful if the designer is faced with "structured uncertainty", for example if he knows that disturbances acting on certain outputs are "high frequency" while those acting on others are "low frequency" . Interaction in the closed-loop design, as well as robustness in the face of partial or complete loop failures, can also be assessed from suitable Nyquist arrays.

The return-ratio of a multivariable system depends on the point of the loop at which it is calculated. Thus, if the plant transfer function is $G(s)$ and the compensator transfer function is $K(s)$, the return-ratios $Q_1(s) = G(s)K(s)$ and $Q_2(s) = K(s)G(s)$ are not the same. Information about robustness in the face of sensor failures is carried in $Q_1(s)$, whereas if actuator failures are of concern then $Q_2(s)$ must be looked at. For this reason CLADP allows all the computations mentioned above to be performed for any point of the loop. Fig.2 shows the Nyquist array of $(I+Q_2)^{-1}Q_2$ for a 2-input, 2-output design, with "generalised Gershgorin circles" [6] superimposed. From these it can be deduced that the designed closed loop will be stable for the gain of the actuator on input 1 lying anywhere between 0 and 2.2 (nominal value = 1), and the gain of the actuator on input 2 simultaneously lying anywhere between 0 and 3.6.

In addition to these frequency-domain tools for analysis, CLADP provides some basic simulation facilities, which can be used to look at the time-domain behaviour of open or closed-loop systems. Responses to step, pulse, sine, triangular, square and random signals can be obtained, with the spectra of the random signals being defined by the user. These signals can be injected at two standard points of the loop, namely the reference signals (i.e. those to be tracked) and the plant outputs (to simulate disturbances). It is also straightforward to inject signals

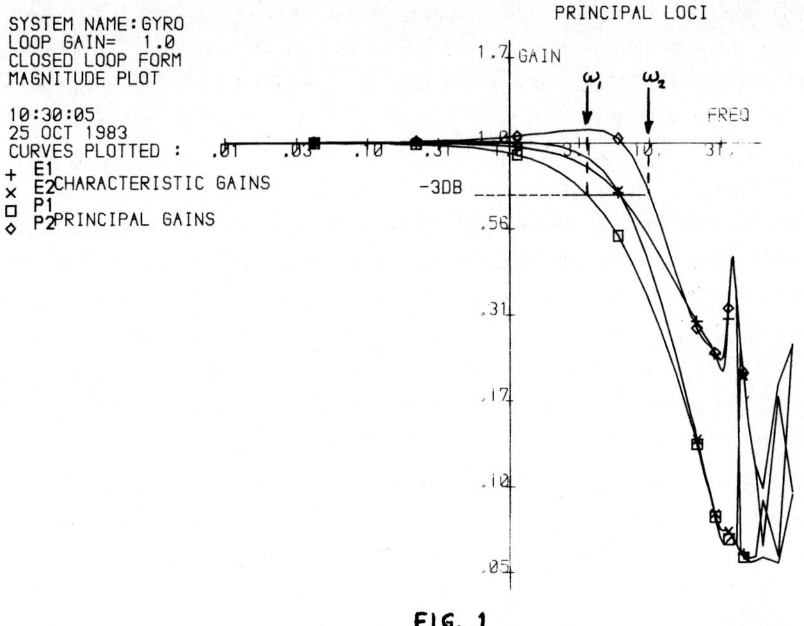

SYSTEM NAME: GYRO
LOOP GAIN= 1.0
CLOSED LOOP FORM
MAGNITUDE PLOT
10:30:05
25 OCT 1983
CURVES PLOTTED :
+ E1
× E2 CHARACTERISTIC GAINS
□ P1
◇ P2 PRINCIPAL GAINS
PRINCIPAL LOCI
GAIN
FREQ
-3DB

FIG. 1

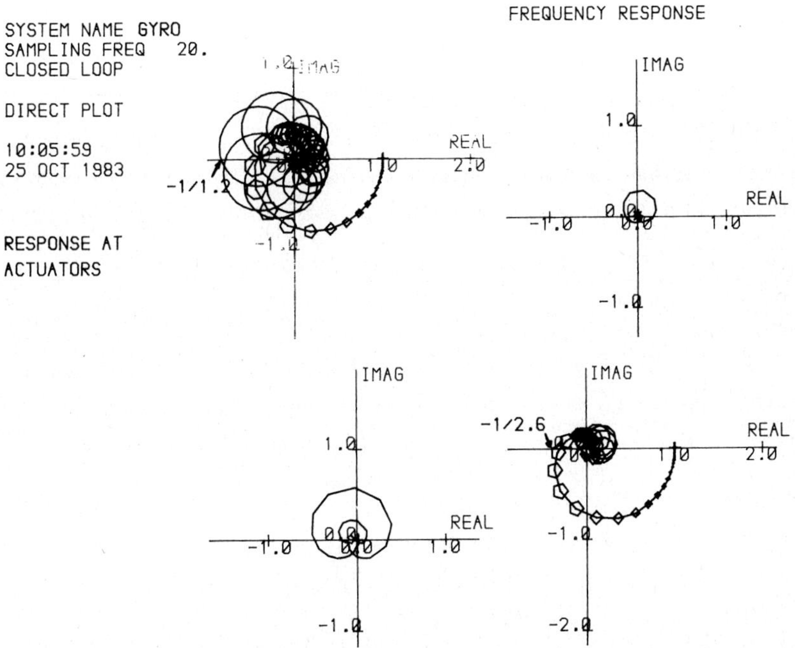

SYSTEM NAME GYRO
SAMPLING FREQ 20.
CLOSED LOOP
DIRECT PLOT
10:05:59
25 OCT 1983
RESPONSE AT
ACTUATORS
FREQUENCY RESPONSE
IMAG
REAL
-1/1.2
-1/2.6

FIG. 2

at any other point of a loop, by using CLADP facilities to rearrange the 'block diagram' which defines the system. The simulation assumes the same linear model as is used elsewhere in CLADP, so a final design usually needs to be checked on a separate nonlinear simulator.

Synthesis Facilities

The synthesis facilities available in CLADP range from very simple manipulative aids to quite sophisticated algorithms. At the lowest level, there are facilities for entering and modifying compensators from the keyboard, and for arranging these compensators into simple networks around the plant. These facilities alone are of significant assistance when using design techniques such as Inverse Nyquist Array [8] or Sequential Return Difference [9], which put almost all the load of synthesis onto the designer. They also allow classical procedures for SISO systems to be followed.

At the next level of sophistication comes the 'ALIGN' algorithm, which computes a real approximation to the inverse of a complex matrix. This is useful for decoupling the forward path in the region of the desired crossover frequency, and for achieving diagonal dominance if one is using the INA design technique. It is also an essential tool if one is designing an Approximate Commutative Controller [10], which is a technique for manipulating characteristic loci. Here the designer is still left with the problem of choosing compensators for each characteristic locus (which he does using classical frequency-domain methods), but the tedious and complicated task of computing real approximations to complex eigenvector frames is taken over entirely by the program.

Much closer to a complete synthesis facility are two algorithms for tuning compensator parameters. One of these [5] requires the designer to specify the closed-loop transfer function matrix which he would like to achieve. The designer must also specify the structure of the compensator, namely the poles of each element, the order of the numerator of each element, and whether any elements are constrained to be zero. The numerator coefficients are tuned by obtaining a least-squares fit to the desired closed-loop transfer function at a number (typically 50) of specified frequencies. This algorithm is very flexible, particularly when combined with some of the other facilities of CLADP. For example, by augmenting the plant model suitably, one can impose constraints on plant input variations. On the other hand, the algorithm is not very robust if the designer demands too much: he must use the analysis facilities carefully to deduce achievable closed-loop performance - in some cases this will involve him in performing a preliminary design using some other technique.

The other algorithm, which we have called the 'Quasi-Nyquist' method, requires the designer to specify the required open-loop behaviour. This is based on singular-value and generalized polar decompositions of transfer- function matrices, and enables one to aim at a simultaneous satisfaction of specifications on stability, performance and robustness. In this approach, particular emphasis is given to the robustness aspects of closed-loop behaviour. After a singular-value decomposition of the transfer-function matrix, phase information is transferred from the singular-direction frames to the singular values to generate a set of what have been called Quasi-Nyquist diagrams. A careful analysis of robustness behaviour then gives a structure for the controller which uses the singular-direction frames of the plant (in reversed order) but with appropiately-specified Quasi-Nyquist diagrams. The usefulness of this approach stems from the fact that it enables one to specify the compensating controller in a way which handles all three key aspects of behaviour: stability, performance and robustness. A further advantage of this quasi-classical approach is that it is well suited to the computer-synthesis phase of design. The controller is handled in the form of a general matrix-fraction decomposition whose parameters are optimised using a double (i.e. two nested loops) weighted least-squares procedure. Plants with different numbers of inputs and outputs can be handled in this controller synthesis approach, which is described in detail in [7].

A key feature of CLADP is a very powerful matrix manipulation facility. More will be said about this later, but here we note that this facility can be used to perform steady-state LQG design. Plant models can be augmented with disturbance dynamics, and advanced procedures such as the'robustness recovery' advocated by Doyle and Stein [3] can be implemented very easily. Controllers designed in this manner tend to have a high dynamic order. This is often not a problem nowadays, in view of the advanced technology available for implementing controllers. But if it is a problem, a very effective order reduction algorithm, based on the theory of balanced realisations [11] [24], is available.

'Utilities'

A control system designer spends much of his time performing routine tasks such as factorising polynomials, inverting matrices, combining connected sets of system equations into single system descriptions, converting state-space descriptions to transfer-function descriptions and vice-versa, and so on. Facilities for all these and other similar tasks are provided by CLADP.

The most powerful of these 'utilities' is the matrix manipulation facility, which allows the manipulation of algebraic and other expressions involving matrix names. Its capabilities are best demonstrated by an example. Suppose the steady-state Kalman filter gain is to be calculated for a system with disturbance covariance Q and measurement noise covariance R, and the system equations are

$$\dot{x} = Ax + \omega$$

$$y = Cx + v .$$

One way of computing the Kalman filter involves finding the eigenvalues and eigenvectors of the matrix:

$$\begin{bmatrix} -A^T & C^T R^{-1} C \\ Q & A \end{bmatrix}$$

In CLADP this can be done by the simple statement:

```
W = EIG((-TR(A)!(TR(C)*INV(R)*C))'(Q!A),VAL).
```

Here operators ! and ' are used to assemble partitioned matrices, so that X!Y = [X Y], and X'Y = $\begin{bmatrix} X \\ Y \end{bmatrix}$. TR(·) is the transposition operator, while INV(·) denotes inversion. The eigenvalues of the assembled matrix are stored in the (complex) vector VAL, and its eigenvectors are stored in the (complex) matrix W.

Sequences of statements of this kind can be run from "batch files" which provide a kind of macro facility, with some conditional control of flow. Indeed, batch files containing <u>any</u> CLADP statements can be run non-interactively. This is useful for example for repeating the same computations for a number of different models.

Another vital utility is the provision of appropriate prompts for the user who is not sure of the options which are available to him at any point. For such a "help" facility to be effective, a balance has to be struck between restricting the user's options excessively, and swamping him with so many options that he becomes confused. The latter possibility poses a serious problem, since most users are either beginners or occasional users. In CLADP this is avoided by the provision of hierarchically organised "menus" of options, so that the designer can choose to examine only the "display manipulation" options, or only the "computation" options, and so on.

Discrete-Time

The analysis and design of discrete-time systems in the form of either state-space descriptions or z-transform transfer function matrices is fully supported in CLADP. All facilities exist in parallel for continuous and discrete-time systems. There are also utilities for converting from one form to the other, and it is possible to apply a bilinear transformation to the frequency variable, which enables the designer to work in the w-domain rather than the z-domain, if he prefers.

Other Facilities

Since CLADP has been developed in a university research group, it contains some facilities which aid theoretical investigations, but are not directly useful in the design process - or at least, no direct use has yet been found for them. An example of these is the capability of displaying individual sheets of the Riemann surface which is the domain of a characteristic gain or frequency function.

Of more practical use, but still unproven in design, are facilities for analysing systems described by irrational transfer function matrices.

A convenient facility is provided by the capability of treating several models as the description of a single system. This is used to represent several linearizations of a system about various operating points, or to represent uncertainty in a model.

Macros

Macros of CLADP commands (known as 'batch' files) can be kept on file, and executed by calling them from the keyboard. These are useful for tasks such as LQG design, solution of matrix equations, and for 'tuning' CLADP for particular applications. For example, a recent student project was to write macros which provide an interface between a novice user and CLADP, and lead him through the design of a 1-input, 1-output feedback system.

Macros cen be called from within other macros, so a useful library of macros can be rapidly established by each user. Prompts to the screen can be generated from macro files, as can annotations of graphic displays.

3. ALGORITHMS

CLADP generally uses reliable numerical algorithms, although it does not make use of some of the latest advances, since it has been under development since 1976. Some of the algorithms are outlined below.

The key decision which was taken in CLADP was to avoid analytical evaluation of the resolvent matrix $(sI-A)^{-1}$, but to use pointwise evaluation instead. Considerable use is also made of curve fitting. This approach has resulted in the ability of CLADP to handle the complex models which usually arise in real design studies. Successful designs have been performed for a 40-state, 2-input, 3-output continuous-time model, and for a 17-state, 5-input, 5-output discrete-time model, both of these models being given to CLADP in state-space form. The pointwise evaluation of the frequency response allows the user to spot immediately any dubious results, since these are usually revealed by discontinuities in the displayed loci. The user can also change the set of frequencies at which the evaluation is performed, and can therefore make these frequencies more dense in the region of a resonance or other important feature, and less dense elsewhere.

The option is provided of transforming the 'A' matrix of a state-space model to either Hessenberg ot tridiagonal form before computing the frequency response,which can save substantial amounts of computing time if the model has many states. Computation from the tridiagonal form is faster than from the Hessenberg form, but this is offset by the possible numerical instability of the transformation to tridiagonal form.

Eigenvalue-eigenvector computations are performed frequently in CLADP, particularly for finding the characteristic loci. These are again evaluated pointwise, and then sorted so that a continuous set of loci is displayed. The algorithm used for finding eigenvalues (and eigenvectors when required) is that used in EISPACK, namely reduction to upper Hessenberg form, followed by a 'modified LR' algorithm [12].

Principal gains are the singular values of the frequency response matrix, also evaluated pointwise. The Golub and Reinsch algorithm for finding the singular value decomposition is used [12].

The transmission zeros of a model given in state-space form are the eigenvalues of the matrix $(A-BD^{-1}C)$. However, the matrix D is often singular, and even if it is regular, computation by this route can be numerically unstable. Nevertheless, if the model is square (i.e. D is square), then a bilinear transformation of the frequency variable usually leads to a transformed state-space representation in which the 'D' matrix is regular. The eigenvalue calculation can then be performed, and the zeros obtained by inverting the bilinear transformation [13]. By repeating this procedure with a different transformation, the results can be checked. However, this check can on occasions be misleading, and an alternative approach is available. This is to search iteratively in a region of the complex plane for a point at which an eigenvalue of the frequency response matrix is zero.

A third way of calculating zeros in CLADP is to use the matrix manipulation routine, which includes algorithms for solving generalised eigenvalue problems for pairs of matrices [14], using 'QZ' techniques [15]. (This is a good example of a task which is best solved by use of a macro file.)

If a transfer-function description of a model is needed, it is obtained from a state-space description by pointwise evaluation of both det(sI-A) and $(sI-A)^{-1}$, followed by fitting polynomials to the values of det(sI-A) and the elements of C adj(sI-A)B, where the adjoint matrix is obtained by multiplying values of $(sI-A)^{-1}$ by values of the polynomial which approximates the determinant. (Of course, CLADP also allows models to be specified directly by their transfer function matrices.)

Conversion from continuous-time to discrete-time state-space models is performed by first using 'scaling and squaring' to obtain an approximation of exp(AT) (where T is the inter-sample interval). This is the least unsatisfactory of the methods reviewed in [15]. The input matrix $(\exp(At)-I)A^{-1}B$ is then obtained by solving the linear equation AX = (exp(AT)-I)B for X.

When a potentially precarious computation is performed in CLADP, some check is usually made on the correctness of the result, and a message is output to the user if there are indications that the result is unreliable. However, there are a few cases, such as the computation of zeros, in which the user is not alerted to possible problems. The range of facilities available in CLADP is so wide that almost any computation can be checked by some means, given sufficient ingenuity on the part of the user. This makes CLADP not only powerful, but also reliable if used intelligently, but it undeniably falls short of the ideal situation, in which the designer could proceed confidently at every step without needing to know the details of the algorithms he is executing. It remains to be seen how closely such an ideal can be approached.

4. STRUCTURE AND PORTABILITY

CLADP is written entirely in FORTRAN, and contains its own libraries of linear algebra and graphics subroutines. It consists of a main program and about 30 major subroutines, one of which is the 'supervisor'. The main program is little more than a multi-position switch, which calls one or other of the major routines. Initially the user enters a one-word command which is interpreted by the supervisor: the supervisor sets up a sequence of up to 10 major routines which are to be called, in the appropriate order. For example, the command 'NYQUIST' causes the 'Nyquist Calculation' routine to be run first, followed by the 'Nyquist Display' routine. Each of these routines writes messages to the terminal screen and accepts input from the keyboard (via input subroutines). Extended command lines with arguments are not used.

Altogether there are about 400 subroutines, and the source code runs to about 10^5 lines of FORTRAN. Although the user interacts directly with subroutines at various levels, input from the keyboard is processed by a limited number of machine-dependent routines. Consequently CLADP is reasonably portable between machines. To date it has been installed on GEC 4070, GEC 4090, Prime 550 and Vax 11/780 computers. Earlier versions have been installed on a PDP10, PDP 11/45, and a Honeywell 6000. On the GEC machines the compiled and linked object code occupies about 800 kbytes, and a further 200 kbytes are taken up by data arrays.

5. APPLICATIONS

Purchasers of CLADP include the (US) General Electric Company, the Australian Ministry of Defence, and, in the UK, Imperial Chemical Industries, the Central Electricity Generating Board, the Royal Aircraft Establishment, Dowty Electronics, and Cambridge Consultants.

A number of applications of CLADP for control system design has been described in the literature. Foss etal [17] report its use for the control of a two-bed catalytic reactor, while Grimble and Fotakis [18] use CLADP for the design of shape control systems for a Sendzimir steel rolling mill. Kouvaritakis and Edmunds [19] and Foss [20] describe the use of CLADP for the design of gas-turbine control systems. Limebeer and Maciejowski [21] use CLADP to design controllers for a large turboalternator, and for a two-gimbal gyroscope. Foss [22] and Thompson [23] describe the integration of CLADP with other CACSD facilities, in a commercial environment.

6. COMMERCIAL AVAILABILITY

Marketing of CLADP is being undertaken by Cambridge Control Ltd., of Madingley Road, Cambridge, CB3 OH8, UK., to whom all commercial enquiries should be addressed.

7. REFERENCES

[1] MacFarlane, A.G.J., and Postlethwaite, I., 'The generalised Nyquist stability criterion and multivariable root loci'. Int. J. Cont. 25, 81-127, (1977).

[2] MacFarlane, A.G.J., and Scott-Jones, D.F.A., 'Vector gain', Int. J. Cont., 29 65-91, (1979).

[3] Doyle, J.C., and Stein, G., 'Multivariable feedback design: concepts for a classical/modern synthesis'. IEEE Trans. Auto. Contr., AC-26, 4-16, (1981).

[4] Postlethwaite, I., Edmunds J.M., and MacFarlane, A.G.J., 'Principal gains and principal phases in the analysis of linear multivariable feedback systems', IEEE Trans. Auto. Contr., AC-26, 32-46. (1981).

[5] Edmunds, J.M., 'Control system design and analysis using closed-loop Nyquist and Bode arrays'. Int. J. Contr., 30 773-802, (1979).

[6] Limebeer, D.J.N., 'The application of generalised diagonal dominance to linear system stability theory', Int.J. Contr, 36, 185-212, (1982).

[7] Hung, Y.S., and MacFarlane, A.G.J., 'A quasi-classical approach to multivariable feedback systems' Lecture Notes in Control and Information Sciences, Springer-Verlag, (1982).

[8] Rosenbrock, H.H., 'Computer-aided control system design', London: Academic Press, (1974).

[9] Mayne, D.Q., 'The design of linear multivariable systems, Automatica, 9, 201-207, (1973).

[10] MacFarlane, A.G.J., and Kouvaritakis, B., 'A design technique for linear multivariable feedback systems' Int. J. Control.25, 837-879, (1977).

[11] Moore, B.C., 'Principal component analysis in linear systems, controllability , observability, and model reduction', IEEE Trans. Auto. Contr.AC-26, 17-32, (1981).

[12] Wilkinson, J.H., and Reinsch, C., 'Handbook for automatic computation, vol. II: linear algebra', New York: Springer, (1971).

[13] Kouvaritakis, B., and Edmunds, J.M., 'Multivariable root loci: a unified approach to finite and infinite zeros'. Int. J. Contr. 29, 393-428, (1979).

[14] Laub, A.J., and Moore, B.C., 'Calculation of transmission zeros using QZ techniques', Automatica, 14, 557-566, (1978).

[15] Moler, C., and Stewart, G.W., 'An algorithm for generalised matrix eigenvalue problems'. SIAM J.Num. Anal., 10, 241-256, (1973).

[16] Moler, C., and van Loan, C., 'Nineteen dubious ways to compute the exponential of a matrix', SIAM Review, 20, 801-836, (1978).

[17] Foss, A.S, Edmunds, J.M, and Kouvaritakis, B., Multivariable Control System for Two-Bed Reactors by the Characteristic Locus Method'. I & EC Fundamentals, 19, 109-117, (1980).

[18] Grimble, M.J., and Fotakis, J., 'The design of strip shape control systems for Sendzimir mills', IEEE Trans. Auto. Contr, Ac-27, 656-666, (1982).

[19] Kouvaritakis, B., and Edmunds, J.M., 'The characteristic frequency and characteristic gain design method for multivariable feedback systems', in: Sain, M.K., Peczkowski, J.L., and Melsa, J.L., (eds), Alternatives for linear multivariable control, (Chicago: National Engineering Consortium), 229-246, (1978).

[20] Foss, A.M., 'A practical approach to the design of multivariable control strategies for gas turbines', ASME Paper No. 82-GT-150, presented at the 27th International Gas Turbine Conference, Wembley, (1982).

[21] Limebeer, D.J.N., and Maciejowski, J.M., 'Two tutorial examples of multivariable control system design', Camb. Univ. Eng., Dept. Technical Report CUED/F-CAMS/TR-229 (1982).

[22] Foss, A.M., 'DICAST: Integrated use of CACSD', Proc. Workshop on Computer Aided Control System Design, Brighton, September, (London: Institute of Measurement and Control), (1984).

[23] Thompson, M.A., 'Computer-aided control system design software in the Central Electricity Generating Board', Proc. Workshop on Computer Aided Control System Design, Brighton, September, (London: Institute of Measurement and Control), (1984).

[24] Glover, K., 'All optimal Hankel-norm approximations of linear multivariable systems and their L^{∞}- error bounds' Int. J. Control, 39, 1115-1193,(1984).

DELIGHT. MIMO: An Interactive, Optimization-Based Multivariable Control System Design Package*

E. Polak, P. Siegel, T. Wuu and W. T. Nye

Department of Electrical Engineering and Computer Sciences
University of California
Berkeley, CA 94720

D. Q. Mayne

Department of Electrical Engineering
Imperial College
London, SW7 2BT, U.K.

Abstract

This paper describes an interactive, optimization-based multivariable control system design package which is currently under development at the University of California, Berkeley. The package will combine a number of subroutines from the Imperial College Multivariable Design System and the Kingston Polytechnic SLICE library with DELIGHT, the University of California, Berkeley, general purpose, interactive, optimization-based CAD system.

1. Introduction

Optimization, either heuristic or algorithmic, is an integral part of engineering design. Heuristics are most frequently used in selecting a system configuration, while algorithmic optimization is used to determine parameter values which satisfy design specifications or optimize a performance function. A new generation of semi-infinite optimization algorithms, see e.g., [G1, P1, P2] enable the multivariable control system designer to satisfy complex specifications, some involving constraints on time responses, others involving constraints on closed loop system eigenvalues or on frequency dependent singular values of various system matrices, see e.g., [D3, T1 Z1]. For best results, these new algorithms must be implemented in an interactive computing environment.

The DELIGHT system [N1] was conceived as an interactive computing environment for multidisciplinary, optimization-based engineering design. It incorporates a high level language (RATTLE) which simplifies the programming of algorithm; standard FORTRAN numerical analysis programs; a modular RATTLE optimization library; as well as highly flexible color graphics and interaction capabilities. DELIGHT.MIMO is a multivariable control system design package which was constructed by incorporating into the DELIGHT system routines for control system definition, response evaluation and graphical display. It permits the designer to use interactively both modern multivariable system design tools as well as semi-infinite optimization algorithms.

Section 2 of this paper describes the "basic" DELIGHT system, section 3 presents the control system design specific enhancements, while section 4 illustrates the use of DELIGHT.MIMO, in the optimization of a multivariable control system design, by means of an example.

2. The DELIGHT System

The DELIGHT system [N1] is an interactive computing environment which was developed by W. T. Nye, E. Polak, A. Sangiovanni Vincentelli and A. L. Tits as part of a broad project dedicated to optimization-based computer-aided design. It provides the user

This research was supported by the National Science Foundation under grant ECS-79-13148 and the U.K. Science and Engineering Research Council.

with a number of features which are present in the UNIX C-shell [U1] as well as with a number of others that facilitate interactive optimization-based design and optimization algorithm development. It can be extended into an application specific design package by the addition of system definition and simulation programs. DELIGHT.MIMO is such a package.

The major features of DELIGHT are (i) simple command and algorithm execution, (ii) a capability to rescale or modify either the design problem being solved or the optimization algorithm being used, without recompilation and reloading of the programs or reinitialization of the algorithm, (iii) powerful, terminal independent color graphics which can be used to display information stored in arrays, in a number of ways, in colors and windows designated by the user, (iv) a high level language which permits the writing of compact computer programs closely resembling the mathematical description of the algorithms being implemented, eliminating most of the usual coding errors and shortening programming time, (v) provisions which facilitate the addition of new built-in FORTRAN functions, simulation subroutines, utilities, or other FORTRAN application dependent features.

The high level, structured, interactive programming language in the DELIGHT system is called RATTLE (an acronym for RATfor Terminal Language Environment) and was evolved from the structured language RATFOR. The structured constructs include "while", "repeat-until", "if-then-else", etc. It allows matrix and graphical commands, eliminates the need for dimension statements, common block declarations and time consuming load/linkages.

It is possible in RATTLE to create new language constructs or new commands from existing ones by means of *macros* and *defines*. Macros are used to call highly complex FORTRAN procedures by means of very simple commands. For example, to solve a linear program one uses the following RATTLE code, which uses the macro '**1*p***':

$$1pz = argmin\ \{c'\ *x \mid x>=0,$$
$$x<=d,\ A*x<=b\}$$

where the array z is assigned the minimizing value of x, and the matrix A and vectors b, c, d have been defined previously. Macros not only enhance readability, they also relieve the programmer of the need to create work arrays and to master the other requirements of the library routines. Both binary and unary matrix computations are carried out by means of the *matop* macro. The table on this page gives a sample of operations available to the user through macros.

An important use of *defines* is in the creation of simple commands for invoking complex, terminal independent graphics procedures that are written in RATTLE. For example, there exists a define which allows the command "window name" to be substituted for the specification of the particular set of world coordinates, and viewport coordinates which are associated with the window [N2].

The RATTLE language permits incremental program development [W1], so that one can execute by just typing it in, a single statement, a procedure, or a section of an algorithm, without having to write and load/link a whole program. The following is a complete RATTLE statement, implementing Newton's method, which would execute when the closing braace and carriage return are typed in:

```
while (f(x)>eps){x=x-f(x)/derf(x)print x}
```

Operation	Macro Syntax
add matrices	matop A = B + C
multiply matrices	matop A = B * C
eigenvalues	matop lambda = eigen (A)
inverse of matrix	matop Ainv = inv(A)
solve linear eqns.	lineq A*x = b
inner product	<<x,y>>

An important RATTLE feature, both from the designer's and algorithm developer's point of view, is that execution of a program can be interrupted by the user or by the program and later resumed after modifying variable values or even recompiling an entire subprocedure. While execution is suspended, the values of both global and local variables can be displayed and modified by appropriate commands.

The nice relationship between the mathematical description of an algorithm and its implementation in RATTLE is illustrated by the following code implementing the Armijo gradient method [A1]:

```
procedure Armijo {
  repeat {
    evaluate h = grad (X[Iter])
    lambda = step (X[Iter],h)
    update X[Iter + 1] = X[Iter]
        - lambda * h
    Iter = Iter + 1
    output
    }
 forever
 }
```

where "evaluate" and "update" are defines for calling the appropriate subprocedures, step is a function which computes a step size by the Armijo rule, "output is a procedure that produces a display, while X[Iter] is a vector in a sequence whose last k iterates (typically 20) are stored.

The DELIGHT system contains an ever growing library of RATTLE routines implementing algorithms for solving unconstrained and constrained, both ordinary and semi-infinite optimization problems. This library is organized to exploit the natural modularity off modern optimization algorithms which, in the simplest case, can be assembled from search-direction, step-size and update subalogrithms. In turn, search-direction subalgorithms can be constructed from subprocedures which compute the gradients to be used for search direction construction and from linear or quadratic programs. Similarly, step-size subalgorithms can be built up from constrained and unconstrained step-size blocks. The user may interactively explore algorithms component and output options and select those that suit his needs. Substitutions from the options list may be made at any time, including when execution has been interrupted in the middle of an optimization run (by depressing the break key).

The use of the RATTLE optimization library has been highly mechanized. Thus, the user defines his problem by inserting appropriate lines in a setup file, in a cost file, in a constraint file and in an initial data file. The setup file contains information on the nature and number of constraints to be used and the dimension of the design vector. The cost and constraint files contain code for evaluating the cost and constraint functions and their gradients, which may involve calls to FORTRAN simulation subroutines. Once the problem files have been created, the user may select an algorithm from the optimization library and link it to his problem by a command of the form ***solve probname using algoname***.

3. Control System Design Aids.

As already mentioned in the preceding section, the "basic" DELIGHT system includes the LINPACK]D1] linear algebra and Harwell [H1] linear and quadratic programming subroutines which can be used for all relevant MIMO design computations by means off high level macros. DELIGHT.MIMO was formed by (i) adding to the DELIGHT system an overall control system assembly and graphical display program, (ii) interfacing a number of FORTRAN subroutines for MIMO design from the Multivariable Design System (MDS) [S2], via the built-in function mechanism provided by DELIGHT, and (iii) adding a number of RATTLE routines for interaction and graphical display of results. The control system specific design aids can be grouped into two categories: (a) aids for data entry and manipulation and (b) aids for producing an initial design and evaluating system responses.

Multivariable control systems may be entered in either state-space or transfer function form. State-space descriptions are limited to 40 states, 10 inputs and 10 outputs. Transfer function matrices can be of maximum dimension 10 x 10, where each entry of the transfer function matrix may contain rational functions of maximum order 20. Systems may contain up to 20 blocks; each block may contain parameters to be optimized.

To enter a control system, one must first enter the individual blocks, either from a file or interactively, using the *entblock* command. One may verify if a block has been entered correctly by means of the *check* command. Editing facilities are available for modifying existing blocks. Parameters to be optimized are specified by means of the command *entparam* ; initial values for these parameters are entered via the *entblock* command. The *convert* command can be used to convert a state space description into transfer function form and vice versa. Descriptions can be saved in files by means of the *save* command.

Once the individual blocks have been entered, their interconnection is specified in terms of a reverse polish list. Three operations may be used to connect blocks: * cascade connection; + feedforward connection; < feedback connection. This method of entry allows systems of arbitrary complexity to be specified. See Fig. 1 for an example. The *bd* command may be used to display graphically the block diagram of the system (computed from the reverse polish list) in order to verify that the control system was correctly entered.

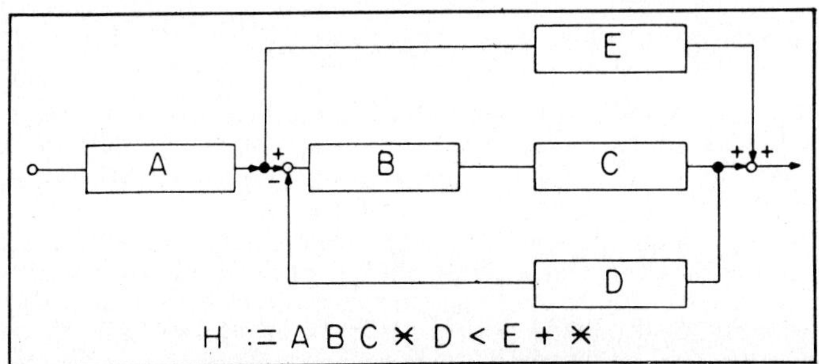

Fig. 1. System interconnection.

The control system definition utility is currently undergoing revision. The polish list method of system interconnection definition is being replaced by a menudriven graphical input facility. The designer will be able to enter the block diagram of the system by "drawing" it on the screen with the aid of a graphics tablet and stylus. It will be possible to edit block diagrams graphically. Input waveforms will be entered either by tracing the waveform on the tablet or by selecting from the menu.

Since optimization requires derivatives of responses with respect to design parameters, we are constructing a symbloc differentiator, as described in [B1]. This requires that control system components containing design parameters be entered only in state-space form, with the elements of the matrices described as quotients of multivariable polynomials in the design variables. The remaining blocks can be entered in state space form, as matrix transfer functions, or as polynomial matrix fractions. The time domain responses which the symbolic differentiator can handle include responses to polynomial, sinusoidal and exponential inputs. Frequently domain responses handled by the symbolic differentiator include distinct singular values of transfer function matrices.

The MDS system [S2] and the SLICE library [D2] contain most of the commonly used subroutines for modern control system design. A number of these have already been incorporated into DELIGHT.MIMO. Additional ones will be incorporated as necessary.

4. Control System Design Using DELIGHT.MIMO.

Control system design by means of semi-infinite optimization is a four-stage process. The first stage consists of sorting the system performance requirements into "soft" and "hard" categories. The "soft" requirements are modeled as a composite cost $f(x)$, to be minimized, while the "hard" requirements are expressed as simple inequalities of the form

$$g^{i}(x) =< 0, \text{ for } j=1,2,\ldots 1, \tag{1}$$

or as semi-infinite inequalities of the form

$$\phi^{k}(x,\ w_k) =< 0 \text{ for } all\ w_k \in [w_k',\ w_k''],\ k=1,2,\ldots,m, \tag{2}$$

where x is the design vector to be introduced in the second stage. The second stage consists of defining a system configuration and a controller structure containing x as a vector of n design parameters to be adjusted by optimization algorithm. The third stage consists of computing an initial value for the design vector x by means of some "classical" method, such as LQR [S1], or multivariable root loci [M1], which usually results in a control system that fails to satisfy a number of the "hard" specifications. The last stage involves the use of a semi-infinite optimization algorithm to adjust the parameter x so that all the "hard" constraints are satisfied and the cost minimized, or at least reduced.

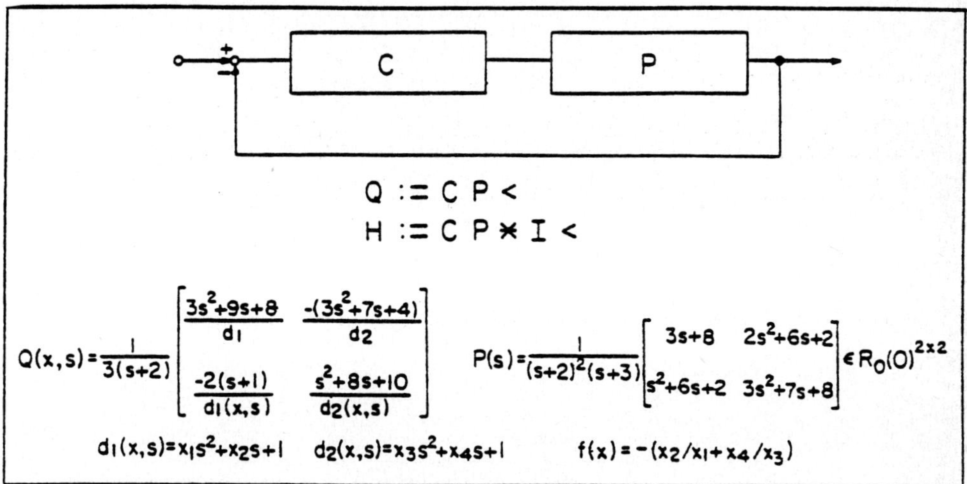

Fig. 2. Design example.

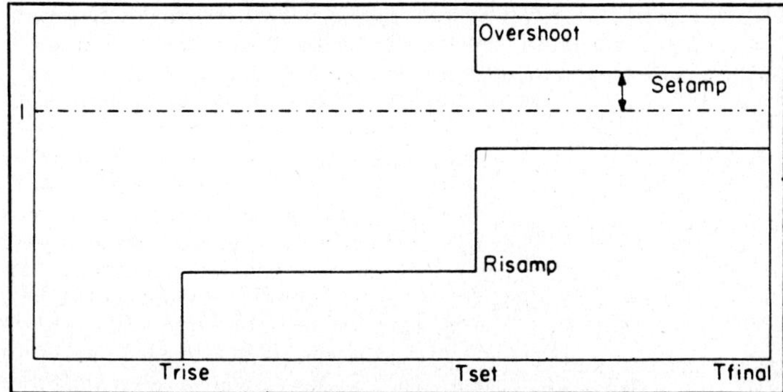

Fig. 3. Step Response constraints.

For example, consider the system in Fig. 2, where the z_i are the design parameters. The structure and initial values for the controller are designed by the method in [G2]. The design objective is to maximize the bandwidth of the closed loop system subject to constraints on the closed loop step response, as in Fig. 3, and constraints on the input amplitude of the plant, which are expressed as bonds on the singular values of the matrix Q, as shown in Fig 4. The routine *opt init* helps the designer to transcribe interactively these requirements into functions and subroutines conforming to the format in (1) and (2). The result is a DELIGHT format problem called optdesign.

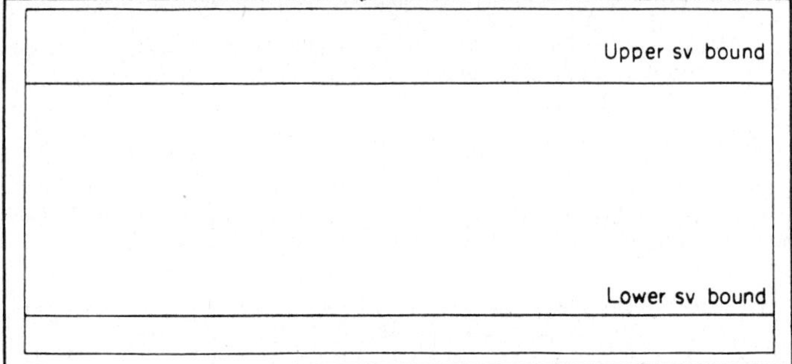

Fig. 4. Singular value constraints.

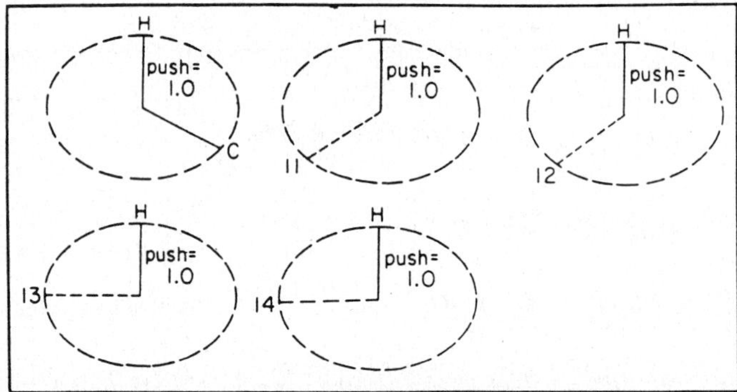

Fig. 5. Gradient clock.

Next an algorithm is selected and coupled to the problem by using the *solve* command. For example, *solve optdesign using Apolwar* selects the Polak-Wardi [P1] nondifferentiable optimization algorithm from the optimization library. This algorithm consist of a phase I phase II search direction finding subprocedure, an Armijo type step size subprocedure and an update subprocedure. The algorithm contains a number of parameters which need to be adjusted for efficient coupling of the problem and algorithm. This adjustment is carried out interactively, with the aid of graphics, as follows. First, the search direction procedure is executed, by means of the command *dir*. At this point the algorithm is suspended, allowing the user to change the problem scaling parameters (push factors). To decide whether this is necessary, the user types in *Gangles* to display the *gradient clock*, see Fig. 5, which show the angles between the search direction and the gradients of the cost and active constraints. In Fig. 5 the gradient clock indicates that the gradients of inequality constraints 3 and 4 are almost perpendicular to the search direction h due to poor initial scaling of the design problem and, possibly, inadequate precision in the search direction calculation. The user attempts to remedy this by changing the values of the precision parameter. To verify that the problem has been eliminated, the user repeats the dir and Gangles commands, to obtain Fig. 6. Since this is satisfactory, the user decides to examine the step-size parameters. For this purpose, the user types in the command *Garmijo* to obtain Fig. 7 which show all the information that governs step size com-

putation and active constraint selection. Since the algorithm is not spending an excessive amount of time in the step-size loop, the user decides to carry out five iterations of the algorithm: he types in *reset carriage return run 5.* When these are completed, the user examines the results by typing in *response green 0 0 1, svplt green 0 1, response sky 0 5 2, svplt sky 0 2,* to display the appropriate step and singular value responses in the color indicated. These commands produce the display shown in Fig. 8, in which the colors were converted to dash patterns. This figure shows in solid curves the responses corresponding to the initial design, while the dashed curves correspond to the design parameters at the end of the 5th iteration. Note that after the 5th iteration the design satisfies all of the constraints (indicated in the figure by solid lines). The box marked *pallet* gives the color in which the results of each iteration were plotted.

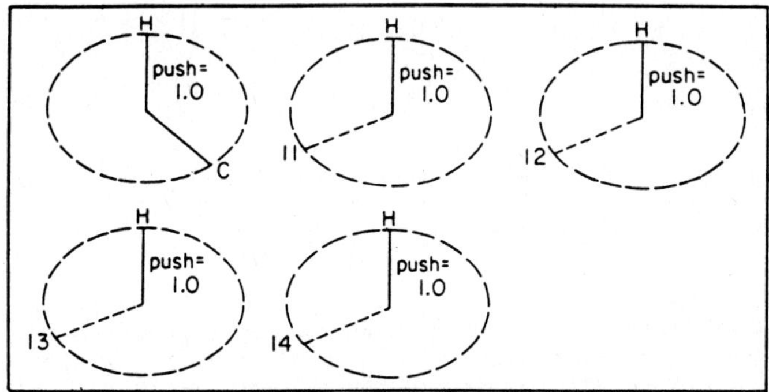

Fig. 6. Gradient clock.

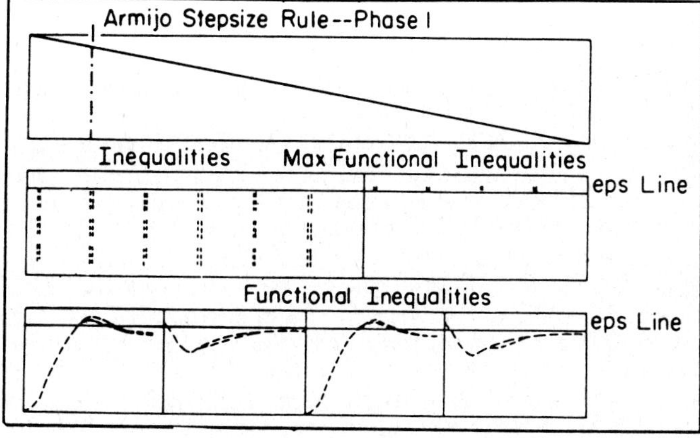

Fig. 7. Step size computation.

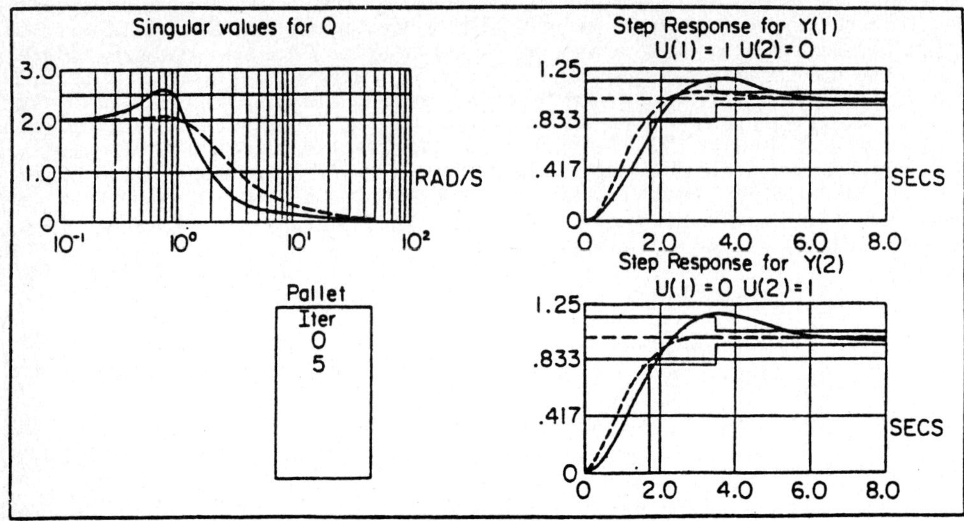

Fig. 8. Design results.

5. Conclusion.

The DELIGHT.MIMO control system design package is intended both as a practical design tool and as a test bed for concepts to be used in interactive control system design. It is hoped that it will prove to be highly obsolescence-resistant and that it will eventually evolve into a comprehensive facility.

6. References

[A1] L. Armijo, "Minimization of functions having continuous partial derivatives", *Pacific J. Math.*, Vol 16, pp 1-3, 1966.

[D1] J. J. Dongarra, et al., LINPACK *Users' Guide*, SIAM, Philadelphia, PA., 1979.

[D2] M. J. Denham, and C. J. Benson, "SLICE: a subroutine library for control system design", Internal Report 01/82, School of Electronic Engineering and Computer Science, Kingston Polytechnic, Kingston upon Thames KT1 2EE, England, 1982.

[D3] J. C. Doyle, and G. Stein, "Multivariable feedback design: concepts for classical modern synthesis", *IEEE Trans.*, Vol. AC-26, No. 1, pp. 4-17, 1981.

[G1] C, Gonzaga, E. Polak, and R. Trahan, "An Improved Algorithm for Optimization Problems with Functional Inequality Constraints", *IEEE Trans.*, Vol. AC-25, No. 1, 1980.

[G2] C. L. Gustafson, and C. A. Desoer, "Controller design for linear multivariable feedback systems with stable plants, using optimization with inequality constraints", University of California, Electronics Research Laboratory Memo No. UCB/ERL M81/51, 1981.

[H1] Harwell Subroutine Library, Harwell, England.

[M1] A. G. J. MacFarlane, and I. Postlethwaite, "Generalized Nyquist Stability Criterion and Multivariable Root Loci", *Int. J. Control*, Vol. 25(1), 1977.

[N1] W. Nye, E. Polak, A. Sangiovanni-Vincentelli, and A. Tits, "DELIGHT: and Optimization-Based Computer-Aided-Design System", Proc. IEEE Int. Symp. on Circuits and Systems, Chicago, Ill, April 24-27, 1981.

[N2] W. M. Newman, and R. F. Sproul, *Principles of Interactive Computer Graphics*, 2nd Edition, McGraw-Hill, NY, 1979.

[P1] E. Polak, and D. Q. Mayne, "An Algorithm for Optimization Problems with Functional Inequality Constraints", *IEEE Trans.*, Vol AC-21, No. 2, 1976.

[P2] E. Polak, and Y. Y. Wardi, "A nondifferentiable optimization algorithm for the design of control systems subject to singular value inequalities over a frequency range", *Automatica,* Vol. 18, No. 3, pp. 267-283, 1982.

[S1] Special Issue on LQG Problems, *IEEE Trans. Autom. Control,* 1975.

[S2] B. R. Shearer, and A. D. Field, "Multivariable Design System (MDS): An interactive package for the design of multivariable control systems", Publication No. 75/29, Department of Computing and Control, Imperial College, London, SW7 2BZ, 1975.

[T1] O. Taiwo, "Design of a Multivariable Controller for a High Order Turbofan Engine Model by Zakian's Method of Inequalities", *IEEE Trans.* Vol. AC-23, No. 5, 1978.

[U1] UNIX programmer's manual, Computer Science Division, Department of Electrical Engineering and Computer Science, University of California, Berkeley, CA 94720, 1979.

[W1] J. Wilander, "An interactive programming system for Pascal", BIT

[Z1] V. Zakian, "New Formulation for the Method of Inequalities", *Proc. IEE,* 126(6), 1979.

COMPUTER-AIDED CONTROL SYSTEMS ENGINEERING
M. Jamshidi and C.J. Herget (Editors)

A STRUCTURAL APPROACH TO CACSD

Magnus Rimvall
Institute for Automatic Control
Swiss Federal Institute of Technology (ETH)
CH-8092 Zurich, Switzerland
01/256'28'42

François E. Cellier
Department of Electrical and Computer Engineering
University of Arizona
Tucson, AZ 85721, U.S.A.
(602) 621-6192

In this paper different structural aspects of the new CACSD-package IMPACT are presented. In a first chapter, the different data structures needed in a general control package are presented using examples from IMPACT. In a second chapter, the need for a structured command interface is discussed. In a last sector, we elaborate on the advantages of using well structured implementation languages like Ada for CACSD-applications.

1. INTRODUCTION

Many CACSD-packages perform their operations on one single data structure: the complex matrix [4], [10]. As long as we want to treat linear systems in the time domain, this structure is adequate as each system can be described by four such matrices. On the other hand, if we for example work in the frequency domain, we would like to describe our systems by transfer-function matrices. This four-dimensional structure can not easily be represented by two-dimensional matrices. Therefore, the new CACSD-Package, IMPACT (Interactive Mathematical Program for Automatic Control Theory), supplies the user with several data structures common in control theory, e.g. polynomial and transfer-function matrices, system descriptions, domains and trajectories. Moreover, IMPACT differs from other packages not only through the supported data-structures, it also offers an extremely versatile user interface. From a computer engineering point of view, IMPACT gives a new dimension to CACSD by being the first package to be implemented in Ada [2].

IMPACT is presently being implemented at the Swiss Federal Institute of Technology (ETH), Zurich, Switzerland [3], [8]. At this time, a kernel (controlling the interactive user dialogue) and a data administrator (handling the dynamically used data structures) exist. In the present phase, the necessary control algorithms are developed/collected and included into IMPACT. The package is already internally used at ETH and will soon be generally available.

2. DATA STRUCTURES IN IMPACT

One of the most serious drawbacks of many control packages is their lack of adequate data structures, many CACSD-packages support the complex matrix as their only data structure. On the other hand, control scientists usually work with structures like polynomial matrices, transfer-functions and linear system descriptions. For this reason, and because the absence of proper data structures in a large software package cannot easily be remedied afterhand (as such a remedy would require extensive changes to the central data structures, and thereby a recoding of large sections of the package), great attention has been given to the initial design of the data structures in IMPACT. In this chapter, these structures will be presented together with a description on how such structures are interactively created (see also [7]).

After calling the program IMPACT, the user will find himself in an interactive environment, where he can create variables of different kinds and enter commands to be executed. The available data structures range from simple scalars over matrices and polynomial matrices to complex system descriptions. The form of the commands used to create these structures is similar to that of MATLAB [5]. If we wanted to create a 2 by 2 matrix, we would write:

```
TWO_TWO =  <1, 3
            5, 7>;
```

Moreover, for more complex structures not available in MATLAB, a similar syntax is used. For example, the input line

```
Q = < 1^2^1 , 2^1 >
```

will result in the polynomial row vector

```
Q(p)      =
   1. + 2.*p + 1.*p**2                  2. + 1.*p
```

Alternatively, a longer but better readable way of entering the polynomial matrix Q would be to first define the variable P as

```
P = <^1>;
```

Thereafter Q can be entered as

```
Q  =  <1 + 2*P + P**2, 2 + 1*P >;
```

Until now, all polynomial matrices have been entered by specifying all non-zero coefficients of the polynomial elements, that is in non-factorized form. Structures in this form can of course be transformed to a factorized form:

```
QF  = FACTOR (Q)
```

will transform the matrix Q to

```
QF(p)    =
    (p + 1.)*(p + 1.)                    (p + 2.)
```

It is also possible to enter factorized polynomial matrices directly:

```
QF = <-1¦-1, ¦-2>
```

The basic matrix operations addition, subtraction and multiplication may be used on polynomial matrices if the basic dimensional rules are fulfilled. However, only in special cases can the inverse of a polynomial matrix be described in the form of another polynomial matrix. On the other hand, the inverse of any polynomial matrix can be described by another structure very useful in control theory - the transfer-function matrix. For example, an element-by-element division of two polynomial row vectors will result in one transfer-function vector. E.g.

```
NUMER = <P,1>;
TRAFUN = NUMER ./ Q
```

will result in the structure

```
TRAFUN(p) =
         p                              1
    -------------------             ---------
    1. + 2.*p + 1.*p**2             2. + 1.*p
```

which of course also can be obtained by

```
TRAFUN = < ^1/(1^2^1), 1/(2^1)>
```

Whereas polynomial and transfer-function matrices can be used in IMPACT to describe linear control systems in the frequency-domain, a system description containing four matrices in the form

$$\dot{x} = A*x + B*u$$
$$y = C*x + D*u$$

can be used to describe linear systems in the time-domain. Given the component matrices, the function LCSYS will form a continuous linear system description

```
CSYS1 = LCSYS(A,B,C)
```

whereas LDSYS will form a discrete linear system description with a sampling rate of DT:

```
DSYS1 = LDSYS(F,G,H,DT)
```

The D matrix was here assumed to be a null matrix of correct dimensions. However, if the user wants to define a D-matrix, this can be entered through the use of default redefinition:

```
CSYS2 = LCSYS(A,B,C //D=DD)
```

will include the matrix DD as the direct-path matrix.

Mathematical operations on system descriptions have been defined such that the physical meaning of the operation is the same as on transfer-function matrices. For example, in the frequency-domain a multiplication of two transfer-functions correspond to a cascading of the two systems. Similarly, if a system of 2nd order has been defined through the matrices

```
A = <1, 1
     0, 1>;
B = <0
     1>;
C = <1, 0>;
SIMPLE = LCSYS(A,B,C);
```

the operation

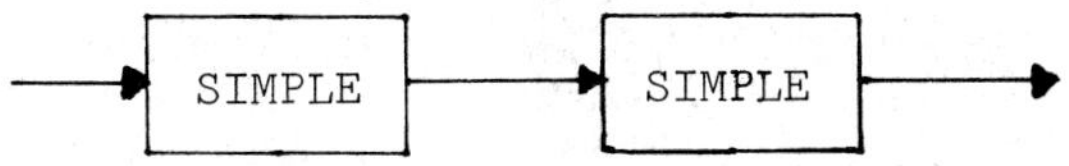


```
CASC = SIMPLE * SIMPLE
```

will result in a system of order 4 with the component matrices

```
CASC.A = <1, 1, 0, 0
          0, 1, 0, 0
          0, 0, 1, 1
          1, 0, 0, 1>
CASC.B = <0
          1
          0
          0>
CASC.C = <0, 0, 1, 0>
```

Note that the dimension of the system matrix is doubled, just as the order of the physical system.

The IMPACT-structure domain contains a sequence of discrete values. A domain can for example consist of increasing discrete values to be used to form the independent variable of a table.

```
TIME = LINDOM(0.,50.,0.1)
```

would thus define a sequence TIME with 501 elements, the first of which has the value 0 and the last the value 50, using an increment of 0.1.

A trajectory is a table of function values which uses a domain as independent variable. Such a table results from a variety of operations performed on domains. E.g. would the operation

```
TRA = SIN(TIME)
```

result in a table where each entry contains an independent variable copied from the domain TIME and the sine-value thereof.

Mathematical operations are defined on trajectories using the same domain, e.g. would the operation

```
TRB = TRA + COS(TIME)
```

once again be a table with one row of values as function of the independent variable TIME.

All graphical functions return a trajectory as result, this trajectory can then be plotted with the command PLOT.

```
PL1 = BODE(1/<9^5^9^1> //DOMAIN=LOGDOM(.1,1000.,100))
```

or a little more verbose but better readable

```
S = <^1>;
G = 1 / (S**3 + 9*S*S + 5*S + 9);
FREQ = LOGDOM(.1,1000.,100);
PL1 = BODE (G//DOMAIN=FREQ)
```

will compute a Bode-diagram of the system

$$G(s) = \frac{1}{s^{**}3 \ + 9.{*}s^{**}2 \ + 5.{*}s + 9.}$$

plot it and store this diagram as a trajectory.

Furthermore, domains and trajectories can be used to simulate system behaviour. If SSYS is any system representation (e.g. a transfer-function matrix or a system description),

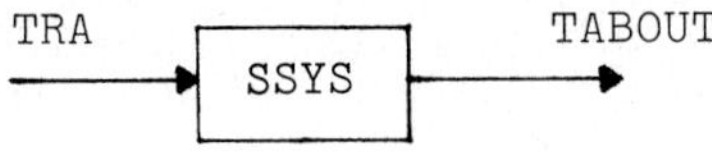

TABOUT = SSYS * TRA

will perform a simulation and store away the values of the output signal at the discrete times of the trajectory TRA, thus making TABOUT another trajectory variable specified over the same domain as TRA.

As we live in an imperfect world, control scientists usually have to use non-linear models to describe real systems. Therefore, IMPACT will include structures to describe non-linear systems as well. From such a structure, it will be possible to get a linearized model which can be used in a linear design of an appropriate controller. Thereafter, this controller can be inserted into the non-linear model, and the behaviour of the controlled, non-linear system can be simulated. If the simulated results are not satisfactory, this design sequence can be repeated, using e.g. another criterion for the controller design. IMPACT also provides several possibilities to transform linear system descriptions to nonlinear system descriptions, and vice-versa. Both continuous and discrete systems can be modeled, and sampled data systems are connections of the two.

3. MODE OF INTERACTION

Many modern interactive CACSD-packages are command driven in the sense that the user controls the flow of action of the package using an (often quite complex) command language. Compared with other means of communication, e.g. question-and-answer or menu-driven interaction, a command driven interface is faster, gives the user a greater flexibility and is better suited for the algorithmic kind of problems found in control theory. However, any developer of such a command language must make compromises: if the language is made too rich, the complexity of the system makes it hard to use. On the other hand, if too few language elements are included, the system will not be flexible enough to let the user solve all his problems. In this chapter, we will discuss the need for a highly structured command language in CACSD, and how at the same time elements can be built into the language to facilitate its use by inexperienced users.

As no general CACSD-package can include every conceivable control algorithm, especially not if it is to be used as a tool in scientific research, the user must be supplied with an interface flexible enough to let him extend the package according to his own needs. In particular, it must be possible for the user to assemble existing base algorithms to form more powerful or more general algorithms. Taking the development in software engineering during the last decade into account [9], this is most adequately fulfilled by a highly structured command language containing the necessary flow control elements (e.g. FOR-, WHILE-loops, IF-statements). Such a command language could be developed from scratch, giving the developer full freedom of design, or it could be derived from any existing structured computer language like Algol, Pascal or Ada. For IMPACT, a command language with a syntax similar to that of Ada had been developed. This similarity to Ada brings three advantages:

- Together with the data structures presented in the previous chapter, the command language can be made rich enough to describe almost any control algorithm.

- It can be expected that Ada will become a widely used language in the near future. Therefore, many users will find a very familiar user interface and will require only a short time to get acquainted with the package interface.

- As the user input, and any functions described in the command language, have to be interpreted rather than compiled, the execution of complex algorithms described by command language is rather slow. During the development of new algorithms, this is offset by the time not spent on compilations. However, to obtain shorter execution times, any complex algorithm should be compiled and incorporated into the package itself as soon as it is developed and tested. As the implementation language (Ada) is similar to the command language of IMPACT, such a transition can be made with a minimum of recoding.

As an example of the IMPACT command language, let us consider the problem of solving the Riccati Equation

$$\dot{x} = A*x + B*u \qquad \int_0^\infty (x'*Q*x + u'*R*u)dt \overset{!}{=} MIN$$
$$y = C*x$$

After defining the matrices A, B, Q and R, this problem can be solved by a simple algorithm described in [6]:

```
a = <...>; b = <...>; q = <...>; r = <...>;

<v,d> = EIG(<a, -b*(r\b'); -q, -a'>);
k=0; n = DIM(a);
FOR j IN 1 .. 2*n
 LOOP
  IF d(j,j) < 0 THEN k=k+1; v(:,k) = v(:,j); END IF;
 END LOOP;
p  = REAL(v(n+1..2*n,1..k)/v(1..n;1..k));
fc = -r\b'*p
```

In this algorithm, we first calculate the eigenvectors and eigenvalues of the Hamiltonian. We store the eigenvalues diagonally in d and the eigenvectors as columns in v. Thereafter we extract the columns corresponding to negative eigenvalues and finally we can calculate the feedback coefficients fc. We note the similarity with Ada, the main differences derive from the notation of MATLAB and include the use of '=' for assignment statements , '<' and '>' to describe the mathematical structures and ':' to form substructures (e.g. to form column vectors out of matrices).

Although the above sequence of statements could be directly entered on the terminal, for repetitive use it should be made into a function. In IMPACT, such a function is as simple to define as to use:

```
Function Riccati(a,b,q,r);
BEGIN -- Riccati
  <v,d> = EIG(<a, -b*(r\b'); -q, -a'>);
  ..
  ..
  p = REAL(v(n+1..2*n,1..k)/v(1..n;1..k));
  RETURN -r\b'*p;
END Riccati;

ffcc = Riccati(aa,bb,qq,rr)
```

In the previous examples, variables have been created without being previously declared. The reason for this disparity to Ada is that in IMPACT command sequences ("algorithms") are entered interactively. It would then be very cumbersome for the user to be forced to define every single variable in the beginning of every session, especially if he does not exactly know which method to use and which intermediate variables will be needed. On the other hand, explicit variable declarations help detecting programming errors in functions and increases the security of the functions by performing run-time type checks on the parameters. Therefore the header of the Riccati function could be complemented with type declarations, in which case all variables used in the function have to be declared:

```
FUNCTION Riccati(a,b,q,r : MATRIX);
  v,d,p : MATRIX;
  k,n   : INTEGER; --j is an implicitly declared loop counter (cf. Ada)
BEGIN -- Riccati
  ..
```

To further enhance the flexibility of IMPACT functions, so called default parameters can be used. In the last chapter, we saw an example on how to produce a Bode plot:

```
BODE(1/<9^5^9^1>)
```

Using the above call, a Bode plot with 100 points over the default frequency range 0.01 to 100.0 is produced. However, if we wanted to magnify this plot over a smaller range, we could specify the optional (defaulted) parameters LOW_BOUND and/or HIGH_BOUND:

```
BODE(1/<9^5^9^1> //LOW_BOUND=0.1//HIGH_BOUND=10.0)
```

The use of defaulted parameters simplifies the standard calls, yet offers an extensive functional flexibility. Moreover, through the use of mutually exclusive qualifiers, different modes and/or input/output specifications can be selected. Hence, the parameters of the Bode function specifying the frequency bounds can, as we have previously seen, be replaced by a //DOMAIN qualifier (helpful for the user producing several Bode plots over the same domain):

```
FREQ = LOGDOM(.1,10.,100);
BODE (G //DOMAIN=FREQ)
```

With the potent data structures and command language of IMPACT, the advanced control scientist is given a very powerful algorithmic environment, which he can further adapt to his own needs. On the other hand, if first time users are directly presented with the full IMPACT package, they will most certainly be stunned by the complexity of the package. Many CACSD-packages try to resolve this problem by including an interactive HELP facility. However, whereas such help is excellent once you have a general overview of the package and only need information on a particular subject, during an introduction to the package it is as pedagogic as a 200 page reference manual. Of course, IMPACT does support an interactive HELP, but to prevent the initial shock it also gives the user a gradual introduction. A tutorial presents only the simplest language elements, e.g. how to create variables and how to call standard functions. Moreover, as even this might be too complicated for a beginner unfamiliar with the standard concepts of control theory, a query-mode has been introduced. In this mode, the initiative is transferred from the user to the system. Through a guided conversation, the system will determine the correct action to take.

Assuming that an inexperienced user wants to use the for him new function BODE. He would then call the function using the //QUERY qualifier, forcing IMPACT to enter the query mode. Thereafter the user will be asked for values of each parameter. Optional (defaulted) parameters need to be specified only when another default value is to be changed (user input has been underlined):

```
BODE (//QUERY)
BODE>>The Bode function produces one or several Bode plots of system(s)
BODE>>described by transfer-function(s).
BODE>>SYSTEM    (NO DEFAULT): 1/<9^5^9^1)
BODE//LOW_BOUND (DEF=0.01)  : 0.1
BODE//HIGH_BOUND(DEF=100.0) : 10.0
BODE//POINTS    (DEF=100)   :
```

Especially for functions with many parameters, this facility is very useful. Moreover, if the user is uncertain on the meaning of a particular parameter, he can enter a HELP for further information.

```
BODE>>SYSTEM: HELP
BODE>>Please enter a transfer-function (transfer-function matrix)
BODE>>describing the system.
BODE>>
BODE>>When the system is given by a transfer-function matrix,
BODE>>one Bode plot is produced for each transfer-function of the
BODE>>matrix (can be overridden by the //SELECT_COMPONENT qualifier).
BODE>>SYSTEM:
```

If at this point the users description of the system is in the form of a system description and not a transfer-function, one of the options available to the user would be to open another interactive session through the command SPAWN, there use the general HELP facility to find out how a transfer-function is obtained from a linear system description, perform the transformation and return the result to the BODE command:

```
BODE>>SYSTEM: SPAWN

%IMPACT-MESSAGE, Global session 3 is started.
%IMPACT-MESSAGE, All variables are imported as local copies.

>> HELP
 ..
>> RETURN TRANS(LG)

%IMPACT-MESSAGE, Global session 3 is closed.
%IMPACT-MESSAGE, All local variables are deleted.
%IMPACT-MESSAGE, 1 parameter is passed back to queried function BODE.

BODE//LOW_BOUND (DEF=0.01)  : 0.1
```

The possibility to start a new, "virtual" session can also be used by the more advanced users to open up local workspaces for intermediate calculations (scratchpads), using an environment similar to the virtual processes/windows found on many modern workstations.

4. IMPLEMENTATION CONSIDERATIONS

Until now, most CACSD-packages and CACSD-libraries have been implemented in FORTRAN. This insures a fair portability, as FORTRAN compilers are available on virtually all computer systems and ANSI FORTRAN standards exist. However, due to the diminishing cost of hardware and the soaring cost of software, software engineering aspects like reliability, portability, and maintenance costs will play a more important role in future implementations of larger application packages. Whereas these aspects certainly do not favour FORTRAN, the new computer language Ada was designed with these particular aspects in mind [2]. Being the first larger CACSD-project to be implemented in Ada, IMPACT therefore gives a new dimension to CACSD.

In Ada (as in other languages, e.g. Pascal) the user can define his own structured data types. This is very useful to CACSD-programs, permitting us to implement program structures directly corresponding to the structures used in control theory. For example, we can define a linear system description type as one record containing four different-sized matrices:

```
TYPE syst_descr_type(n_dim, m_dim, p_dim : positive) IS
        RECORD
          state_matrix  : matrix_type(n_dim, n_dim);
          input_matrix  : matrix_type(n_dim, m_dim);
          output_matrix : matrix_type(p_dim, n_dim);
          direct_matrix : matrix_type(p_dim, m_dim);
        END RECORD;
```

The so called discriminants n_dim, m_dim and p_dim specify the order of the system, the number of inputs and the number of outputs, respectively. We notice that the four matrices are of different sizes. Moreover, as Ada allows us to dynamically create any number of variables of this kind (syst_descr_type) with different discriminant values, we will be able to simultaneously work on several different-sized systems without wasting any storage space. Furthermore, the use of such high-level structures makes the program more readable and enhances the robustness of the program, as it is automatically checked for consistency. During run-time, IMPACT will do a type-check on each operation, guaranteeing that you do not for example multiply a system description in the time domain with a transfer-function.

Through the use of <u>data abstraction</u>, we can hide the details on e.g. data types or algorithm implementations from other parts of the program. This feature is a cornerstone in Ada's aim at robust programming, and should be used to modularize larger software projects. Data abstraction is used quite extensively in IMPACT. For example, procedures containing mathematical algorithms are "hidden" from the programmer not working directly on these algorithms. Such a programmer can of course use all these algorithms to perform the mathematical operations, but he has no possibility to change the algorithms, or to access the internal data structures of the algorithms. Moreover, in some cases the whole implementation of the algorithm is hidden, so that the programmer has no way of knowing how the algorithm internally works, and thereby cannot use this knowledge to adapt ("improve") his program in such a way that it relies on the particular implementation used. This means that, even at a very late stage of development, it is possible to replace internal routines (e.g. for numerically better ones), without the risk of influencing the behaviour of other routines.

Ada is per definition portable, there may not exist any sub- and/or super-set of Ada with that name. Furthermore, the language is rich enough to support CACSD-packages; for example Ada's support of recursiveness allows for a much more elegant coding of the IMPACT expression parser than FORTRAN would do. In this way, IMPACT shall be easier maintainable and updatable than most other CACSD-packages.

Ada provides for a unique means of exception handling. The main difference between the Ada exception handler and most conventional (user defined) error handlers is that Ada can handle user-errors (e.g. erroneous interactive input sequences) as well as system-/programming-errors (e.g. division by zero or array-index out of range) in a portable manner. This is heavily used in IMPACT to return the program to a consistent state in the sequel of an interactive input error and/or program error, guaranteeing that the program is not aborted with a loss of all intermediate results.

A current disadvantage of Ada is that no Ada-libraries of e.g. mathematical algorithms are available. However, such libraries are planned and should emerge on the market in the near future [1]. Moreover, Ada allows for a so called "PRAGMA INTERFACE" to access programs written in other languages.

5. SUMMARY

The keyword of this contribution was "structure". Only through structured data elements it is possible to describe all mathematical entities used in control theory. The incorporation of a highly structured command language in a CACSD-package guarantees the flexibility needed in a research environment. A package implemented in a highly structured computer language like Ada can be expected to be more reliable and require less maintenance than a package using FORTRAN. Examples from the new CACSD-package IMPACT have given illustrations to these "structural" aspects of CACSD.

6. REFERENCES

[1] ACM Ada letters, 4 (1984) Iss.3 p.68.

[2] ANSI/MIL-STD 1815 A, Reference manual for the Ada programming language (January 1983).

[3] Cellier, F.E. and M. Rimvall, Computer Aided Control Systems Design 1984., in Ameling, W. (Ed.), Proc. First European Simulation Conference ESC'83, (Informatik Fachberichte, Springer Verlag, 1983).

[4] Little, J.N. et alia, CTRL-C and matrix environments for the computer aided design of control systems, in Proc. 6'th International Conference on Analysis and Optimization (INRIA), (Lecture notes in Control and Information Sciences 63, Springer Verlag, 1984).

[5] Moler, C., MATLAB, User's Guide, Department of Computer Science, University of New Mexico, Albuquerque, USA, (1980).

[6] Potter, J.E., Matrix quadratic solutions, SIAM Journal of Applied Mathematics, 14 (1966) 496-501.

[7] Rimvall, M., IMPACT, Interactive Mathematical Program for Automatic Control Theory, A Preliminary User's Manual, Institute for Automatic Control, ETH Zurich, Switzerland (1983).

[8] Rimvall, C.M, Cellier, F.E, IMPACT - Interactive Mathematical Program for Automatic Control Theory, in Proc. 6'th International Conference on Analysis and Optimization (INRIA), (Lecture notes in Control and Information Sciences 63, Springer Verlag, 1984).

[9] Scientific American, Special issue on Computer Software, (September 1984).

[10] Walker, R. et alia, $MATRIX_x$, A Data Analysis, System Identification, Control Design, and Simulation Package, IEEE Control Systems Magazine (December 1982).

COMPUTER-AIDED CONTROL SYSTEMS ENGINEERING
M. Jamshidi and C.J. Herget (Editors)

KEDDC - A Computer-Aided Analysis and Design Package for Control Systems

Chr. Schmid

Dept. of Electrical Engineering, Ruhr-University Bochum,
F. R. Germany

ABSTRACT

KEDDC is a comprehensive CAD package designed to cover a wide range of control system engineering tasks. It contains modules for process identification, system analysis, controller design, simulation and controller implementation based on a broad variety of modern as well as classical approaches for SISO and MIMO systems. The user-friendly interface, graphic output and state-of-the-art numerical algorithms give the control engineer more power at his fingertips than ever before. Based on the KEDDC framework, developed at Ruhr-University Bochum, KEDDC is available in different configurations and versions. It can easily be tailored and adapted by the user. KEDDC can be used in the form of complete main programs or in the form of a comprehensive control engineering library. It provides for a system which minimizes engineering and programming resources required for the complete cycle of system identification, control design and design verification.

INTRODUCTION

Modern control techniques are based on more sophisticated process models, controller design and simulation studies. We are observing a growing number of special purpose CACSD systems, which are designed for dedicated tasks in those control disciplines. A few number of systems are available which allow a combination of the control system design including process modelling and simultaneously the on-line application of control techniques. In most cases the problems facing a control engineer are very diverse. Thus he wants to have a set of possible solutions, selecting a special method according to the particular situation and using a large variety of design aids. It seems advantageous to have a comprehensive, easily applicable and user-friendly software system, which is not only restricted to a special class of problems but also supports all classes of methods and algorithms. The control engineer wants to take tools from a toolbox to handle the diverse problems at the particular stages of design and implementation. In addition to this he wants to arrange and adapt these tools to special tasks. A CACSD system which shows the flexibility to provide tools for these problems, was developed at the Department of Electrical Engineering, Ruhr-University Bochum [1]. This system, called KEDDC, was already planned in 1973 to combine works and results of different research projects with the objective to make advanced methods of system modelling, controller design and adaptive control accessible to engineers. The approved concept has been extended by many projects in the last years. To this date it is a still growing system with many capabilities.

At first a survey of the most significant capabilities of KEDDC is given. The system representation forms and the methods are listed. The software structure is presented and finally the user interaction, graphics and documentation are discussed.

COMPONENTS OF KEDDC

Fig. 1 provides a block diagram of the main components of the CAD system and their interfaces with the applications environment to research and development. The major components are a set of interactive non real-time programs for system and signal analysis, a database, an extensive program library and real-time programs interfacing with experimental hardware. Access to a common database improves the communication between research groups working on a common project. In particular, access by all users to modelling and design procedures can help to improve the interfacing between system analysts and control designers. KEDDC also forms a common software base reducing duplication of program development. It is organized as an open system, parts of which may be added, updated or removed at any time. This unlimited extendability is of particular importance in an applied research and development environment with on-going of new programs and methods. Program development is supported by the program library which contains about 1000 routines. Implementation of existing real-time programs will allow immediate implementation of control algorithms, on-line identification and adaptive control.

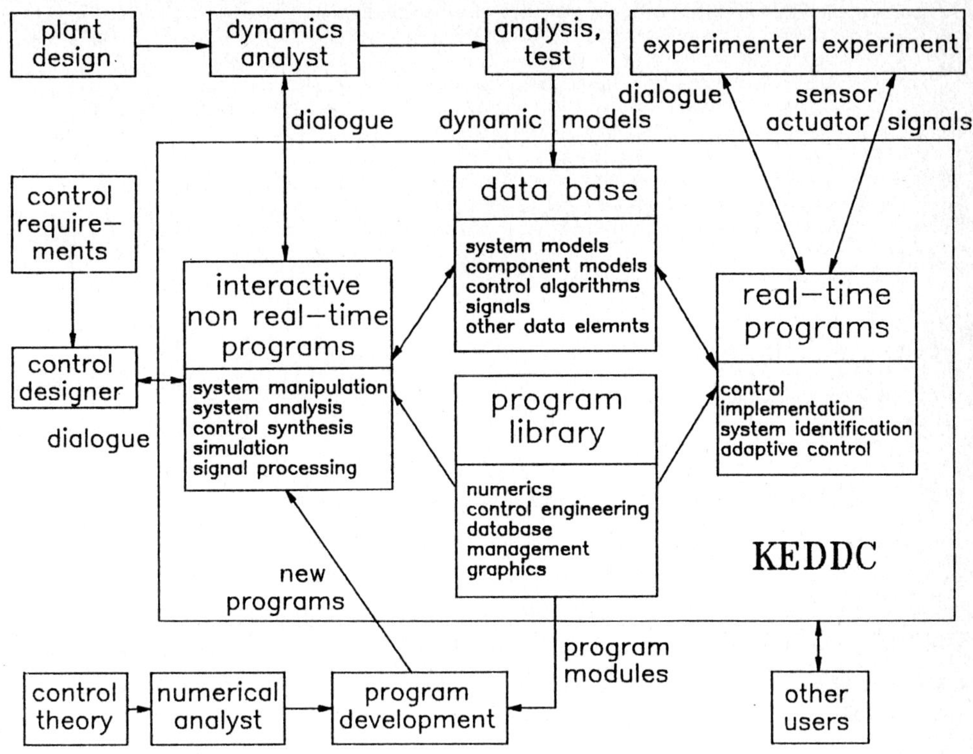

Fig. 1. KEDDC Main Components and Applications Environment

SYSTEM REPRESENTATION FORMS

The package supports a wide variety of operations for system manipulation and analysis, control system synthesis, simulation and signal processing. Systems may be described in any of the 10 different forms shown in Fig. 2 and transformed with ease between time domain, state space and frequency domain. The choice of system descriptions includes deterministic and stochastic input/output signals, continuous and discrete time transfer matrices, continuous and discrete state space representations, as well as matrix fraction descriptions. The solution of system transformations or controller design procedures are usually supported by a number of numerical options in order to ensure efficient and accurate solution of high-order problems.

Fig. 2. System Description Forms and Transformations:
1. series of deterministic or stochastic input and/or output signals,
2. series of auto- or crosscorrelation values,
3. discrete values of the pulse response,
4. discrete values of the step response,
5. transfer function (matrix) in s-domain,
6. transfer function (matrix) in z^{-1}-domain,
7. discrete values of frequency responses, spectra
8. matrices for continuous state space,
9. matrices for discrete state space,
10. polynomial matrices for MFD.

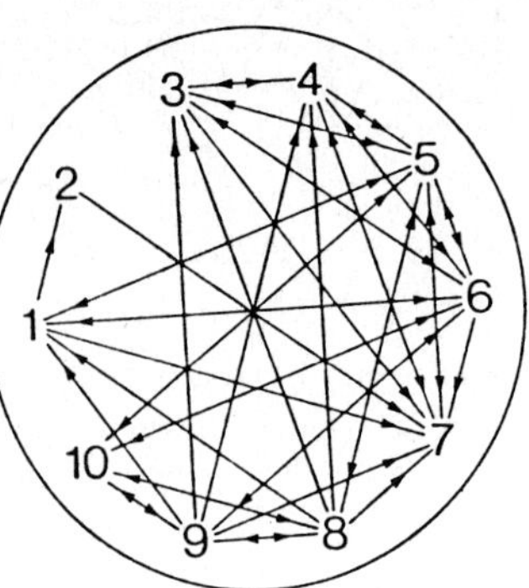

CORE PROGRAMS

Frequently used tasks, like transformations to different representation forms, data base definitions or basic calculations, are separated from other tasks and are concentrated in main interactive system handlers, the so-called Managers. E. g. matrices can be typed in using dialogue and combined to a state space description, which may be tested for controllability or observability. Eigenvalue analysis, minimal realization, calculation of transfer matrices or multivariable zeros and similar tasks are standard tools solved by Managers. The following Managers are available:

Signal Manager
- generation, analysis and display of signals

System Manager
- manipulation and analysis of SISO transfer functions and elements of transfer matrices
- continuous and discrete time

Matrix Manager
- matrix manipulation
- analysis and manipulation of MIMO systems in state space
- continuous and discrete time
- transformation to and from transfer matrix

Frequency Manager
- calculation and display of frequency response data or spectra
- principal gains
- continuous and discrete time

Polynomial Matrix Manager

- manipulation of polynomial matrices
- analysis of MIMO systems described by matrix fraction description (MFD)
- continuous and discrete time
- transformation to all other system representation forms

Graphics Manager
- generating of all graphs in KEDDC
- manipulating of graphical representations.

Using the Polynomial Matrix Manager typical command subtasks for handling systems which are described by the MFD

$$\underline{G}(p) = \underline{Q}(p)\ \underline{P}(p)^{-1}$$

or

$$\underline{G}(p) = \underline{P}(p)^{-1}\ \underline{Q}(p)$$

are explained here in extracts:

Input/Output Operation

Type in $\underline{P}(p)$, $\underline{Q}(p)$ or System ($\underline{P}$, $\underline{Q}$) using dialogue.
Read in $\underline{P}(p)$, $\underline{Q}(p)$ or System ($\underline{P}$, $\underline{Q}$) from database.
Save $\underline{P}(p)$, $\underline{Q}(p)$ or System ($\underline{P}$, $\underline{Q}$) to database.
Edit $\underline{P}(p)$, $\underline{Q}(p)$ or System ($\underline{P}$, $\underline{Q}$).
List $\underline{P}(p)$, $\underline{Q}(p)$ or System ($\underline{P}$, $\underline{Q}$).
Show polynomial matrices in database.

Polynomial Operation

Perform arithmetic operations (*, +, -) using polynomial matrices.
Calculate determinant, determinant degree.
Calculate inverse.
Test properness of polynomial matrix.
Generate a column or row proper matrix from a MFD.

System Operation

Calculate transmission poles and zeros.
Calculate transfer matrix.
Calculate relatively prime MFD from transfer matrix.
Calculate relatively prime MFD from state space description.

Calculate minimal realization from MFD.
Calculate left MFD from right MFD and vice versa.
Calculate greatest common divisor.
Calculate inverse or dual system.

There are about 250 such subtasks in the core programs, which play the role of elementary service functions.

METHODS

KEDDC supports a wide variety of operations for off-line system manipulation and analysis, control system synthesis, simulation and signal processing, as well as on-line automatic controller implementation, system identification and adaptive control. The methods are essentially limited to finite-dimensional linear multivariable systems, which may be described in any of 10 different forms and transformed with ease between time domain, state space and frequency domain. Some extensions to nonlinear systems, e. g. harmonic linearization, are available. KEDDC is not restricted to a specific class of methods. It is a system containing tools which support systematic design strategies, and which help the problemsolver to combine his intuition with the results of high-accuracy numerical solution. In addition to modern analysis and design concepts also classical methods are integrated, which are very simple. This may be seen under the automatization aspect of classical techniques. This is a concession to the process engineering community, where KEDDC is mainly used and where classical analysis and design tasks are interlaced into modern control techniques.

From the theoretical point of view the type of analysis and design procedure depends strongly on the design philosophy, which is applied. This is an idealization, but in practical applications a planning phase is preceded, which includes extensive calculations, e. g. compensator design, simulation studies, modelling of the plant or a signal analysis. For instance an automatic multiobjective optimization procedure or a state-feedback and Kalman filter approach can facilitate the choice of filters in MIMO self-tuning or model reference adaptive schemes. Such options for this phase are supported using some parts of KEDDC. So any part of KEDDC can be treated as a present library of utilities.

A detailed presentation of all implemented methods cannot be given here. But a short survey about the classes of methods is listed here. More informations about modelling can be get from [2], about control from [3] and about special applications from [4].

System identification methods
- Approximation in time domain (6 methods),
- Approximation in frequency domain (2 methods),
- Approximation by exponential functions,
- Approximation by momentum methods,
- Least squares algorithms,
- Correlation and spectral analysis,
- Various parameter estimation methods using different numerical methods and strategies.

Control system design methods
- Continuous and discrete time compensators,
- Finite settling time methods,
- Various pseudo-compensator methods,
- Optimization of PID-type controllers,
- State-feedback controllers with or without PI-action designed by pole placement or LQG procedures,
- Inverse-Nyquist-array technique and systematic compensator design,
- Reduced and full-order observers by pole placement or LQG procedures,
- diverse utilities to support design procedures, e. g. root-locus-plots, return difference magnitude plots.

Adaptive control system design methods
- Adaptive PI-controller using periodic test signals,
- Adaptive compensator using various parameter estimation methods,
- Various self-tuners,
- Various model reference adaptive control approaches,
- Adaptive observers.

SIMULATION FEATURES

Standard structures for linear control systems are simulated by a standard simulator. The overall system to be simulated may be composed of different subsystems of different representation forms. Each block in this structure may be of the form 5, 6, 8, 9 or 10 of Fig. 2. A second type of simulator performs block-oriented continuous-time simulation using elementary CSMP-like blocks. In a third type of simulator the user can describe the simulation problem using the vector differential equation $d\underline{x}(t)/dt = \underline{f}[\underline{x}(t), \underline{u}(t), t]$. The user has to specify the right-hand side by primitives like summers, signal generators, or by superblocks like hysteresis or transfer functions. For direct digital control operation different special simulators are available. They simulate the environment for testing algorithms in real-time applications.

SOFTWARE DETAILS

The KEDDC package is organized in a four-level hierarchy of separate program modules (Fig. 3). All basic functions are concentra-

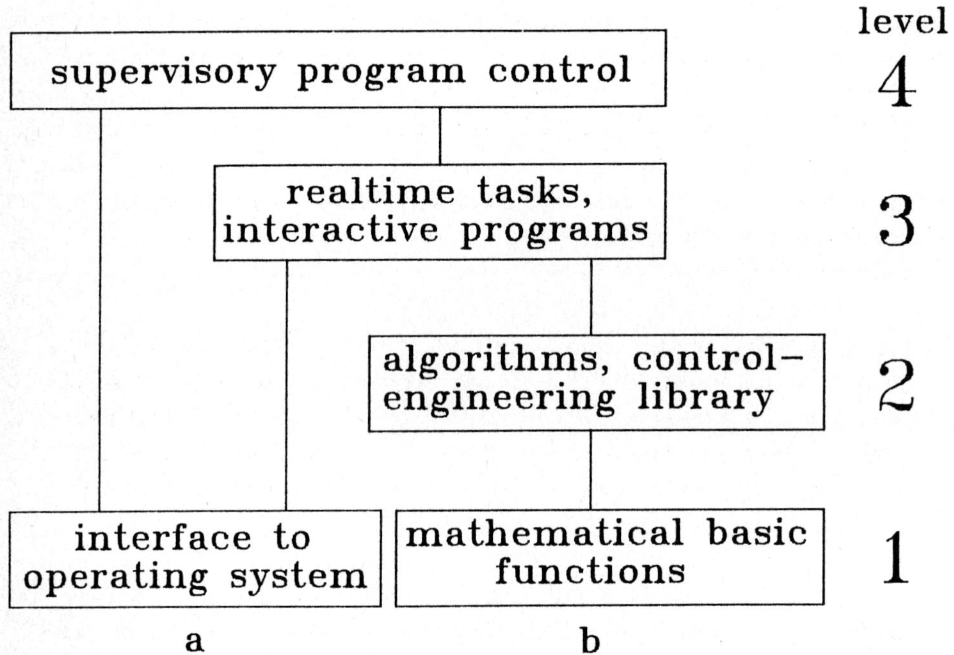

Fig. 3. Software structure of KEDDC

ted within the first level. These functions include all numerical algorithms (e. g. for matrix or polynomial operations). The second level represents the control engineering library containing all relevant specialized algorithms as subroutines which make extensive use of modules of proven numerical software (EISPACK, LINPACK, MINPACK, etc.). Level 1 and level 2 subroutines together form the complete library system. This is one form in which KEDDC can be used. Modules of the third level are designed for dedicated main tasks, in which a set or a class of methods is grouped together. They are complete, interactive or real-time programs which can run as stand-alone programs or which can run supervised by a central monitor which resides in the top level. Data exchange between the modules in facilitated through data files. Any choice of level 3 programs is possible. With a proper choice from a large set of programs (some are prerequisite) a KEDDC system can be tailored according to the user's request.

KEDDC provides a high degree of portability, which results from the interface of KEDDC to the operating system and to hardware-dependent functions. Generally the level 1a routines build an interface to the world outside of the programs. The strict separation of the core system and the interactive code in this level makes this possible. For the implementation in an other computer environment only level 1a modules have to be modified. That level contains a number of device dependent functions also supporting the user interface. For a minimal configuration (e. g. no graphics, no help functions), dummy routines are available to replace non-essential functions. Program development may take place on other than the target machine as the core system is portable.

Level 1b and level 2 routines are portable to any FORTRAN ANSI X3.9-1966 environment, but are fully upward compatible to FORTRAN ANSI X3.9-1978. The level 1a routines are machine dependent, and are predesigned for any computer using the FORTRAN ANSI X3.9-1978 standard. This is valid also for level 3 and 4 programs. An exception is the Graphics Manager, for which ANSI PASCAL is used.

All vector, matrix and polynomial primitive operations are assembled as routines in dedicated libraries. These subroutines can be simply exchanged by subroutines, which are optimized for the user's machine, e. g. using assembly or micro programming. These routines speeds up processing.

The entire package consists of about 10 Mbyte of source code. Level-3 programs can be configured to run in about 56 kbyte partitions. Dimensions like max. order, number of inputs and outputs etc. are not restricted to fixed values. The maximum dimensions in source code of level 3 programs can be automatically changed using a source preprocessor, which is an integral part of KEDDC. Level 1 and level 2 routines are free of fixed dimensions.

NUMERICAL ALGORITHMS

The control engineering library which contains the basic design and analysis procedures in level 2 is based on the level 1b numerical software. Subsets of LINPACK, EISPACK, MINPACK and other similar reliable software sources are included in the lower level. Most control-related algorithms (e. g. parameter estimators, Riccati equation solver) are realized using different approaches and numerical algorithms. They are accessible by the user. He can select a specific algorithm, but the default is that algorithm which has been proved to be the most reliable one over some years, and which is suitable for general purpose application. Due to the library structure this strategy allows to include new algorithms drawn from recent research in numerical analysis and to remove obsolete parts from library. In addition to this KEDDC can be used as a test bed for new algorithms, which are tested by the KEDDC community under real conditions. Most hints at unreliable parts and the impact to improve design and analysis procedures are coming from this group. It is the authors opinion that those design and analysis procedures are the successful candidates, which are robust enough as to be used as black boxes by the practical working engineer.

INTERACTIVE OPERATION

The user interacts with KEDDC through a command-driven dialogue combined with a question-and-answer dialogue. He can perform operations in arbitrary sequence, access the implemented methods and algorithms like tools in a toolbox. Each interactive program has its own set of commands, which can be enabled by supervisory scheduling of the programs using the monitor. An active program prompts with its name. A two-character command initiates a subtask or a local question-and-answer dialogue, where values can be entered in free format. Commands are natural, simple and powerful. The local question-and-answer dialogue consists of questions sent by the computer, which ends with '=?', where the user has to specify one or some values by typing a character string. If the question ends with '?' only a yes/no answer is requested. Reasonable defaults make this dialogue easy to use. On zero input default values are assumed.

Menu, status and help facilities normally enable casual users to master the system without reference to manuals. At any stage the user may get 'help' information. This is essential for error recovery. The user is provided with a detailed analysis of his errors and an explanation about what happened. In addition he will get some hints, how he should proceed. The command interpreter handles both local commands and central commands which can also be tailored by the user. Level-3 programs are interruptible at any stage, and new subtasks from other level-3 programs can be interlaced into the running dialogue if program cloning is supported by the operating system. This feature allows a very free and flexible use of KEDDC. A KEDDC session does not require any pre-planning or programming as forgotten steps can be interlaced in the dialogue at any time, such that a "hang-up" cannot occur.

DATABASE

KEDDC is organized as a number of stand-alone program modules. Data exchange between the various program modules and permanent data storage is affected through data files. In order to pass data from one module to another, a file has to be created or appended with the first module and this file then has to be read by the destination module. The user is responsible to specify and memorize file names. In order to keep track of the file content, the user may enter descriptive informations when creating a file. Permanent data files are used to save data for continuation of a KEDDC session at a later point in time and to build a database for a project. All modules allow to save and retrieve the data necessary to restore a session.

In order to store data for all system description forms from Fig. 2, files for

- signals, correlation data,
- transfer functions and transfer matrices,
- frequency tables,
- frequency response tables and spectra,
- matrices and systems in state space,
- polynomial matrices and system in MFD,
- nonlinear characteristics

must be handled. The database can only be accessed in terms of those system description forms through level 1a interface routines. For example to read model data from file first an open request must be issued, which includes checking of the validity of the description form to be read. In the next step a record can be get from that file, which may be a transfer function or a matrix. For other database operations this procedure is similar.

GRAPHICS CAPABILITIES

KEDDC provides a control engineer with tools for design, but computer graphics can further aid in the design with visual representations. Control system engineers are accustomed to evaluating designs in terms of pole zero patterns, frequency response or time response data. Because he may want to compare various results, the graphical output should allow several functions on a single plot and multiple plots in a single output. In addition to this, many attributes such as line styles, labels, markers, titles may vary from application to application.

In KEDDC graphics operations are performed by a central Graphics Manager (GRMGR), which is an autonomous subpackage. The GRMGR can be implemented using a graphics coprocessor, or on an intelligent graphics terminal or as an internal graphics task on the same computer. Fig. 4 shows the Graphics Manager, which is a more extended concept of that, which is discussed in the CASCADE project [5]. Here we can also find two interfaces, one between the application modules and the Graphics Manager - called GRMLB [6] - and a second between

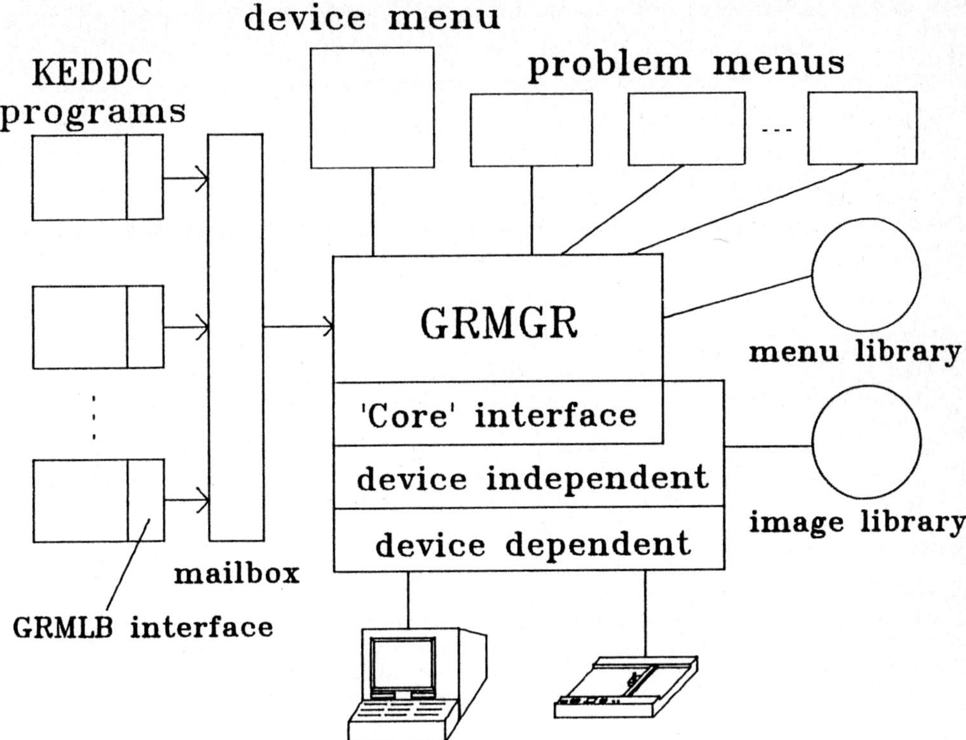

Fig. 4. The Graphics Manager in KEDDC

GRMGR and the graphics system, which is device independent. The first interface is part of the level 1a modules and the second is based on the SIGGRAPH 'Core' standard. A version using GKS is in preparation. By separating application modules from graphics tasks, KEDDC is also portable, expandable and maintainable from the graphics point of view.

Data including destination attributes are put by GRMLB into a mailbox. GRMGR maps these mailbox data into graphical form, which is described by problem menus. Data destination attributes can be given as direct viewport addresses on a virtual screen or as indirect addresses, which are described by a character string, like 'map data to viewport which is assigned to a problem menu for Bode plot' or 'which has y(t) as label for y-axis'. For addressing multiple plots the virtual screen is split into some viewports of arbitrary size.

To each viewport one problem menu is assigned, which specifies the type of diagram. Mapping to the destination devices is controlled by a device menu, which contains the information about physical size and special device characteristics.

Problem menus specify all informations which are necessary to generate a diagram. The header contains a viewport address and a type identifier (see Table 1). Text labels for axes, scales, grids,

type identifier	type of output graphs
NYQUIST	Nyquist diagram
BODE-MAG	Bode plot for magnitude, principal gain
BODE-PHA	Bode plot for phase
NICHOLS	Nichols chart
TY	time response
POLE-S	root locus plot and poles and zeros in s-domain
POLE-Z	root locus plot and poles and zeros in z-domain
MATRIX	3D-mesh surface plot for matrices
XY	general xy-diagram, nonlinear characteristic
XLOGYLOG	general x-logarithm-y-logarithm diagram
XLOGY	general x-logarithm-y-linear diagram
XYLOG	general x-linear-y-logarithm diagram

Tab. 1. Type of Output Plots in KEDDC

tic space values, text font, label orientation, isotropic mapping, curve characteristics and many other things can be specified. Application modules have full control over the characteristics of the graphs using the GRMLB interface. For standard cases prepared menus can be called from a menu library, which can be individually tailored by the user under control of GRMGR. In addition to remote operation, GRMGR can be used interactively. This is for editing menus, generating menu libraries, toggling-off or on of operating modes, rescaling of particular viewports or zooming or locating data in a diagram.

Using the GRMGR concept the user application software only controls the position of a plot, the data to be plotted, and which type of diagram is desired. Standard diagrams are available. If the user wants to have a special design for a plot, he can tailor his private menu without any programming. This shows a great deal of flexibility in use and in interfacing to different graphics configurations and ties the parts of KEDDC to an unified system.

REAL-TIME OPERATION

KEDDC supports the real-time implementation of all control schemes, which are designed using KEDDC. A multi-purpose real-time suite serves as a development tool in basic studies or pilot experiments. It consists of a command-driven central task monitor, where the user has to specify his personal configuration of hardware channels, the class of controllers to be implemented, attributes for monitoring signals on graphs or for writing them on a mass storage etc.

A second component is a set of class monitors, which is offered to the user by the task monitor. A class monitor, once selected from a menu, is responsible for one class of control algorithms, e. g. for parameter-adaptive controllers using parameter estimation methods. It serves as an intelligent interface between the user and the related real-time task. Typing chained commands the user can specify all data which are necessary to implement the controller and to run the real-time task. He can change or list parameters, call components from the data base, or monitor the real-time response. A class monitor is only active during an implementation or monitoring phase.

The communication between the class monitor and the real-time task is performed by a global data area, where all real-time relevant controller data reside. The real-time task contains redundant code which is necessary to realize the entire class of controllers. In addition options for start-up procedures, anti-windup, default controller or other facilities are available. The detailed structure of a real-time task is not unified, since it depends on the specific class of controllers and on the control problem of the plant. However, some types of standard frames are prescribed.

This development tool is supplied with a set of evaluation facilities. The user can both monitor and store the real-time behaviour of any internal or external signal. At the same time or at a later date he can evaluate and presentate results in some form. If the development tool is implemented on the same computer as the design part of KEDDC, at any instant the user has the entire power of KEDDC at his fingertips. It is possible to interlace a redesign into the running start-up or monitoring phase of a controller. If real-time signals are stored on a mass storage, during the same time an off-line evaluation can be started. Actual plant input and output signals of an adaptive closed loop can be fed to an off-line identification task. The result could be a model basis for a redesign of an adaptive controller, which might be tested in a simulator with the same specifications as the actual real-time task. After extensive verifications the new controller can be called by the class monitor for exchange with the actual one.

DOCUMENTATION

The comprehensive documentation consists of a user's manual (1200 pages) and a programmer's reference manual (1400 pages). Self-explanatory code (40 % of the source code is comments) supports the maintainability of the package. The programmer's documentation is contained in the source code. Each library routine is documented by a description of the purpose, by complete specification of input parameters, output parameters, data types, workspaces, remarks on methods, hints, references to literature, list of subroutine calls, and of the revision level. Using these informations the subroutines can be easily integrated into any software system. In addition, self-explanatory code supports the documentation. For most subroutines and sublibraries extended documentation in form of reports is available. A KEDDC interface guide, tree diagrams, module crossreference lists, and reports related to special control engineering methods and topics are available. Detailed run-time test documentation listings with many examples including ill conditioned cases are available.

SOFTWARE IMPLEMENTATION

KEDDC is organized as an open system, parts of which may be updated or added at any time. The development of new programs and methods is supported by the program library which contains about 1000 routines for basic numerical functions, control engineering, database management and graphics. At Ruhr-University, recent program development is focussed on nonlinear system identification, multivariable adaptive control and state-feedback design techniques. These efforts were complemented by the development of other techniques at industries and at research institutes. The flexible and uncomplicated organization permits an implementation on small computers. The package is quite large of about 10 Mbyte source code. Due to its structure it can be divided in subsystems of any size. Small subsystems are installed at many places in industry. Maior implementations are listed in Table 2.

location	computer	operating system
Ruhr-University Bochum	HP2100	RTE-II
	HP1000	RTE-IVB,-VI,A
	INTEL86/310	iRMX/86
Bayer AG Leverkusen	PDP11/23	RT11
		RSX11M
	VAX/750	VMS
CRC Ottawa	DPS8/52	CP6
	INTEL86/380	iRMX/86
University of Sussex	VAX/780	VMS
Philips Hamburg	P800M	MAS
CSDL Cambridge USA	HP1000	RTE-IVB
IBM Germany	IBM/370	VM(CMS)
KHD Koeln	PDP11/23	RSX11M
Voest-Alpine Austria	VAX/780	VMS
Anschuetz Kiel	VAX/750	VMS
Honeywell Germany	Series 60	GCOS 6

Tab. 2. Maior Implementations of KEDDC

EXAMPLE

To illustrate the user dialogue of the package, a hardcopy of some sections of a longer KEDDC session has been copied into this chapter. The following is showing the original dialogue. Operator input is underlined, command lines are printed in bold letters and command menus are framed. Graphics output is not exactly placed in chronological order with the dialogue. The hardcopy is printed when the image is complete.

In order to understand this demonstration, section numbers are added to the left margin. Section (1) starts with Frequency Management. The command menu is shown which appears in a special window. Principal gain plot is requested but rejected as explained using 'help' at (2). The principal gains for a system read from file TGM are plotted at (3). At (4) Matrix Management is interlaced to generate a 3D-mesh surface for the system given in state space description. Simulation is interlaced at (6) and then a second Matrix Management at (7) to display the pole zero pattern of the discrete-time system to be controlled. At (8) Simulation is continued. After some configuration dialogue, which is not shown here, the screen shows the results of all sections at (9). The graphic viewports are numbered from left to right beginning at the bottom of the screen. The viewport number 5 is a small one which is inside of number 3.

```
(1)KEDDC
   KED66: FR
   FRM66: ??

   ####################### Frequency Management #######################Page 1
      SYSTEM:                                       FURTHER CALCULATIONS
   RE Read system from file                       PO  Popov-frequency response
      G(s),G(z),u(t),y(t),G(jw)                   IV  Inverse frequency response
      g(t),h(t),(A,B,C)                           WC  Magnitude/phase for spec. freq.
   LI List system datas   | WH Write step         RA  Magnitude or phase margins
      FREQUENCY TABLE:   response on file         TT  additional deadtime
   GO Generate frequency table                    FH  Step response from freq. resp.
   EO Edit frequency table                            CONNECTIONS:
   RO Read frequency table from file              S2  Read 2nd system from file
   IO Input frequency table using dialog          ++  Parallel connection
   LO List frequency table                        **  Series connection
   WO Write frequency table on file               //  Negative feedback connection
      FREQUENCY RESPONSE:                             GRAPHICS:
   FG Calculate frequency response                    BO Bode plot complete
   RF Read frequency response from file               BM Bode plot magnitude
   IF Input frequency response using dialog           BP Bode plot phase
   EF Edit frequency response                         NY Nyquist plot
   LF List frequency response                         NI Nichols plot
   WF Write frequency response on file                DH Step response
   FM Create frequency response matrix                TR Toggle curve tracking on/off
      on file                                         NP Change no. of dots per dia.
   ST Status protocol                                 PG Principal gain plot
   Standardfilenames see next page

   FRM66: PG
   /FR-004 THERE IS NO FREQUENCY TABLE !
(2) FRM66: HE
   /FR-004 THERE IS NO FREQUENCY TABLE !
   You have initiated an operation which requires a frequency
   table. This table isn't defined.
   Read table from file with RO, or specify table interactively
   with IO, or generate it automatically with GO. The last one is
   the most comfortable way if the frequency range is initially
   unknown.
   FRM66: GO
   RECOMMENDED FREQUENCY RANGE FROM    1.79416      TO    899.2309
   FREQUENCY RANGE FROM, TO    = ? 1,1E3
   NUMBER OF FREQUENCY VALUES = ? 100
   FRM66: PG
   FREQUENCYFACTOR  = ?
   FILENAME         = ? TGM::DA
   ~~~~~~~~~~~~~~~~~~~~~~~~~~~~~~~~~~~~~~~~~~~~~~~~~~~~~~~~~~~~~~~~~~~~~
   file=TGM    ::DA, type=UE
   created using SMGR  by SCHMID on  4. 9.1983 at 11.35
   TURBOGEN. MODEL 1 MIMO
   ~~~~~~~~~~~~~~~~~~~~~~~~~~~~~~~~~~~~~~~~~~~~~~~~~~~~~~~~~~~~~~~~~~~~~
   IT IS A TRANSFER MATRIX
   NUMBER OF INPUTS =  2, NUMBER OF OUTPUTS =  2
```

```
(3) MAXIMUM PRINCIPAL GAIN IS MAPPED TO VIEWPORT  4, CURVE  1
    MINIMUM PRINCIPAL GAIN IS MAPPED TO VIEWPORT  4, CURVE  2
(4) FRM66: @@
    KEDDC: MM
    MMG66: RS
    FILENAME = ? VFELDE::DA
    ~~~~~~~~~~~~~~~~~~~~~~~~~~~~~~~~~~~~~~~~~~~~~~~~~~~~~~~~~~~~~~~~~~~~~~~
    file=VFELDE::DA, type=MA
    created using MMGR  by SCHMID on 31.10.1980 at 13.25
    TURBOGEN. STATE SPACE MODEL A2Y3
    ~~~~~~~~~~~~~~~~~~~~~~~~~~~~~~~~~~~~~~~~~~~~~~~~~~~~~~~~~~~~~~~~~~~~~~~
     2 BLOCKS,  3 MATRICES IN 1ST BLOCK
     BLOCKNO., MATRIXNO. = ? 2
    MMG66: MS
    DO YOU WANT GRAPHICS ? Y
    MATRIX A
    + 0 0 0
    0 + 0 0
    0 + + 0
    + 0 0 +
    MATRIX B
    + 0
    0 +
    + +
    + +
    MATRIX C
    0 0 1 0
    0 0 0 1
(5) MATRICES MESH SURFACE IS MAPPED TO VIEWPORT  2, CURVE  1

    MMG66: EX
(6) FRM66: @@
    KEDDC: DG
    DIG66: @@
(7) KEDDC: MM
    MMG66: RS
    FILENAME = ? TZM::DA
    ~~~~~~~~~~~~~~~~~~~~~~~~~~~~~~~~~~~~~~~~~~~~~~~~~~~~~~~~~~~~~~~~~~~~~~~
    file=TZM   ::DA, type=MA
    created using MMGR  by SCHMID on  9. 9.1983 at 11.46
    TURBOGEN. STATE SPACE MODEL A2Y4
    ~~~~~~~~~~~~~~~~~~~~~~~~~~~~~~~~~~~~~~~~~~~~~~~~~~~~~~~~~~~~~~~~~~~~~~~
     3 BLOCKS,  3 MATRICES IN 1ST BLOCK
     BLOCKNO., MATRIXNO. = ? 2
    MMG66: DM
    SAMPLING INTERVAL = ? .2
    MMG66: EI
    DO YOU WANT GRAPHICS ? Y
    ------------ EIGENVALUES -----------
     NO  REALPART       IMAGINARYPART  S/I
      1   .385317         .285704       -
      2   .385317        -.285704       -
      3   .462243         .308177       -
      4   .462243        -.308177       -
      5   .644391         .250971       -
      6   .644391        -.250971       -
      7  2.131015E-02    .437934       -
      8  2.131015E-02   -.437934       -
      9   .715286        0.00000       -
     EIGENVALUES ARE MAPPED TO VIEWPORT  5, CURVE  1
```

```
MMG66: NU
DO YOU WANT GRAPHICS ? Y
-------------- ZEROS ---------------
 NO  REALPART      IMAGINARYPART  S/I
  1  -2.03490        0.00000       +
  2  -1.73257        0.00000       +
  3    .770414       0.00000       -
  4    .368101        .463010      -
  5    .368101      -.463010       -
  6    .449659        .360620      -
  7    .449659      -.360620       -
ZEROS ARE MAPPED TO VIEWPORT  5, CURVE  2
MMG66: EX
```

(8)
```
DIG66: RE
FILENAME = ? TGM::DA
     .
     .
     .
DIG66: SI
SIMULATION STEP SIZE = ? .1
SIMULATION TIME      = ? 10
SIGNAL U ( 1) IS MAPPED TO VIEWPORT  1, CURVE 1
SIGNAL U ( 2) IS MAPPED TO VIEWPORT  1, CURVE 2
SIGNAL Y ( 1) IS MAPPED TO VIEWPORT  3, CURVE 1
```

(9)
```
SIGNAL Y ( 2) IS MAPPED TO VIEWPORT  3, CURVE 2
```

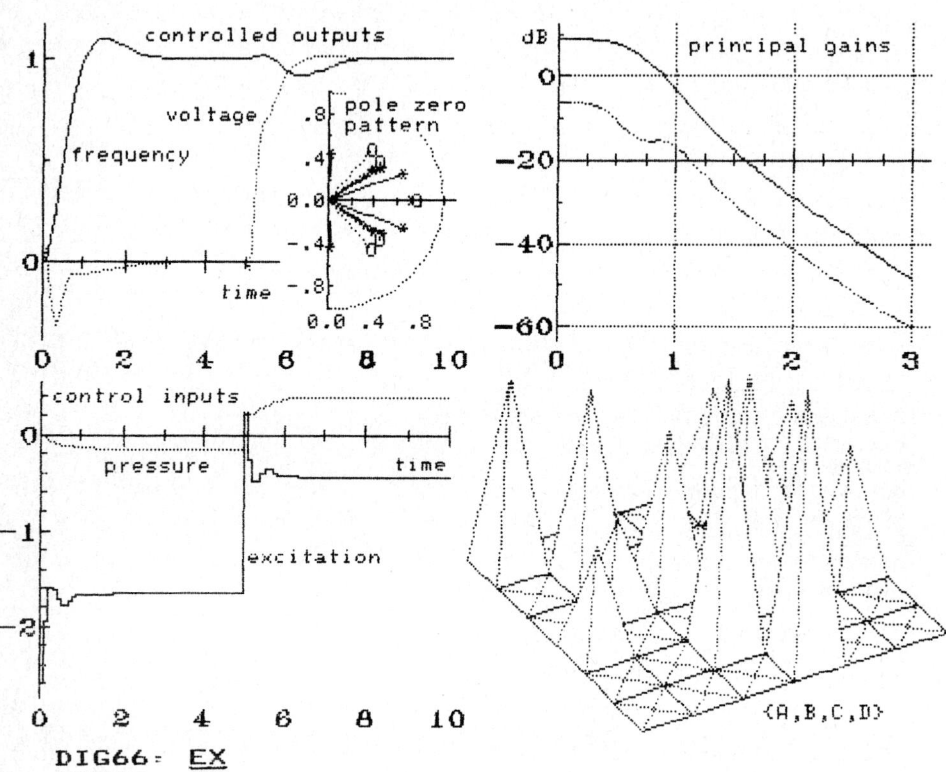


```
DIG66: EX
FRM66: EX
KED66: EX
```

This example is a small abstract of a longer demonstration which was performed to show, how KEDDC can be used for modelling, control design, simulation and controller implementation for a turbogenerator plant in an active experiment. This demonstration starts using the knowledge of a rough model of the turbogenerator plant. Some time and frequency response data had been measured, and on the basis of these measurements the model is completed. After model validation a design model is evaluated. LQ design technique is applied. The controller is simulated in different versions and implemented in real time. Finally adaptive control operation is demonstrated. A complete documentation about this example can be requested from the author.

CONCLUSIONS

The availability of the CAD system KEDDC has improved the quality of control engineering designs and greatly enhanced the productivity of the control engineer in industries and research institutes. It has been found a very efficient and versatile, almost indispensible tool for the reliable analysis of high-order systems [4]. Its architecture and user interface are particularly well suited for an applied research and development environment. It serves not only as an efficient tool for control system analysis and design but also is intended to play a role as an interface and stimulus for the communication between the various research groups within industries and universities.

REFERENCES

[1] Schmid, Chr. and H. Unbehauen: KEDDC, a general purpose CAD software system for application in control engineering. Prep. 2nd IFAC/IFIP Symposium on Software for Computer Control (SOCOCO), Prag (1979), paper C-V.

[2] Schmid, Chr. and H. Unbehauen: Identification and CAD of adaptive systems using the KEDDC package. Prepr. 5th IFAC Symposium on Identification and System Parameter Estimation. Darmstadt (1979), paper C1.

[3] Schmid, Chr.: CAD of adaptive systems. Proc. IFAC Symposium on computer aided design of control systems, Zürich (1979), 625-630.

[4] Stieber, M. E.: Design and Evaluation of Control Systems for

Large Communication Satellites. IFAC-Workshop of Identification and Control of Large Space Structures. San Diego, CA, June 4-5, 1984.

[5] Bly, S. A.: Graphics capabilities. In: M. Marietta, Issues in the design of a computer-aided systems and control analysis and design environment (CASCADE). Oak Ridge National Laboratory, Report ORNL/TM-9038, Part V, August 1984.

[6] Schmid, Chr.: GRMLB-Graphics Manager Library Users' Reference Manual. Ruhr-University Bochum, 1984.

COMPUTER-AIDED CONTROL SYSTEMS ENGINEERING
M. Jamshidi and C.J. Herget (Editors)
© Elsevier Science Publishers B.V. (North-Holland), 1985

MATRIX$_x$: CONTROL DESIGN AND MODEL BUILDING CAE CAPABILITY

Sunil C. Shah
Michel A. Floyd
Larry L. Lehman

Integrated Systems, Inc.
101 University Avenue
Palo Alto, CA 94301-1695

OVERVIEW

MATRIX$_x$ is a computer-aided-engineering (CAE) package that can be used to efficiently complete the entire control design cycle. The package includes features to graphically and interactively develop nonlinear simulation models from analytical equations or test data. These models can be linearized and simplified for use in control design. Both classical and modern control design tools are available. The simulation capability allows for rapid validation of control laws under off-design and failure conditions.

The package features a powerful command interpreter, a flexible command language, interactive on-line graphics, state-of-the-art numerical algorithms, and user-transparent model catalog and data management facilities.

This paper describes the capabilities of Version 4.0 of MATRIX$_x$ and SYSTEM_BUILD (released Nov. 1984).

INTRODUCTION

Computer-aided engineering (CAE) tools will strongly direct future control design practice. Each aspect of related analysis in the control design process (Fig. 1) from system identification, data analysis, model building, model reduction, control design, simulation, to implementation and validation can be done more efficiently with good software tools. Achieving higher performance in increasingly complex systems, in spacecraft printing control to servomechanical control of flexible disc drives, requires powerful CAE technology. Analog and digital single/multi-rate controller designs require analysis and evaluation under a wide range of conditions.

MATRIX$_x$ provides an extensive set of capabilities in a single integrated package with uniform data and file formats. "Bookkeeping" chores are performed by the software, leaving the control designer free to concentrate on design issues. Integrated graphics allow the designer to visualize information rapidly.

The SYSTEM_BUILD feature in MATRIX$_x$ allows graphical development of simulations for systems represented by block diagrams. Such simulations are the basis for simplified models from which control laws can be synthesized, and are at the same time a test bed for evaluating those control laws.

Both classical and modern tools are provided for control design and system identification in MATRIX$_x$. This allows engineers to use the tools which are appropriate for the job. Experience has shown that MATRIX$_x$ users use a larger repertoire of tools because of the ease with which tools can be accessed.

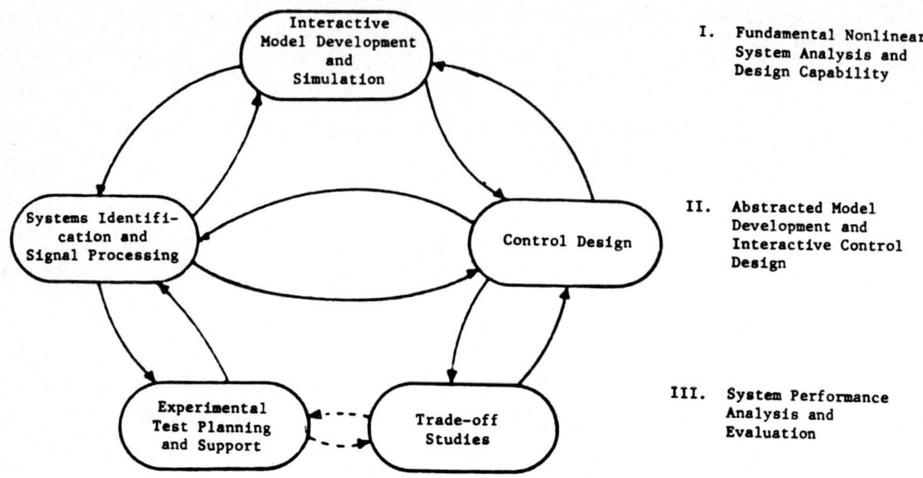

Figure 1
System Design Cycle

Students of the "modern" school tend to become slightly more "classical" and vice versa. This in itself has dramatically improved the quality of control designs.

$MATRIX_x$ is built on reliable numerical algorithms drawn from LINPACK, EISPACK, and recent research in numerical analysis. Numerical stability and robustness are important, particularly for high order systems. The designer can handle large systems in $MATRIX_x$, relying on numerical software which comprehensively reports and controls numerical conditioning.

Commands in $MATRIX_x$ have a natural and simple syntax, yet can be combined to give powerful hierarchical structures. An internal editor allows correction of the last command line. On-line HELP, extensive, direct diagnostics, and reasonable defaults make $MATRIX_x$ easy to use. A working level of expertise can be acquired in several days.

Flexible tools are available to the user for use in expanding the $MATRIX_x$ command language. Functions can be written as a series of $MATRIX_x$ commands, with local variables and parameter passing. New functions can also be added in FORTRAN if computation speed is critical.

$MATRIX_x$ inherited many of its capabilities from its predecessor MATLAB [1], which was developed by Cleve Moler. The package has also incorporated ideas from the pioneering work of Professors Karl Astrom [2] and Dean Frederick [3] in the use of a high level command languages and the utilization of block diagram model representations.

The paper is organized as follows: Section 2 provides an overview of $MATRIX_x$ capabilities. A simple system is analyzed in Section 3. Section 4 illustrates system identification, adaptive control and optimization using $MATRIX_x$. Future work in related areas is summarized in Section 5 followed by a discussion of the availability of $MATRIX_x$.

OVERVIEW OF $MATRIX_x$ CAPABILITIES

The overall capabilities of $MATRIX_x$ may be divided into the following broad categories:

1. Matrix, vector and scalar operations
2. Graphics
3. Control design
4. System identification and signal processing
5. Interactive model building (SYSTEM_BUILD)
6. Simulation and evaluation

Figure 2 shows the structure of $MATRIX_x$. The details on how each of these capabilities can be used are shown in subsequent sections. A brief summary is given here.

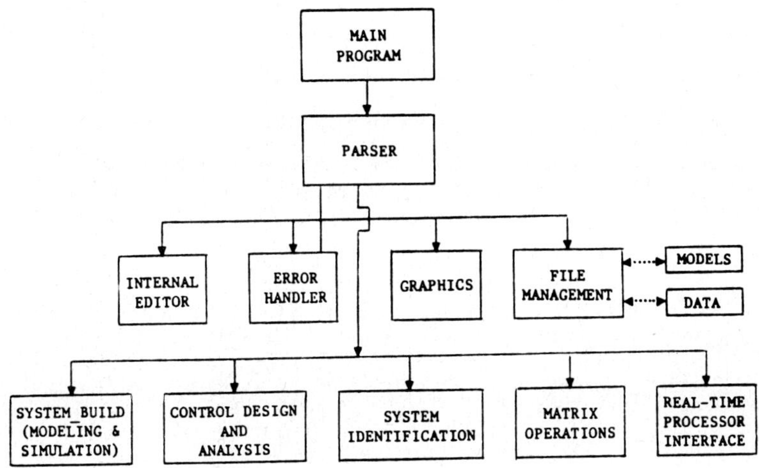

Figure 2
$MATRIX_x$ Software Architecture

Matrix, Vector and Scalar Operations (Table 1)

Basic arithmetic operations on compatible matrices are performed using standard symbols for addition, subtraction, multiplication, division, and exponentiation. For example, the product of the matrix A with the inverse of B (AxB^{-1}) can be computed by typing A/B or A*INV(B).

Basic algorithms for solving linear systems of equations, eigensystem decomposition (including reliable determination of the Jordan form), singular value decomposition (SVD), QZ decomposition, and matrix algebraic operations are implemented as language primitives. Many of the primitives were inherited from MATLAB. Most commonly used matrix operations are available.

TABLE 1
EXAMPLES $MATRIX_x$ CAPABILITIES: MATRIX ARITHMETIC

Data Entry, Display and Editing
Addition, Subtraction, Multiplication and Division
Absolute Value, Real Part, Imaginary Part and Complex Conjugate of a Matrix
Sum and Product of Matrix Elements
Element-by-Element Multiply and Divide
SIN, COS, ATAN, SQRT, LOG, EXP of Matrix Elements
Eigenvalue, Singular Value, Principal Values, Schur, LU, Cholesky and QR Decomposition of Matrices
Random Vector and Matrix Generation and Manipulation

Graphics (Table 2)

$MATRIX_x$ includes a powerful and flexible interactive graphics capability. The command PLOT(x,y) plots y against x with automatic scaling and axis labeling. Part of the screen is left for alpha-numeric output and command inputs.

The size of the plots, scaling, the location of plots on the screen, axis labels, titles, grid lines, tic marks and other variables can easily be changed by the user at the command level.

The Apollo Workstation implementation of $MATRIX_x$ uses windowing capability to enhance user-friendliness (Figure 3).

TABLE 2
$MATRIX_x$ CAPABILITIES: GRAPHICS

Flexible Commands
Multiple Plots
Axis Labels and Plot Title
Symbols and Lines
Tics and Grids
Log Scales
Bar Charts
Plot Location and Size
Personalization
Report Quality
3-D Graphics
Parallel & Perspective Projections
Surfaces
Curves
Viewing Transformation

Control Design (Table 3)

Control design in MATRIX$_x$ can be based on any of the following:

(a) Classical methods including root locus, Bode, Nyquist, and Nichols (single-input/output or multivariable plants)

(b) Linear-Quadratic-Gaussian (LQG)

(c) Methods based on A-B invariant subspaces

(d) Eigenstructure assignment and zero placement

(e) Adaptive control using self-tuning regulators and other techniques

These capabilities are available for use with both continuous and discrete systems.

For the LQG problem, the algebraic Riccati equation is solved from extended Hamilton equations, avoiding inverses which are troublesome in the singular case. The equations are row compressed with an orthogonal transformation followed by the QZ pencil decomposition and a backward stable ordering of the eigenvalues.

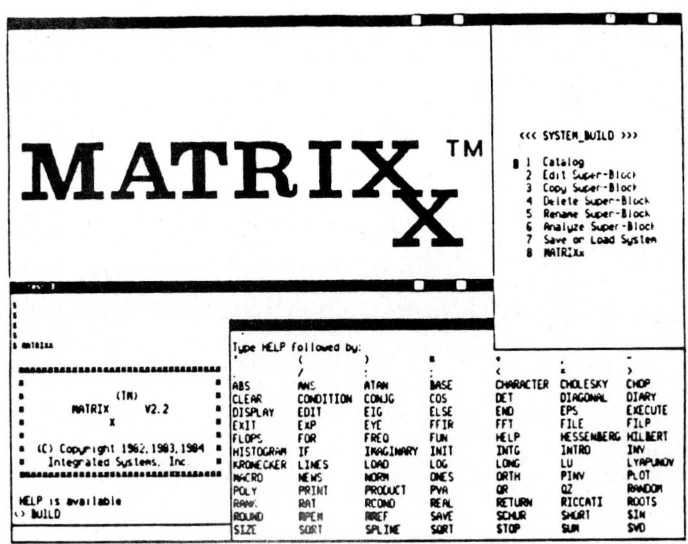

Figure 3
MATRIX$_x$/Apollo Workstation Screen

TABLE 3
MATRIX$_X$ CAPABILITIES: CONTROL DESIGN AND SYSTEM ANALYSIS CAPABILITIES (APPLICABLE TO CONTINUOUS, DISCRETE AND HYBRID SYSTEMS)

Classical Tools

- Root Locus
- Bode Plots
- Nyquist Plots
- Nichols Plots

Modern Tools

- Optimal Control Design, Discrete and Continuous
- Optimal filter Design, Discrete and Continuous
- Frequency-Shaped LQG Design
- Singular-Value Decomposition of the Return-Difference
- Eigensystem Decompositions Including the Jordan Canonical Form
- Model Following Control
- Model Reduction
- Linearization of Nonlinear Systems
- Minimal Realization and Kalman Decomposition
- Geometric Control Algorithms
- Multivariable Nyquist Plots

Meaningful extensions to LQG methods require inclusion of the dynamics of the reference inputs, disturbances, sensors, and actuators. Appending dynamics in frequency-shaped control design or model-following techniques involves forming augmented equations. This is easily accomplished with MATRIX$_X$ primitives. Use of frequency-shaped cost functionals, with singular value plots for robustness evaluation, allow incorporation of engineering judgment in control design.

Evaluation tools for linear systems include frequency response and power spectral density plots, time responses, and determination of transmission zeros. The principal vector algorithm (PVA) primitive for numerically reliable extraction of the Jordan Form (with discriminatory rank deflation of root clusters) is useful in modal analysis. PVA enables computation of residues or partial fraction expansions of multivariable systems.

Data Analysis and System Identification (Table 4)

Data analysis and identification can be performed very efficiently and easily in MATRIX$_X$. With its flexible graphics, MATRIX$_X$ provides a productive environment for batch and recursive identification methods. A universal interface is provided to ease data transfer. Data can be censored, detrended, and analyzed. Batch procedures include the standard regression methods with analysis of variance and step-wise regression. State-space and nonlinear batch maximum likelihood procedures are also available. Recursive algorithms such as recursive least squares, maximum likelihood, and extended Kalman filter with Ljung's modification are available. All covariance factorizations and updates are in U-D form for numerical reliability. Non-parametric batch and semi-batch methods using the FFT are provided for auto/cross covariances/spectras.

Adaptive control algorithms for multivariable systems using U-D updates can be designed using simple commands.

Filter design facilities include Finite Impulse Response design in Chebyshev norm, Wiener Filters, window-based designs, and Infinite Impulse Response design.

TABLE 4
$MATRIX_X$ CAPABILITIES: SYSTEM IDENTIFICATION, SIGNAL PROCESSING AND DATA ANALYSIS CAPABILITIES

Data Display
Time-History Plots
Multichannel Cross-Plots
Scatter Plots
Frequency Plots
Histograms
Data Transformations and Spectral Analysis
Multiplexing/Demultiplexing
Detrending
Censoring
Digital Filtering
Discrete Fourier Transform
Inverse Fourier Transform
Autocorrelation
Cross Correlation
Autospectrum
Cross Spectrum
Decimation and Interpolation
Maximum Entropy Spectrum Estimation
System Identification
Step-Wise Regression and Model Building
Maximum Likelihood Identification of State-Space Models and Nonlinear Models (generated by SYSTEM_BUILD)
Recursive Maximum Likelihood Identification
Extended Kalman Filter Algorithm
Filter Design
Window-Based Methods
Wiener Filter
REMEZ Exchange Algorithm for Finite Impulse Response Filters
Elliptic, Chebyshev, Butterworth Infinite Impulse Response Design

Interactive Model Building (SYSTEM BUILD)

The interactive model building facility called SYSTEM_BUILD is a tool for building models of complex systems for use in simulation, control design, and trade-off studies. The user can develop multi-input/multi-output (MIMO) system models from models of individual parts of the system. Transfer function descriptions can be combined with nonlinear functions and state-space models. It is also possible to connect an externally defined FORTRAN module to models defined in SYSTEM_BUILD. Models can be placed in catalogs for future use. Systems defined using SYSTEM_BUILD can be linearized and simulated with arbitrary inputs. Modules or parts can be changed or replaced without recompiling and relinking FORTRAN code.

A hierarchical structure allows models to be developed "top-down" or "bottom-up." In the top-down approach, the designer specifies an overall system in terms of its major subsystems. Each major subsystem can be defined as an interacting interconnection of lower level subsystems. The lowest level subsystems are finally specified using basic elements, which might consist of nonlinearities, table look-ups, transfer functions, state-space models, and summing junctions. Nonlinearities can include saturation, absolute values, hysteresis, general piecewise linear functions-quantization and general algebraic nonlinearities. Transfer functions can be written as numerator/denominator polynomial coefficients, zeros/poles, or natural frequencies and damping ratios.

In the bottom-up building approach, the lowest subsystem models are developed first. Major subsystem and complete system models may then be assembled from lower-level system models. The basic building blocks available in SYSTEM_BUILD are shown in Table 5.

TABLE 5
SYSTEM_BUILD CAPABILITIES:
CLASSES OF BUILDING BLOCKS

Gain
Nested Super-Block (Continuous of Discrete)
7 Different Algebraic Equation Blocks
8 Different Piecewise Linear Functions
7 Different Classes of Dynamic Systems
8 Different Trigonometric Functions
FORTRAN Blocks

Simulation and Analysis

$MATRIX_x$ provides capabilities for efficient linear and nonlinear simulation. Linear simulation is performed using a discrete representation and is structured to fully use sparseness in system matrices.

Table 6 shows the classes of systems treated within $MATRIX_x$. Table 7 shows the simulation and analysis capabilities.

A variety of integration algorithms is available for dealing with various classes of simulation models (Table 8).

TABLE 6
CLASSES OF SYSTEMS COVERED BY SYSTEM_BUILD

Classes
Linear
Nonlinear
Continuous
Single-Rate Discrete
Multi-Rate Discrete
Single-Rate Hybrid
Multi-Rate Hybrid

TABLE 7
SIMULATION AND ANALYSIS CAPABILITIES

Capabilities
Simulate general nonlinear continuous models
Simulate general nonlinear discrete models
Simulate general nonlinear multi-rate models
Simulate general nonlinear hybrid multi-rate models
Simulate to show transitions due to sampling
Study sampling and intersample behavior of hybrid and multi-rate system
Linearize nonlinear continuous models about initial conditions
Linearize nonlinear continuous models at a point along the trajectory
Linearize nonlinear discrete models about initial conditions (single sample rate)
Linearize nonlinear discrete models at a point along the trajectory

TABLE 8
INTEGRATION ALGORITHMS AVAILABLE IN $MATRIX_x$

Algorithm	Description
Euler	Euler
Rk2	Runge-Kutta (2nd order)
Rk4	Runge-Kutta (4th order)
Kutta-Merson (fixed step)	
Kutta-Merson (variable step)	
DASSL	implicit stiff predictor-corrector

MODEL BUILDING AND SIMULATION WITH SYSTEM_BUILD

The following example illustrates the use of SYSTEM_BUILD in designing a simple multi-rate digital control system.

Consider the following linear dynamic system expressed in state-space form:

$$\dot{x} = \begin{bmatrix} 0 & 1 & 0 & 0 \\ -4 & -.04 & 0 & 0 \\ 0 & 0 & 0 & 1 \\ 0 & 0 & -16 & -.16 \end{bmatrix} x + \begin{bmatrix} 0 \\ 1 \\ 0 \\ 1 \end{bmatrix} u \qquad y = \begin{bmatrix} 3.2 & 0 & 3.2 & 0 \\ 0 & 3.2 & 0 & 3.2 \end{bmatrix} x \tag{1}$$

This system consists of two lightly-damped second-order poles with a single input and two outputs. The first output is position and the second is velocity. $MATRIX_x$ uses a matrix called a "system" matrix to store the A, B, C, and D matrices that make up a linear state-space system. Given

$$\dot{x} = Ax + Bu \quad \text{and} \quad y = Cx + Du \tag{2}$$

the system matrix is defined as:

$$S = \begin{bmatrix} A & B \\ C & D \end{bmatrix} \tag{3}$$

The linear dynamic system represented by Eq. (1) is entered in $MATRIX_x$ as:

```
S = [  0    1    0    0   0
      -4  -.04   0    0   1
       0    0    0    1   0
       0    0  -16  -.16  1
      3.2   0   3.2   0   0
       0   3.2   0   3.2  0 ]
```
(4)

(D is zero in this case)

Fig. 4 shows the block diagram representing the plant. The dynamic system given by Eq. (4) is entered as the state-space system block S. A super-block called ACT, which contains the dynamics of the actuator, is connected to the input of the state-space system block S. In the hierarchical structure of SYSTEM_BUILD, the block diagram show in Fig. 4, which represents the plant, is a super-block. The super-block ACT is nested within PLANT. Since super-blocks may be nested to an arbitrary depth, ACT can also contain super-blocks. Furthermore, since models can be built from the bottom-up or from the top down, ACT does not have to exist for it to be included in PLANT. The existence of ACT will only be resolved when PLANT, or a super-block containing plant, is analyzed. Note that the connections shown in the figure represent vectors, so that the single line going to the output actually contains both the position and velocity signals.

The super-block ACT is shown in Fig. 5. The dynamics of the actuator include a first order lag and a deadband nonlinearity of ± 0.2. The lag is entered as a dynamic system in numerator-denominator polynomial form:

$$H(s) = \frac{5}{s + 5} \tag{5}$$

The ANALYZE option of SYSTEM_BUILD is used to analyze the super-block PLANT. ANALYZE returns the user to the $MATRIX_x$ command level after creating an internal simulation model. The system represented by PLANT is linearized by typing:

```
[SL,NSL] = LIN(1);
```

SL is the linearized system matrix, NSL is the number of states in the linear model, and the argument of LIN is the state and control perturbation to be used in the numerical linearization.

A Bode plot of the open loop system can be obtained by typing

```
BODE(SL,NSL,0.1,10);
```

This produces the plots shown in Fig. 6.

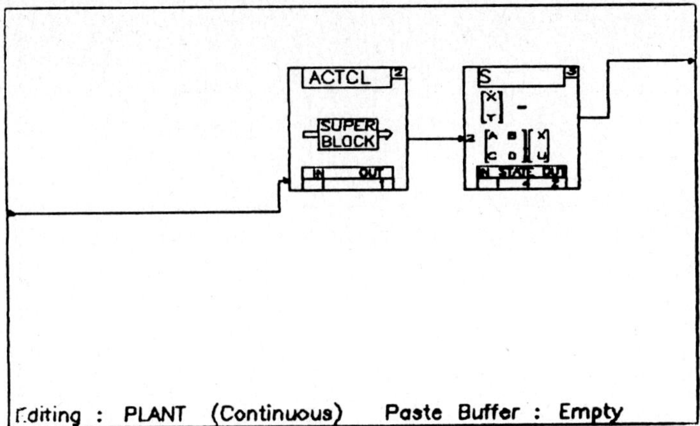

Figure 4

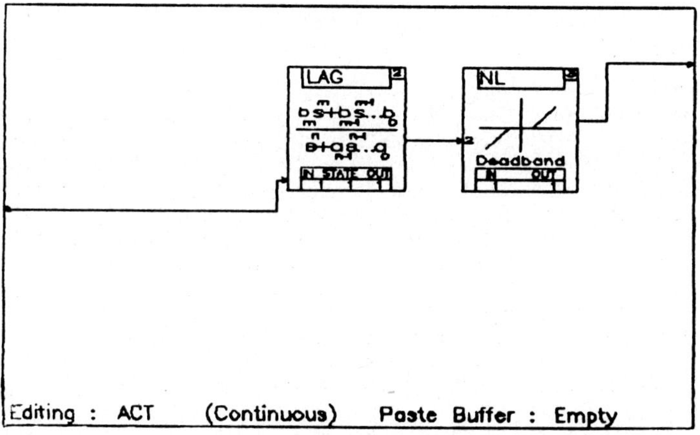

Figure 5

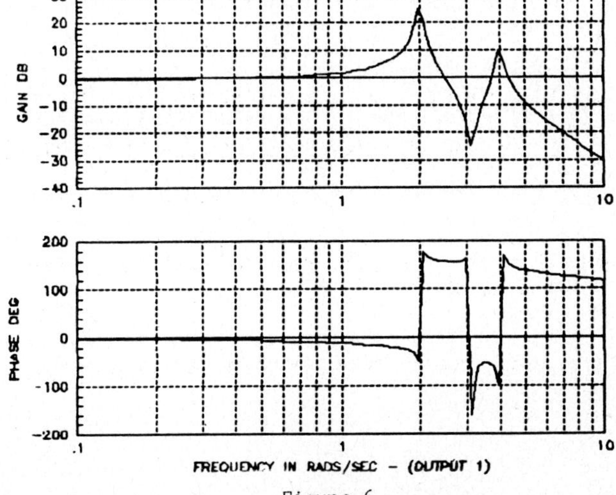

Figure 6
A Bode Plot

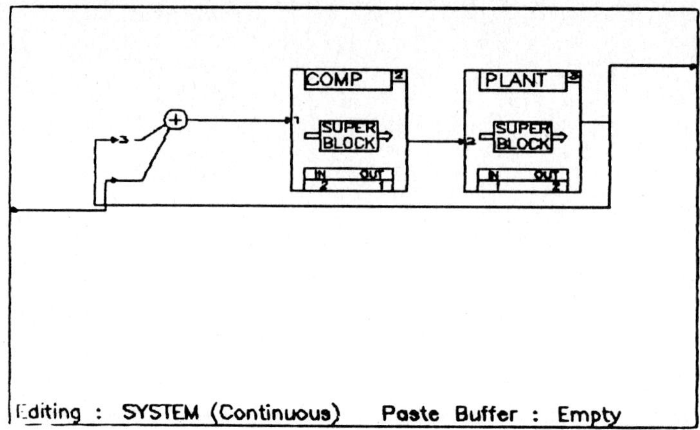

Figure 7
SYSTEM Represents the Compensated PLANT

A control system is required to maintain the system's position. Fig. 7 shows the super-block SYSTEM which contains the plant with a super-block called COMP (representing a compensator) in the forward loop. The inputs to the compensator are the velocity and position errors. The super-block SYSTEM has a single input, which is the reference position, and a single output, which is the actual position. One of the inputs to the summing junction, which should represent the reference velocity, has been left unconnected. This input will automatically be set to zero when the system is analyzed.

The super-block COMP is defined as a discrete-time block with a sampling period of 0.1 sec. It contains a single gain block. The gain block multiplies both the position and velocity errors by 2.5 and sums them. These gains can easily be changed, as can other block parameters and the initial conditions on dynamic systems.

The ANALYZE option of SYSTEM_BUILD is used with the super-block SYSTEM, again returning the user to the MATRIX_x command level. At this level, the MATRIX_x commands

```
TIME=[0:.1:10]'; U=ONES(TIME);
```

create the vector TIME, which contains all the times at which output is desired, and the vector U, which consists of a unit step with the same number of points as TIME. A simulation is then run by typing:

```
[T,Y]=HSIM(TIME,U);
```

HSIM is used for systems which contain discrete and continuous blocks whereas SIM is used for systems which only have continuous blocks. The outputs of the HSIM command are the optional vector T, and the output vector Y. Specifying T causes the simulation to produce results immediately before and immediately after every discrete-time transition in the system.

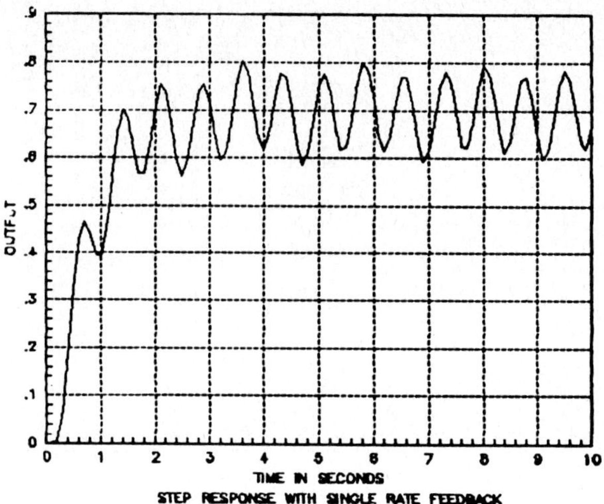

Figure 8
Step Response with Single Rate Feedback

The simulation results, shown in Fig. 8, were plotted by typing:

```
PLOT(T,Y,'TITL/STEP RESPONSE WITH SINGLE RATE FEEDBACK/ ...
XLAB/TIME IN SECONDS/ YLAB/OUTPUT /')
```

The response exhibits a limit cycle oscillation with an amplitude of ±0.2, which is the size of the deadband. The steady-state error is also quite large (.3). A high-gain, high-rate loop is placed around the existing actuator. The resulting super-block, ACTCL, is shown in Fig. 9. The super-block HRGAIN is defined with a sampling period of 0.02 secs. It contains a single gain block with a gain of 10. The super-block ACT in SYSTEM is replaced with ACTCL and a new simulation is run, the results of which are shown in Fig. 10. The high-rate loop around the actuator has eliminated the limit-cycling and has speeded up the dynamics. The steady-state error still remains, however.

A new compensator block is designed by copying the COMP block to INTCMP. An integrator is added which integrates the position error with a gain of 0.15. The resulting compensator is then of the PID type. The performance of the new compensator is evaluated by replacing the super-block COMP in SYSTEM with INTCMP. The step response is shown in Fig. 11. Adding the integral compensation has eliminated the steady-state error without slowing down the response. Damping remains quite good without any traces of limit-cycling.

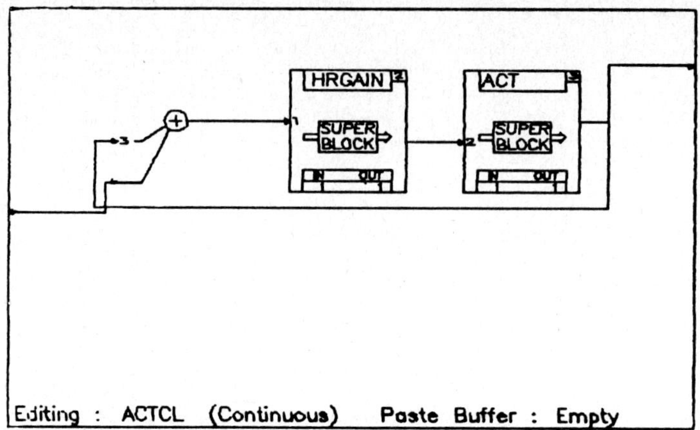

Figure 9
A High Gain, Fast Discrete-Time Compensation Around the Actuator Reduces the Deadband Limit Cycle

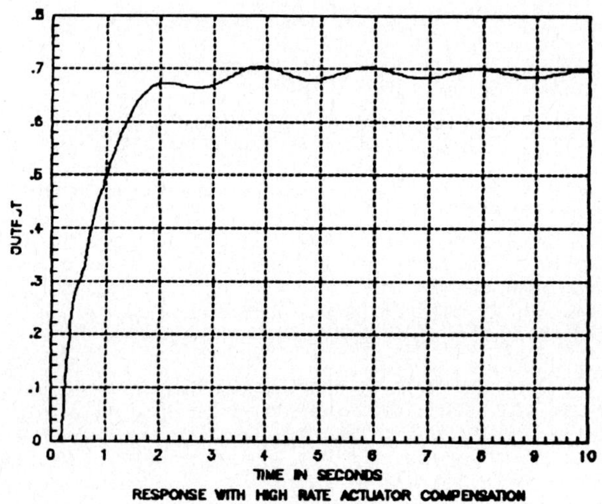

Figure 10
Response wit High Rate Actuator Compensation

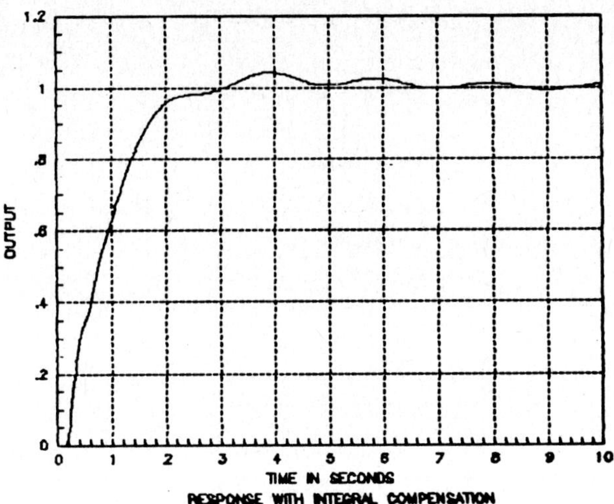

Figure 11
Response with Integral Compensation

A discrete LQG compensator could also be designed based on the linearized system (SL,NSL). Typing

```
SD = DISCRETIZE (SL,NSL,TAU);
```

discretizes the system with a sampling period of tau (previously entered as 0.1). The SD matrix is separated into its component parts by typing:

```
[AD,BD,CD] = SPLIT (SD,NSL);
```

An optimal quadratic regulator with state weighting matrix RXX and control weighting matrix RUU can be designed by typing:

```
[EVC,KC] = DREGULATOR (AD,BD,RXX,RUU);
```

EVC contains the closed-loop eigenvalues and KC contains the optimal gain matrix. A Kalman filter with state driving noise covariance QXX and sensor noise covariance QYY is designed by typing:

```
[EVF,KF] = DESTIMATOR (AD,CD,QXX,QYY);
```

The LQG compensator is assembled from the regulator and estimator designs by typing:

```
[SC,NSC] = LQGCOMP (SD,NSL,KC,KF);
```

The discrete LQG compensator represented by SC could be inserted into the COMPblock as a linear state-space system (with appropriate sign conventions).

ANALYSIS OF A LASER POINTING EXPERIMENT WITH MATRIX$_X$

Analysis results from a laser pointing experiment have been used to demonstrate major steps in a typical control design cycle (see Figure 12).

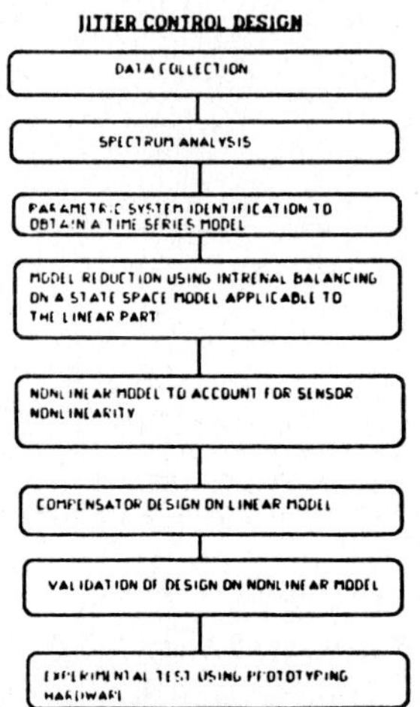

Figure 12
Illustration of a Typical Control Design Process

The experiment objective (see Figure 13) is to control the jitter of a laser beam. The single actuator consists of a proof-mass which exerts a reaction force on the flexible beam when the proof-mass is moved by an applied armature current input. A rate-sensor is pivoted on the actuator to provide a velocity measurement at actuator location. As the flexible beam vibrates, the laser beam changes its angular direction. A second sensor, a quad detector, mounted on the structural support picks up the position of the beam, as long as it is in its field-of-view.

The laser beam strikes a mirror on the flexible beam and then is reflected back by another mirror mounted on the proof-mass actuator. The resulting beam is split by a beam splitter in two rays, one going to the quad detector and the other going to the screen where the jitter is magnified. The proof-mass actuator controls the flexible beam vibrations and hence can reduce the laser beam jitter. The mass of the actuator is greater than the flexible beam and therefore the interaction between the modes of the flexible beam and the actuator is significant.

Obtaining the Model for Control Design

Jitter control was desired in the region of 4Hz to 20Hz. Consequently a sine sweep was applied to the actuator lasting about sixteen seconds sweeping from 4Hz to 20Hz. The sampling rate was 51.2Hz. Figure 14 shows that the magnitude of the input signal between 0Hz and 4Hz and between 20Hz and 25.6Hz is very low and therefore the model will not be accurate in those frequency ranges.

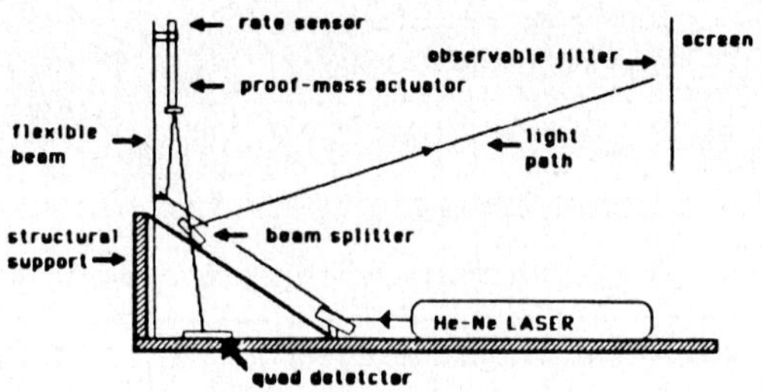

Figure 13
Laser Pointing and Jitter Control Experiments

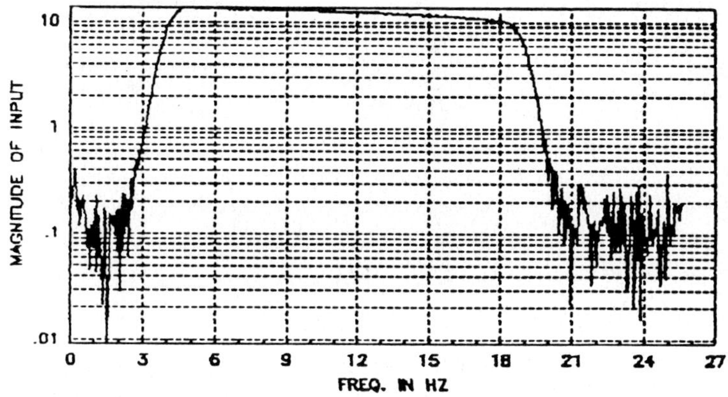

Figure 14
Magnitude of the Input (Note Low Magnitude in Frequency Regions 0-4Hz and 20-25.6Hz)

Figure 15 shows a transfer function estimate obtained by simply taking ratios of the complex Fourier transforms of the output and the input. The estimates are very noisy as expected.

A parametric model was obtained using a recursive least-squares algorithm on an eighth order time series model. The time series model was converted to a discrete state-space model. The state-space model was then converted to its internally balanced form (Moore [4]) and reduced to a fourth order model based on the observation that there were only two modes in the frequency region of interest. The balanced discrete state-space model was converted to its continuous form to evaluate the overall nonlinear system and perform control design. Figure 16 shows a comparison between the transfer function estimate obtained from the parametric state-space model and the estimate obtained by Fourier analysis (as in Figure 16).

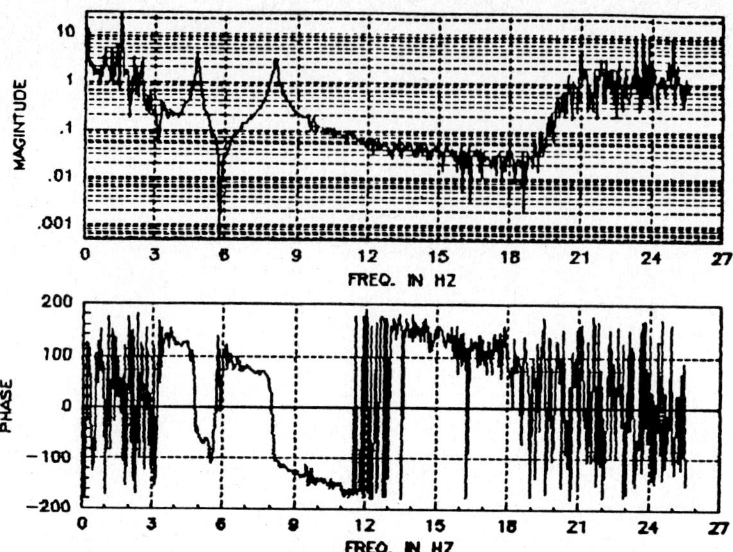

Figure 15
Transfer Function Estimate Obtained by Fourier Analysis from Actuator to Rate-Sensor

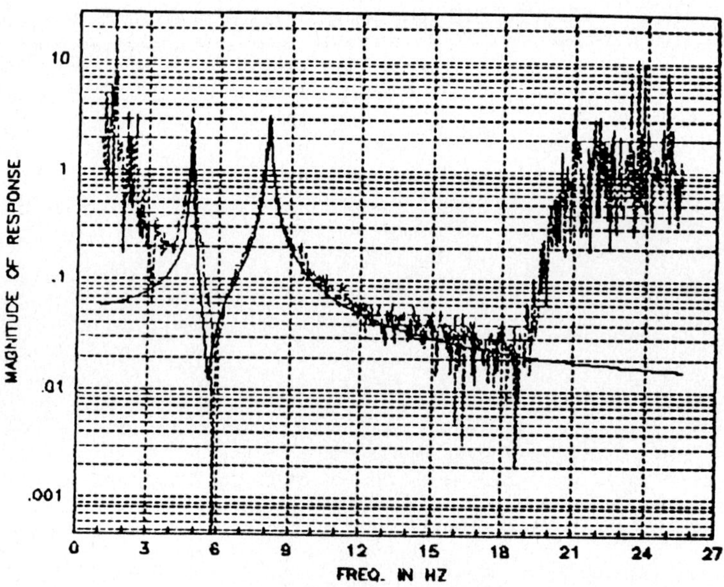

Figure 16
A Comparison of the Transfer Function Estimates Obtained Parametrically (smooth solid line) and by Simple Fourier Analysis Without Averaging (broken noisy line). Actuator to Rate-Sensor Transfer Function.

The primary nonlinearity in the system is due to the narrow field of view of the quad detector. Fig. 17 shows the nonlinearity. A compensator was designed for the linear continuous-time state-space system obtained from system identification using linear-quadratic-Gaussian theory (Kwakernaak and Sivan [5]). The resulting compensator is in the state-space modal form.

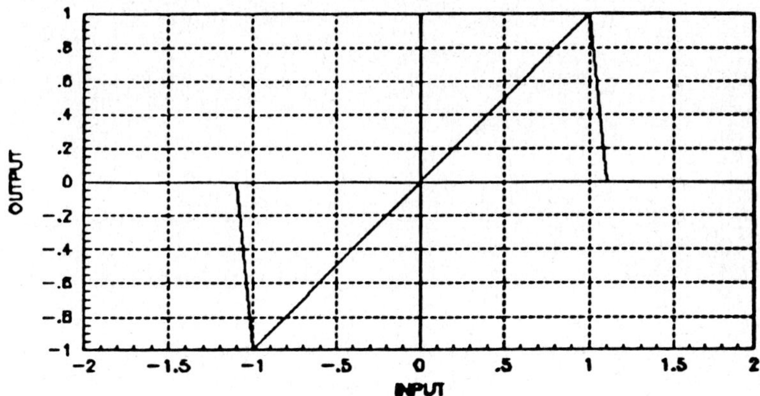

Figure 17
Input/Output Behavior of the Quad Detector

Fig. 18 shows the block diagram representing the linear beam model in its modal form using only the quad detector channel. The overall nonlinear system is shown in Fig. 19. The quad detector nonlinearity is in the block called SENSOR. The compensator is in the block called COMPEN.

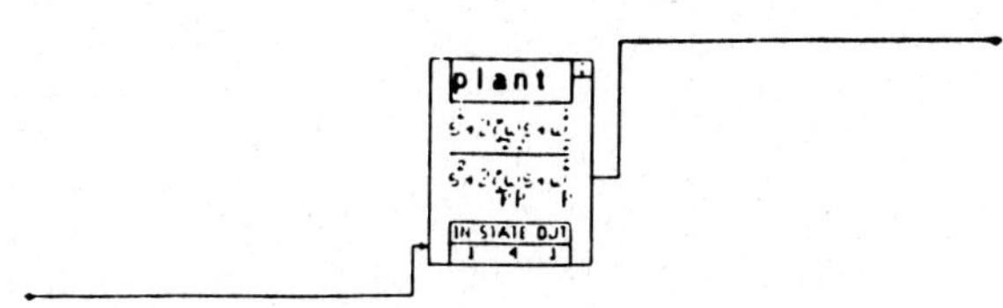

Figure 18
Super-Block Diagram "BEAM"

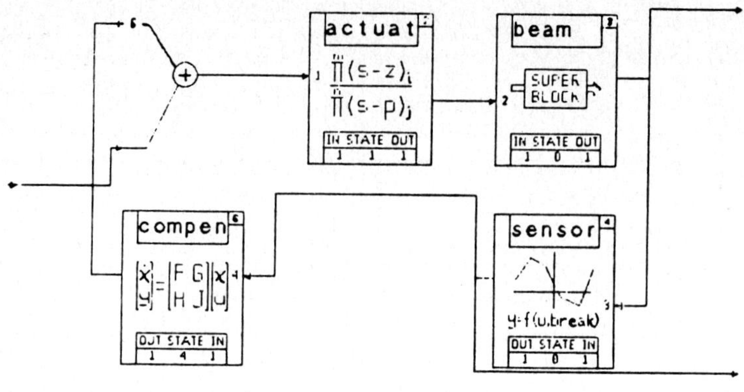

Figure 19
Completed Super-Block Diagram "SYSTEM" for the Jitter Control Experiment

Simulating the closed-loop system, with a trapezoidal input shown in Fig. 20, results in the output of Fig. 21. The impulse response of the linearized closed-loop system is shown in Fig. 22. Notice that if sensor nonlinearity is ignored, the jitter settles down much more rapidly.

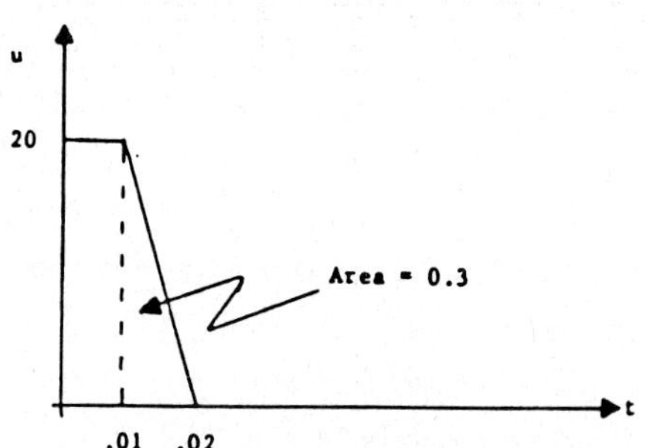

Figure 20
Disturbance Input to be Applied to "SYSTEM"

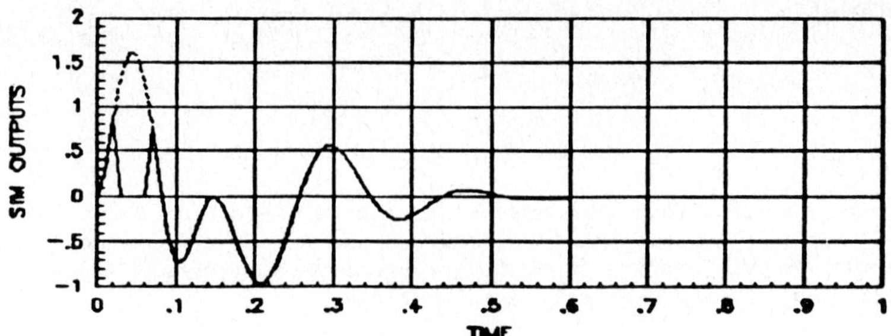

Figure 21
Nonlinear Simulation Output Showing the Effect of the Narrow Field of View of the Quad Detector. The Dotted Line is the Output of the Block Beam. The Solid Line is the Output of the Sensor.

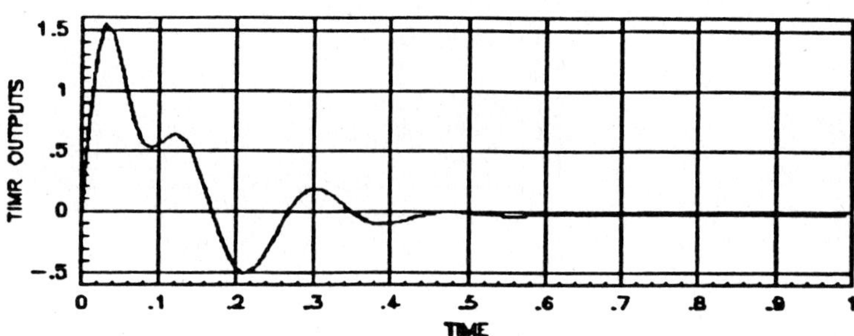

Figure 22
Impulse Response of the Linearized System

The final step is to transfer the compensator design to prototype hardware compensator and test the control system in the physical experiment. The overall conclusion we want to emphasize is that having a single CAE environment for the entire design cycle considerably reduces the menial effort required, and enables efficient design iterations necessary for a good control law design.

AN ADAPTIVE CONTROL EXAMPLE

The adaptive control of the beam experiment from the previous section is described below.

The parameters of the model are estimated using a U-D factored recursive least squares update. A one-step ahead optimization criterion with a weighting on control activity is used. Three aspects of the adaptively controlled system are explored by simulation:

(i) Intersample behavior

(ii) The effect of the quad detector nonlinearity

(iii) The effect of changing the sample rate of adaptive control

SYSTEM BUILD Model

Fig. 23 shows the overall adaptively controlled beam experiment. The input is a disturbance at the beam input. The three outputs are the laser beam deviation, the quad detector output, and the input to the proof mass actuator. Note that the super-block ADAPT is a discrete-time super-block.

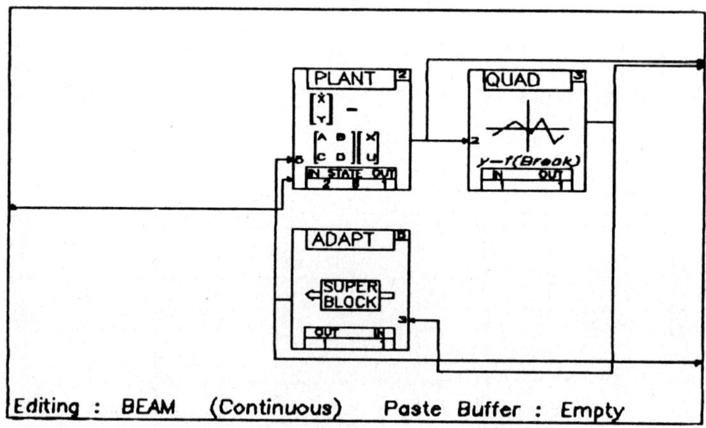

Figure 23
Overall Adaptively Controlled Plant

Fig. 24 shows the super-block ADAPT generating the new output $\bar{y}=y+z^{-1}*gain*u$ that is fed to the super-block AD1.

Fig. 25 shows the discrete-time adaptive control algorithm with a minimum variance control law.

DEL: Generates the regression vector to be used in the exponentially weighted recursive least-squares update.

GAIN: Computes the $\Delta\theta$ to be used for updating the parameter θ.

INT: Discrete-time integrator for θ.

DIV: Computes the minimum variance control based on the most recent output.

SAT: Puts a hard limit on the magnitude of the input.

The super-block GAIN, shown in Fig. 26, computes the parameter update using the regression vector and the prediction error as inputs. The deadband generates a flag that can freeze the parameter update if the prediction error is too small. The FORTRAN block in the upper right corner generates the Kalman gain used for

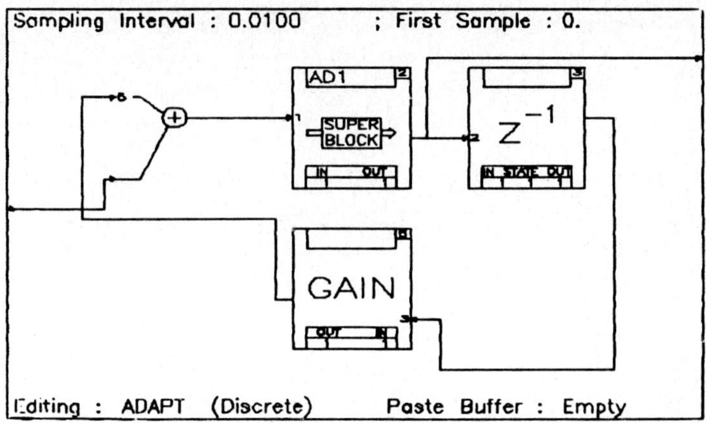

Figure 24
Discrete-Time ADAPT Controls the Beam.

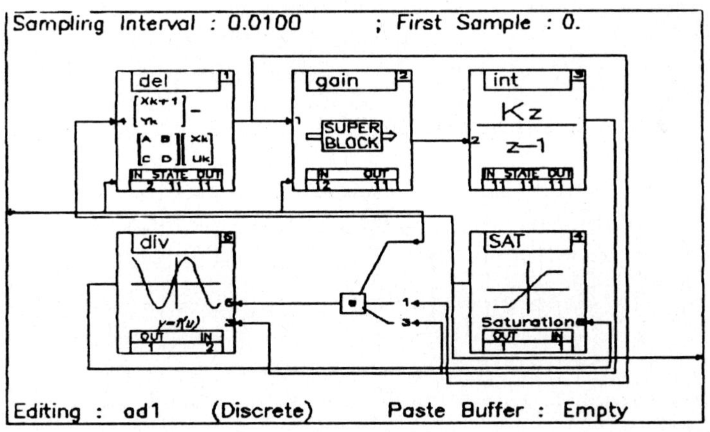

Figure 25
A Minimum Variance Adaptive Controller

parameter updates. It uses a U-D factored covariance update, using exponentially weighted past data. The regularization parameter to suppress reduction in the covariance in those directions that could lead to nearly singular covariance (i.e., an obvious estimator).

The algebraic expression $y = f(u)$ in the upper central part generates the exponential weighting and the assumed measurement covariance. The element-by-element product in lower left corner generates $\Delta\theta$ = Kalman gain * prediction error.

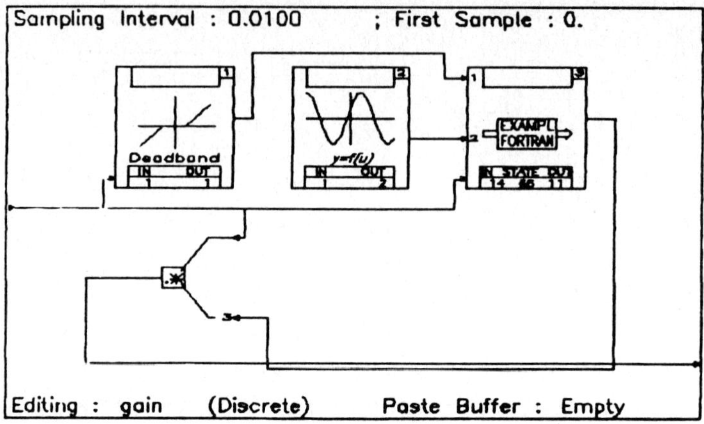

Figure 26
An Exponentially Weighted Recursive Least Squares Parameter Update

Fig. 27 shows the adaptive control behavior of the continuous-time plant with an impulsive input.

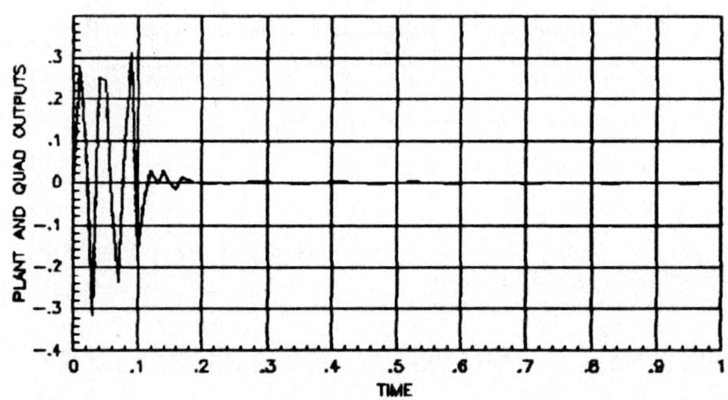

Figure 27
Impulse Response of the Plant with Adaptive Control

Note that the quad sensor field of view is sufficient for complete adaptation to a reasonable control. Increasing the input size by 10% causes the plant output to go outside the quad sensor field of view, resulting in unstable behavior.

Changing the sample rate from .01 sec to .08 sec causes the residual oscillations to increase considerably, as shown in Fig 29.

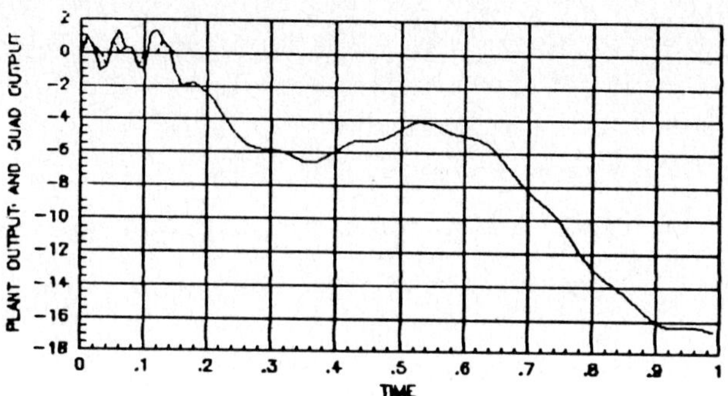

Figure 28
Unstable Behavior Due to the Quad Sensor Nonlinearity

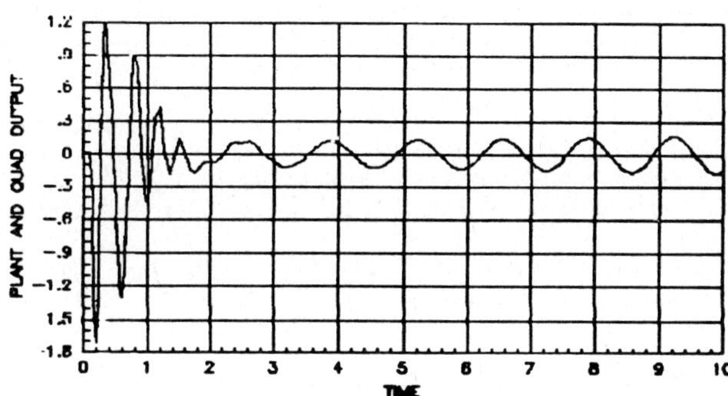

Fig. 29
The Effect of Reducing the Sample Rate of Adaptive Control

The discrete plant input is shown in Fig. 30.

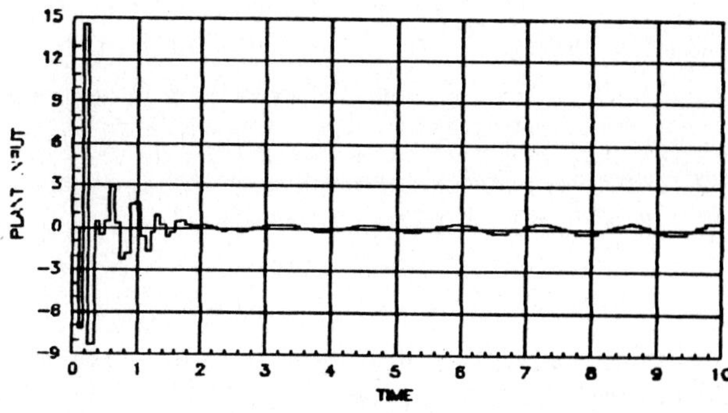

Figure 30
The Discrete-Time Plant Input

FUTURE WORK

Future work in computer-aided-control system design (CACSD) will be directed at the following areas:

1) Enhancing current capabilities and ease of use

2) Providing the capabilities in a distributed processing environment

3) Extension of the capability to do real-time control design implementation and testing

4) Use of artificial intelligence and expert systems

Each of these areas is described in this section.

Enhancement of Current Capabilities

Though $MATRIX_x$ and SYSTEM_BUILD provide a very comprehensive set of control design tools, future developments are essential. These developments are aimed at continuous inclusion of new tools as they mature, more flexible input formats, comprehensive data base management and automatic report writing.

Much research is directed to new control design methods for multivariable systems. Of greatest interest are algorithms for direct design of robust and high performance compensators. Significant advances still need to be made in numerical algorithms. New algorithms will be included as they become available.

$MATRIX_x$ requires input in a combination of menus and commands. The input format needs to be extended so that it can more readily accommodate both naive and expert users.

An extension of current data and model catalog management capabilities will make it easier for a group of engineers or engineering departments to interact more effectively.

Automatic report writing is essential because the documentation process takes a significant portion of most analysts' time. The report writer will produce complete reports in flexible formats and would continuously update the reports as the design status changes.

Distributed Processing Environment

To enhance user friendliness, it is necessary to perform some computations on the user's workstation and some computations on the central computing node. In the current configuration, all computations are performed by the central processor. The trend towards distributed processing will accelerate as more personal workstations are introduced in engineering groups.

Real-Time Control Implementation

A key area of future development is the capability to rapidly implement control laws designed by $MATRIX_x$ and SYSTEM_BUILD. The basic approach is automatic code generation. The code could be generated in a standard portable language like Ada, or the code could be specialized to a particular processor. Both approaches are currently being followed. A key element of the real-time code generation is also the documentation of control code specifications.

Tools that address this area are expected to become available in the near future.

Artificial Intelligence and Expert Systems

Artificial intelligence (AI) and expert system technology could provide natural language interfaces, expert advisors, intelligent searches, and other tools in CAE applications. The integration of AI and expert systems will be evolutionary.

IMPLEMENTATION AND AVAILABILITY OF MATRIX$_x$

Most of MATRIX$_x$ is implemented in ANSI FORTRAN 77. Several features are machine dependent because FORTRAN 77 does not allow certain constructs necessary to enhance user-friendliness. The software is currently available on VAX VMS, VAX UNIX, IBM MVS/TSO, IBM VM/CMS and Apollo Aegis.

Typical MATRIX$_x$ implementation requires virtual memory. Most algorithms access memory in a linear fashion, so that page swapping is minimized. System orders up to 150 can be handled easily within the storage capacity of the computer and the numerical accuracy of the algorithms. Whereas MATRIX$_x$ allows complex operations and arithmetic, all real algorithms such as RICC use real arithmetic and do not use up extra memory. The BLAS subroutines in LINPACK have been modified so that efficiency is not sacrificed in working with complex data with zero imaginary part.

MATRIX$_x$ also allows "chopped arithmetic" operations with smaller effective word-length. This feature can be used to provide performance valuation of on-board small word-length control systems.

MATRIX$_x$ documentation includes the User's Guide, Training Guide, Reference Guide, Command Reference List and on-line HELP.

CONCLUSIONS

MATRIX$_x$ provides a system which minimizes engineering and programming resources required for the complete cycle of the control design process, including modeling, simulation development, system identification, control design and validation. It is fully supported. MATRIX$_x$ and SYSTEM_BUILD are being enhanced and updated at regular intervals to ensure that they represent the state-of-the-art in control system design CAE technology.

REFERENCES

[1] C.B. Moler, "MATLAB User's Guide," Technical Report CS81-1, Department of Computer Science, University of New Mexico, Albuquerque, Nov. 1980, Rev. June 1981.

[2] K. Astrom, "Computer-Aided Modeling, Analysis and Design of Control Systems: Perspective", IEEE Control Systems Magazine, Vol. 3, No. 2, May 1983.

[3] D.K. Frederick, R.P. Kraft, T. Sadeghi, "Computer-Aided Control System Analysis and Design Using Interactive Computer Graphics", IEEE Control Systems Magazine, Vol. 2, No. 4, Dec. 1982.

COMPUTER-AIDED CONTROL SYSTEMS ENGINEERING
M. Jamshidi and C.J. Herget (Editors)
Elsevier Science Publishers B.V. (North-Holland), 1985

THE FEDERATED COMPUTER-AIDED CONTROL DESIGN SYSTEM

H. Austin Spang, III

General Electric Research and Development Center
Schenectady, New York 12345
U.S.A.

The structure and operation of the Federated Computer-Aided Control Design System is discussed. The system is termed "Federated" to indicate that it consists of several independently developed subsystems tied together by a unified database. The objective is to provide the user with a unified system that spans the entire control design problem: modeling, design, and simulation. The paper discusses how the management of the database is handled and the operation of an overall system supervisor. A set of supervisor commands is given which allows considerable flexibility to add subsystems and to tailor commands to the individual installation.

INTRODUCTION

With microprocessors, it is currently possible to implement a wide variety of control algorithms ranging from simple traditional proportional plus integral to sophisticated multivariable or adaptive algorithms. Traditional pencil and paper approaches to control design are no longer cost effective and limit what strategies will be used. Fortunately, computer hardware, software, and interactive graphic terminals are currently available to provide an interactive control design system which allows the engineer to answer the "what if" questions and to cost effectively design even the most complex control algorithm.

In creating a control system, the engineer must construct models for the process to be controlled, analyze their behavior, design an appropriate control strategy, and evaluate its overall performance. Eventually, he will implement the design in appropriate hardware such as a microprocessor. A brief, noninclusive summary of the design procedure and some of the current techniques which might be used are shown in Figure 1. The goal of a computer-aided control design system should be to provide an engineer with a broad spectrum of alternative design approaches.

To meet the need for such a broadly based design system, the Federated Computer-Aided Control Design System has been developed. The system is termed "Federated" to indicate that it consists of several independently developed subsystems tied together by a unified database. In this manner, one takes advantage of existing software while providing the user with a unified system that spans the entire control design problem: modeling, design, and simulation. While numerous computer-aided control design packages exist, as Frederick[1] summarizes, most are

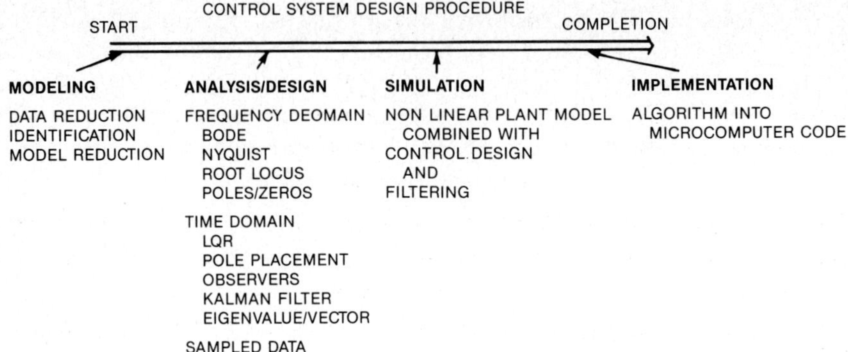

Figure 1. Summary of techniques used in the different stages of a control design.

focused on a particular aspect of the design problem. The Federated System is unique in the way it ties diverse packages together into a unified system. In this paper, the structure and operation of the Federated Computer-Aided Control Design System will be discussed.

THE FEDERATED APPROACH

The structure of the Federated System is designed to meet the following objectives:

1. Each subsystem can be operated as a stand-alone program.
2. Subsystems and other programs can be added to the system easily.
3. Subsystems or programs can be modified without affecting other parts of the system.
4. Federating adds a minimum amount of overhead to each subsystem.
5. User commands can be added easily and are valid for specified subsystems.

To meet these objectives, the Federated System is organized in a hierarchical structure of stand-alone subsystems connected by a supervisory program. A block diagram of the system is shown in Figure 2. The supervisor primarily serves as an operating system interface, translating

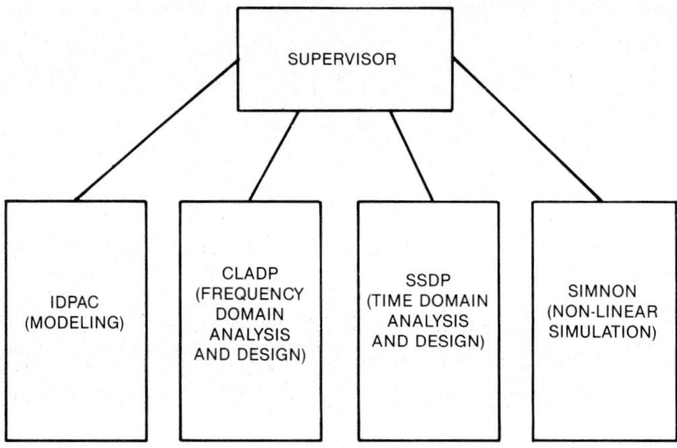

Figure 2. The Federated System.

user commands into the names of programs that will be run. It also passes initialization and file information to allow correct startup of programs. Once a subsystem is entered, control is not returned to the supervisor until the user enters a command that is handled by another subsystem. Thus, the federating of the subsystems generates no overhead except during the transition from one subsystem to another.

It should be recognized that any subsystem may also be organized as a series of stand-alone programs connected by a supervisory program. This approach further enhances the modularity and maintainability of the system. Two of the subsystems, the Cambridge Linear Analysis and Design Package (CLAPD) and the State Space Design Package (SSDP), have been organized in this manner.

AVAILABLE SUBSYSTEMS

The control design software shown in Table 1 are the initial major subsystems within the Federated structure. These subsystems provide most of the desired design capability: IDPAC

Table 1

COMPUTER-AIDED CONTROL SYSTEM DESIGN PROGRAMS
USED IN THE FEDERATED SYSTEM

Source	Name	Author	Function of Program
Cambridge University, England	Cambridge Linear Analysis and Design Program (CLADP)	Prof. A.G.J. MacFarlane J.M. Edmunds[3]	Multivariable control system design by frequency domain methods
Lund University, Sweden	IDPAC	Prof. K.J. Astrom J. Wieslander[4]	System identification
Lund University, Sweden	SIMNON	Prof. K.J. Astrom H. Elmquist[5]	Nonlinear simulation
General Electric	State Space Design Package (SSDP)	H.A. Spang, III[7]	Multivariable control system design by state space and time domain methods

for modeling, CLADP and SSDP for analysis and design, and SIMNON for nonlinear simulation. Recently MATLAB[2] has been added to provide a flexible means of handling matrix equations. In addition, the Federated system includes programs to go from one subsystem to another. These currently include programs for generating SIMNON equations from numerical data files and format conversion from IDPAC to CLADP. As will be discussed in more detail later, the Federated System allows additional software to be added on-line. Thus, the user can customize the system to meet his specialized requirements with, for example, additional graphic display programs or additional design algorithms.

In this section, the capabilities of these initial major subsystems will be summarized. Each of the packages is fully interactive command driven. A listing of the major commands for each subsystem is given in Appendix B.

CLADP[3]

The Cambridge Linear Analysis and Design Package provides single input-output and multivariable extensions of classical frequency design methods. These include Bode, Nyquist, Nichols and root locus diagrams as well as time simulation of the resulting linear system. Multivariable techniques include the characteristic locus and Nyquist array methods. Robustness of the multivariable system can be determined through singular value plots. A number of conversion commands are available to allow the user to change from state space to Laplace transform and vice versa. The package handles equally well discrete or continuous systems. W plane plots and transformations between discrete state space systems and z transforms are available.

It has turned out that an important feature of the package is the ability to describe a system as a series of subsystem blocks. Each of these blocks can be described in state space or Laplace form. The form of the blocks can be intermixed. Most calculation are done in state space form for numerical accuracy with the conversion from Laplace done automatically.

IDPAC[4]

IDPAC is a package for data analysis and identification of linear systems. It provides time series analysis of ARMA and ARIMA models. Its strength is in its capability for manipulation and plotting of data, correlation analysis, spectral analysis and parameter identification. The least squares and maximum likelihood methods of identification are available.

SIMNON[5]

SIMNON provides interactive nonlinear simulation of continuous and discrete interconnected systems. A system is described as an interconnected set of subsystems. These subsystems can be either discrete or continuous. Thus it is particularly well suited for simulation of digital control systems. A full range of functions, noise generation and table lookup functions are available. A more detailed overview of both SIMNON and IDPAC is given by Astrom.[6]

SSDP[7]

The State Space Design Package provides the time domain state space design techniques. It is structured very similarly to CLADP. Included are design programs for the time invariant linear quadratic regulators and Kalman-Bucy filters. Testing controllability and observability and the easy modification of the performance indices are provided. The singular value and loop transfer recovery procedure of Doyle and Stein[8] allows the user to design systems which are robust to plant uncertainties.

Command Interaction

As indicated, each of the packages is fully interactive. The desired operations are specified by commands whose names are easily identified with the desired results. The packages differ on how options are specified. In IDPAC and SIMNON various options are determined by additional subcommands specified on the command line. However, in CLADP and SSDP a question and answer sequence is used to specify options. This has been shown to be helpful to the novice user. To minimize the restrictions to the expert user, there is a type ahead feature which allows one to give the answers to several questions on the same line as the initial command.

Normally commands are entered from the terminal. However, all packages have the important capability of receiving a series of commands stored in a file. In this paper we will follow the terminology of IDPAC and SIMNON and call these command sequences "macros". Such macros provide a convenient simplification of the command dialogue. The user can extend the normal command structure by developing his own sequences of commands which can be activated by a single name. The utility of macros can be considerably enhanced by providing commands to control program flow and passing of parameters.

The details of command decoding and macro handling have been implemented by a front end series of subroutines. In SIMNON and IDPAC this front end is called INTRAC.[9] A summary of those commands handled by INTRAC is shown in Appendix A. CLADP and SSDP have similar routines though not as clearly defined into a separate module.

DATA BASE MANAGEMENT

One of the major problems in tying several subsystems together is to provide a common unified database which describes the user's plant and his associated control system. The user must be able to enter a description of his plant in many ways, either from measurement data or linear or nonlinear models. Once entered, the user must be able to go from one subsystem to another without having to reenter any of the previously entered or generated information.

Each subsystem, however, has its own way of handling the information it needs. Since each subsystem has been developed independently of the others, the way data is handled reflects the developer's insight into the problem and tradeoffs determined by his computer system. Generally, this data handling is integral to the subsystem. Any attempt to force a common database structure would result, essentially, in a complete rewrite of that subsystem.

In the Federated System, this problem of data exchange is handled in one of two ways. The first takes advantage of the similarity of the information, and is shown in Figure 3. This approach is useful in those situations where the information required is the same but the format is different. A common set of files forms the database. Each subsystem reads that information through a set of subroutines forming a 'database manager' for that subsystem. These subroutines read the in-

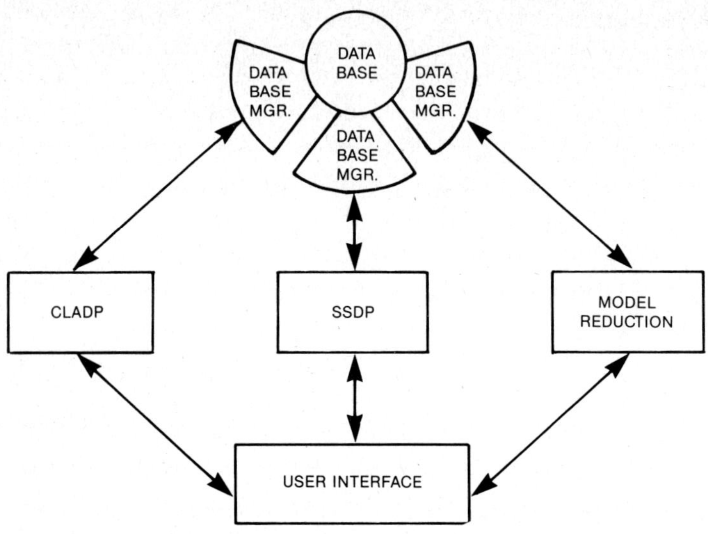

Figure 3. Data file handling common database approach.

formation in the standard format and pass it on in the form required by that subsystem. Thus, the change in the form of the database is invisible to the subsystem. This approach is particularly useful for those subsystems requiring the same matrices.

The second approach to data exchange is by direct data conversion programs. The supervisor calls a conversion program before executing the next subsystem. A block diagram of this approach is shown in Figure 4. The conversion process may be invisible to the user or may interact to determine which files should be converted. The advantage of this approach is that it can handle widely diverse forms of information. The interaction between SIMNON and the other subsystems is handled in this manner. For example, a conversion routine takes the feedback designs from CLADP or SSDP and generates SIMNON code blocks for nonlinear simulation.

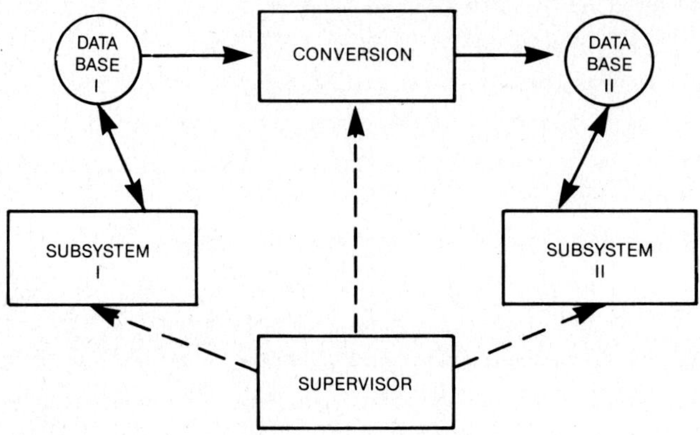

Figure 4. Data conversion.

A block diagram of the necessary data conversions between the initial subsystems of the Federated system is shown in Figure 5. The straight lines indicate that no conversion is necessary. Many are format changes from one numeric form to another. The primary conversions are those between SIMNON and the design subsystems. The finding of equilibrium steady-state solutions of the nonlinear SIMNON models and the generation of linear models requires access to the internal SIMNON database. Therefore, these functions have been made an integral part of the SIMNON subsystem.

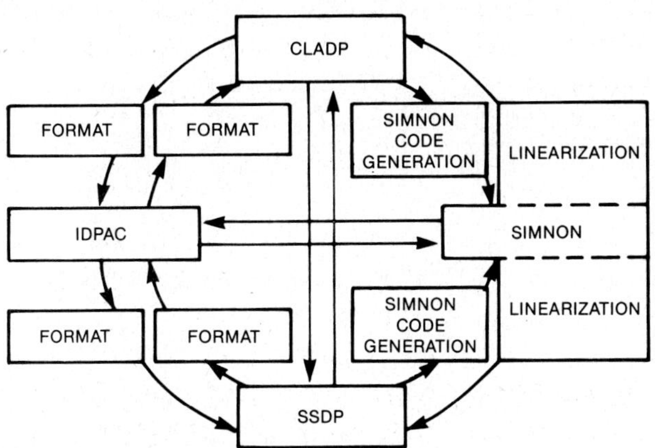

Figure 5. **Data conversions between subsystems.**

INTERCONNECTIONS BETWEEN SUBSYSTEMS

The collection of four subsystems, the supervisor and the conversion programs does not form a useful system by itself. There must be means of passing information between programs. The user should be able to invoke commands in a given subsystem which will provide services from or transfer him to another subsystem. Due to the possible conflict with existing subsystem commands, the same commands cannot in general be available in all subsystems. It would be further desirable to allow the user to customize his commands depending on his particular needs. The Federated System meets these requirements through the interconnection structure shown in Figure 6.

Communications between subsystems takes place through the supervisor using the VAX mailbox. The VAX mailbox is a memory area which can be read or written by two programs in the same manner a device or file is read or written. Prior to executing each program, the supervisor writes information into the mailbox. The program reads the information which generally initializes the subsystem and contains commands recognized by that subsystem.

To provide global commands for interaction between subsystems, the user can "install" commands in the supervisor and make these "known" to specific subsystems. The "known global

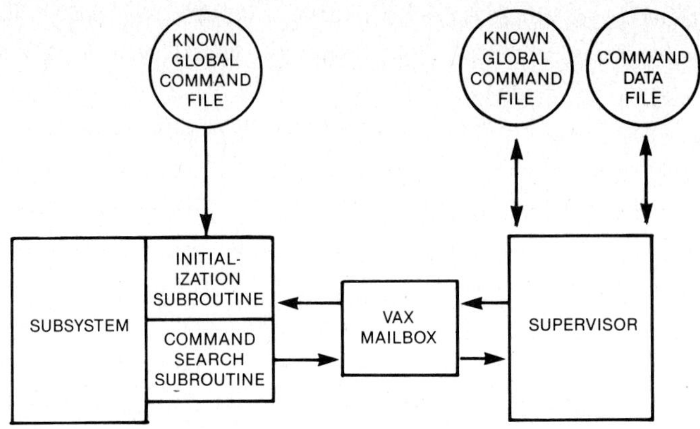

Figure 6. Supervisor and subsystem interaction.

command" file shown in Figure 6 contains the names of the global commands that will be recognized by that subsystem. Each subsystem has its own file. When the user issues a command string, the known global commands are first searched for a possible match, so the user has the capability of redefining any subsystem command by defining a corresponding global command. When a global command is found, control is passed back to the supervisor to initiate the sequence of actions associated with that command.

The capability of installing and removing commands has the potential for one user to interfere with another. One would also like to have a library of already installed commands which determine the basic capability of the system. These both can be handled by dividing each of the known global command files into three separate files: a system library file, a group library file, and a file in the user's current directory. Functionally, these appear to be one file. To avoid interference, commands can be installed or removed only in the user's current working directory.

The changes to each subsystem in order to allow this communication and command handling has been kept to a minimum. The subsystem interface to the Federated System consists of three subroutines: an initialization subroutine, an optional command stack handler, and a command-search subroutine. The initialization subroutine is called immediately after the subsystem is initiated. It first attempts to read the supervisor mailbox. If there is no mailbox, the subroutine assumes that the subsystem has been started stand-alone, and sets a flag to avoid searching the known global commands. If the mailbox is found, the subroutine reads the mailbox and the known global command file for that subsystem. It then initializes the subsystem based on the passed information and inserts the remaining passed information into the subsystems command buffer.

The command stack handler is optional. Commands passed through the mailbox are received as one large block of characters. The function of this stack handler is to pass the characters to the subsystem's own command handler in a manner that they can be understood. This routine may not be necessary if the subsystems command handler can handle an arbitrary length of text.

The command-search subroutine searches the known global command table each time a user command is given. If the given command is found in the known command table, the subroutine returns control to the supervisor by sending the user command string through the supervisor mailbox and terminating execution of the subsystem.

SUPERVISOR

The supervisor primarily serves as an interface between the subsystems and the VAX operating system. It establishes the mailbox, allows commands and subsystems to be installed, updates each known global command file and calls on the operating system to run a program. Its functions are best summarized by discussing its command set.

Supervisor Commands

The INTRAC communication module developed by Wieslander and Elmquist[9] is used to implement these commands. The additional commands provided by INTRAC are given in Appendix A. The Backus-Naur Form (BNF) syntax used to define the commands is also given in the appendix.

INSTALL {<command name>} [<formal argument>|
<delimiter>|<termination marker>]* [<[<subsystem>]*]

The command name is inserted into the known global command file for the supervisor. If a subsystem name is specified, the name is also inserted in that subsystem's known global command file. The supervisor inserts the subsystem name in its SYSTEM table, and creates a known global command file if the subsystem has previously not been specified.

To maintain on-line documentation, the installed command and any FORMAL command lines are optionally followed by comment lines describing the function of the command. These comment lines are displayed by the HELP command.

The install command is usually followed by one or more CONV, PASS, RUN or INTRAC commands. Previously installed user commands or file names may also be included. If a command is not an INTRAC command, a supervisor command or a previously installed user command, the command is assumed to be a filename for a macro containing additional commands. In same manner as INTRAC macros, formal arguments can be specified in installed commands and used in subsequent command lines. When issuing the command, the user provides values which are substituted for the formal arguments wherever they occur. The install command sequence is ended by an END command.

To facilitate the development of general command sequences, two global variables, RUN.PREV and CONV.PREV, are available. These contain the name of the previously executed subsystem and conversion program. These variables can be used to modify a command sequence depending on what was previously executed.

The install command can also be used to add a previously defined command to a subsystem. In this case the install command consists of the single command line followed by the END command.

END

This command indicates the end of an install sequence. The supervisor will return to the previous command sequence, or prompts the user for more commands.

CONV {<filename>}

The CONV command is used to run a program which converts data files from one subsystem to another. The program given by the filename is run. The filename is inserted in the global variable CONV.PREV. Operationally this command operates the same as the RUN command, except that the filename is inserted in the CONV.PREV global variable.

PASS {<string>}1

The PASS command provides a means of conveying information to a program. The given string is passed to the program defined by the subsequent CONV or RUN command. Strings can consist of any ASCII characters. Formal parameters are replaced by their values; otherwise the supervisor does not interpret these strings. The format and meaning of the passed data depend entirely on the subsystem specified in the next RUN or CONV command.

If more than one PASS command is given, the strings are passed in the order defined except if the switch FIRST is set ON by the TURN command. If the switch FIRST is set ON, then any strings from PASS commands in previously INSTALLed commands are inserted in order at the front of the sequence of previously issued passed commands. This allows the user to define commands that contain initialization information for a subsystem and this information will be passed to the subsystem ahead of any other commands.

RUN {<filename>}

The program specified by the filename is run. The filename is inserted in the global variable RUN.PREV.

REMOVE {<command name>} [subsystem]*

If at least one subsystem is specified, the command is made unknown to each of the subsystems specified. If no subsystems are specified, then the command is made unknown to all known subsystems and the supervisor. If a subsystem no longer has any commands associated with it, that subsystem is made unknown to the supervisor. In order to prevent users from interfering with each other, this command applies only to the commands specified in the user's current working directory.

LIST [ALL|FULL [<command>]|<subsystem name>]

This command displays the names of all supervisor and installed commands. If a subsystem name is given, it lists the names of all commands known to that subsystem. If the option ALL is given, all supervisor commands, installed commands and commands known to all subsystems are listed. The option FULL is used to list the full text of all installed commands. The option FULL followed by a command name displays the full text of that command.

HELP [<command name>]

HELP displays the help information associated with that command. If a command name is not given, help information on all the supervisor and installed commands is displayed. The help information for installed commands is obtained from the comment lines that immediately follow the INSTALL and any FORMAL command lines.

TURN FIRST [ON|OFF]

When turned ON, the FIRST switch causes the text strings from PASS commands that are part of previously installed commands to be inserted at the beginning of the PASS string buffer. If the switch FIRST is off, PASS strings are inserted into the buffer in the normal order.

DEV {TTY|PRI|SHOW}

The DEV command changes the device used by the help and list commands:

TTY — Output goes to the users terminal (default).
PRI — Output goes to the printer file.
SHOW — Prints at the user's terminal the current device.

EXIT
STOP
FED

EXIT, STOP, and FED are default user subsystem commands indicating a termination of the subsystem. The supervisor continues execution of the current command sequence. If EXIT or STOP are given at the Federated supervisor command level, the supervisor stops execution and returns to the VAX command level.

<filename> [parameters|delimiters]*

If the filename is not a supervisor or previously installed user command, the supervisor through INTRAC will read the next commands from the specified file. Full INTRAC macro capability is available.

Example of a Command Sequence

The following is an example of a command sequence which is used to call SIMNON from any other subsystem. If the previous subsystem is CLADP or SSDP, then a conversion of their numeric database to SIMNON equations is performed. The user can optionally specify that the conversion not be performed. A second installed command which calls SIMNON directly is used to simplify the code and illustrates the use of other installed commands and macros. This second installed command is known only to the supervisor.

```
INSTALL SIMNON ;NC<CLADP SSDP IDPAC
    "Calls SIMNON. If called from CLADP or SSDP the system files will
    "be converted to SIMNON equations.
    "The command "SIMNON NO" will override the conversion.
DEFAULT NC=YES
IF NC EQ NO GOTO NCONV
    "Perform the conversion
    "if CLADP or SSDP
IF RUN.PREV EQ CLADP GOTO CONV
IF RUN.PREV EQ SSDP GOTO CONV
LABEL NCONV
    "Do not convert the files, go directly to SIMNON
SIMNL
GOTO FINISH
    "Convert the files using program CTOS
    "Initialize CTOS
PASS NEXT=CTOS;MODE=2;
    "Make the coefficients be in the
    "equations rather than separate parameters.
PASS NUM
CONV CTOS
LABEL FINISH
END  "End of command sequence

INSTALL SIMNL
    "Command to directly call SIMNON
```

```
TURN FIRST ON
   "initialize SIMNON indicating where
   "its help library is
PASS LIBLEV2='[SPANG.LUND.COM.SIMNON]'
RUN SIMNON
TURN FIRST OFF
END  "End of command sequence
```

CONCLUSION

The Federated Computer-Aided Control Design System is a means of providing a broad range of design techniques based on a set of existing subsystems. The system is a loosely coupled set of programs thus minimizing the necessary changes. The basic modularity also enhances the maintainability and expandability of the system. It also provides the user with the capability to define his own commands and install subsystems to meet his application. The system is currently in use at 10 General Electric Departments involved in a wide range of aerospace and industrial applications. It has significantly improved productivity. It has also facilitated the transfer of complex control design theory into practice.

REFERENCES

1. Frederick, D.K., "Computer packages for the simulation and design of control systems," Arab School on Science and Technology, 4th Summer Session, Bloudan, Syria (September 1981).

2. Mohler, C., "MATLAB users' guide," Department of Computer Science, University of New Mexico (1981).

3. Edmunds, J.M., "Cambridge linear analysis and design programs," IFAC Symposium on Computer-Aided Design of Control Systems, Zurich (1979) 253-258.

4. Wieslander, J., "IDPAC user's guide, revision 1.," Department of Automatic Control, Lund Institute of Technology, Report 7605, Lund, Sweden (1979).

5. Elmquist, H., "SIMNON, an interactive simulation program for nonlinear systems," Department of Automatic Control, Lund Institute of Technology, Report 7502, Lund, Sweden (1975).

6. Astrom, K.J., "Computer-aided modeling, analysis and design of control systems - a perspective," IEEE Control Systems Magazine, 3, (1983) 4-16.

7. Spang, H.A., III, "State space design program (SSDP) - reference manual," GE Internal Report, Schenectady, NY (1984).

8. Doyle, J.C. and Stein, G., "Multivariable feedback design: concepts for a classical/modern synthesis," IEEE Trans. Automatic Control AC-26, (February 1981).

9. Wieslander, J. and H. Elmquist, "INTRAC, a communications module for interactive programs," Department of Automatic Control, Lund Institute of Technology, Report LUFTFD2/(TFRT-3149)/1-060/(1978), Lund, Sweden, (1978).

APPENDIX A

The material in this appendix is taken from Wieslander and Elmquist.[9]

SYNTAX NOTATION

The following syntax notation is used:

/ or (separates terms in a list from which one and only one must be chosen)
{} groups terms together
[] groups terms together and denotes that the group is optional
{}* denotes repetition one or more times
[]* denotes repetition none or more times

Items are sometimes underlined in the syntax to indicate that an item could be replaced by a variable with the value 'item'.

SUMMARY OF INTRAC STATEMENTS

MACRO <macro identifier>[<formal argument>/
<delimiter>/<termination marker>]*
Begins a macro definition and creates a macro.

FORMAL {<formal argument>/<delimiter>/
<termination marker>}
Declares formal arguments in a macro definition.

END
Ends a macro and ends macro creation mode. Deactivates suspended macros.

LET {<variable> =}* {<number>[{+/-/*//}<number>]
/{+/-}<number>/<identifier>/
[+<integer>]/<delimiter>/
<unassigned variable>
Assigns (allocates) variables.

DEFAULT {<variable> =}* <argument>
Assigns a variable if it is unassigned or does not exist previously.

LABEL <label identifier>
Defines a label.

GOTO <label identifier>
Makes unconditional jump.

IF <argument> {EQ/NE/GE/LE/GT/LT} <argument>
GOTO <label identifier>
Makes conditional jump.

FOR <variable> = <number> TO <number>
[STEP <number>]

Starts a loop.

NEXT <variable>

Ends a loop.

WRITE [([DIS/TP/LP] [FF/LF])]
[<variable>/<string>]*

Writes variables and text strings or displays currently available variables.

READ {{<variable>
{INT/REAL/NUM/NAME/DELIM/YESNO}} /
<termination marker> }*

Reads values for variables from the terminal.

SUSPEND

Suspends the execution of a macro.

RESUME

Resumes the execution of a macro.

SWITCH {EXEC/ECHO/LOG/TRACE} {ON/OFF}

Modifies switches in INTRAC.

FREE {{<global variable>}* / *.*}

Deallocates global variables.

STOP

Stops the execution of the program.

APPENDIX B

CLADP COMMANDS

Utilities

BEGIN	Select type of file for system, ABCD, Laplace, Transition, or z transform
BATCH	Initiate a batch macro file
EXIT	Terminate the program
H	Display brief listing of commands
HH	Display more detailed listing of commands
LIST	List information on the printer file
RET	Return to previous question
SUPER	Return to CLADP supervisor
TYPE	Display information on the terminal

Model Modification

COMB	Combine several blocks into one state space model
MOD	Modify system or compensator coefficients
PRE	Add a precompensator
POST	Add a post compensator
FEED	Add a feedback compensator
PAR	Add a parallel feed forward compensator

Conversions

ATOS	Change state space to Laplace
STOA	Change Laplace to state space
ATOT	Change continuous state space to discrete state space
TTOA	Change discrete state space to continuous state space
TTOZ	Change discrete state space to z transform
ZTOT	Change z transform to discrete state space

Display

AREA	Change size of area being viewed
AXS	Add or delete numbers on axes
CAL	Mark curve with values of dependent variable
CIR	Add Gershgorin circles
COPY	Make a hard copy of the picture on a plotter
NEXT	Draw subset of calculated curves
RED	Redraw the picture
RES n	Rescale by a factor of n
SHIFT	Shift the phase range 180°
TEXT	Add text to the picture

Analysis

ALIGN	Do a high frequency alignment
ARL	Root Locus based on a state space model
RL	Root Locus based on Laplace model
NYQ	Plot a Nyquist or Inverse Nyquist Array
BODE	Draw a Bode plot
MAG	Draw the magnitude only
PHASE	Draw the phase only
NICHOLS	Draw the Nichols plot
ROOT	Obtain poles and zeros of the system
TIME	Do a time response

SSDP COMMANDS

Utilities

See CLADP list

Model modification

QRW	Enter linear quadratic performance measure matrices
COR	Enter correlation matrices for Kalman Bucy filter
MOD	Modify system or compensator coefficients
PRE	Add a precompensator
POST	Add a postcompensator
FEED	Add a feedback compensator
PAR	Add a parallel feed forward compensator
STFB	Add a state feedback block
OBS	Add an observer block

Controllability and Obserability

COTY	Determine controllability
DETY	Determine detectability
OBTY	Determine observability
STTY	Determine stabilizability

Display

See CLADP commands

Analysis

EIG	Calculate the eigenvalues
KBF	Design a Kalman Bucy Filter
LQR	Design an linear quadratic regulator
RIC	Solve the Riccati equation
SVDD	Perform singular value controller design
TIME	Do a time response

SIMNON COMMANDS

Utilities

EDIT	Edit system description
GET	Get parameters and initial values
LIST	List files
PRINT	Print files
SAVE	Save parameter values and initial values in a file
STOP	Stop

Graphic output

AREA	Select window on screen
ASHOW	Plot stored variables with automatic scaling
AXES	Draw axes
HCOPY	Make hard copy
SHOW	Plot stored variables
SPLIT	Split screen into windows
TEXT	Transfer text string to graph

Simulation commands

ALGOR	Select integration algorithm
DISP	Display parameters
ERROR	Choose error bound for integration routine
INIT	Change initial values of state variables
PAR	Change parameters
PLOT	Choose variables to be plotted
SIMU	Simulate a system
STORE	Choose variables to be stored
SYST	Activate systems

IDPAC COMMANDS

Utilities

CONV	Conversion of data to internal standard format
DELET	Delete a file
EDIT	Edit system description
FHEAD	Inspect and change file parameters
FORMAT	Conversion of data to symbolic external form
FTEST	Check existence of a file
LIST	List files
MOVE	Move data in database
TURN	Chagne program switches

Graphic output

BODE	Plot Bode diagrams
HCOPY	Make hard copy
PLMAG	Magnify plot and allow changes of data
PLOT	Plot curves with linear scales

Time Series Operations

ACOF	Compute autocorrelation function

CCOF	Compute cross-correlation function
CONC	Concatenate time series
CUT	Extract a part of a time series
INSI	Generate time series
PICK	Pick equidistant time points
SCLOP	Do scalar operations on a time series
SLIDE	Introduce relative delays between time series
STAT	Compute statistical characteristics
TREND	Remove a trend
VECOP	Do vector operations on a time series

Frequency Response Operations

ASPEC	Compute an auto spectrum
CSPEC	Compute a cross spectrum
DFT	Discrete Fourier Transform
FROP	Operate on frequency responses
IDFT	Inverse Discrete Fourier Transform

Simulation and Model Analysis

DETER	Deterministic Simulation
DSIM	Simulation with noise
FILT	Compute a filter system
RANPA	Pick parameters from a random distribution
RESID	Compute residuals with statistical test
SPTRF	Compute the frequency response of a transfer function

Identification

LS	Least Squares identification
ML	Maximum Likelihood identification
SQR	Least Squares data reduction
STRUC	Least Squares structure definition

COMPUTER-AIDED CONTROL SYSTEMS ENGINEERING
M. Jamshidi and C.J. Herget (Editors)

INTERACTIVE COMPUTER-AIDED CONTROL SYSTEM ANALYSIS AND DESIGN

P.P.J. VAN DEN BOSCH

Delft University of Technology
Laboratory for Control Engineering
P.O. Box 5031
2600 GA Delft, The Netherlands

Simulation and optimization will be proposed as flexible and powerful alternatives to mathematically oriented analysis and design methods. Requirements to be posed on both programs and hardware for computer-aided control system design are discussed and supplemented by several examples.

INTRODUCTION

In this paper the use of computer-aided design programs for system analysis and system design will be discussed. Simulation and optimization will be proposed as a flexible and powerful alternative to mathematically oriented analysis and design methods. Such methods make use of, for example, the linearity of the system so that mathematical solution techniques are available. Then the range of validity of the results may be quantitatively evaluated for both this and other comparable systems.
In general, systems do not satisfy assumptions as to linearity, order information, etc., so that either a mathematical approach cannot be applied or a simplified, linear model has to be used.

An analysis or design approach based on simulation and optimization is much more flexible and can deal with nearly any system description, linear or nonlinear, continuous or discrete or any mixture of differential, difference, algebraic or logical equations. This flexibility has to be paid for by extra calculation time. Suitable software has to be available in order to use such an approach.

In this paper we will describe several analysis and design approaches and requirements to be posed on software and hardware to get a suitable design environment. Moreover, we will illustrate this paper with examples of design programs and computer installations.

SYSTEM ANALYSIS

Identification methods are generally based on an analysis of the input and output signals of the system that has to be identified. Estimates of the parameters of a model, whose structure and order are determined in advance, can be calculated. Depending on the identification method a priori information, based on noise characteristics, can also be included in the algorithm in order to obtain better estimates. Let us briefly consider the Least-Squares Method (LS), as illustrated in figure 1.

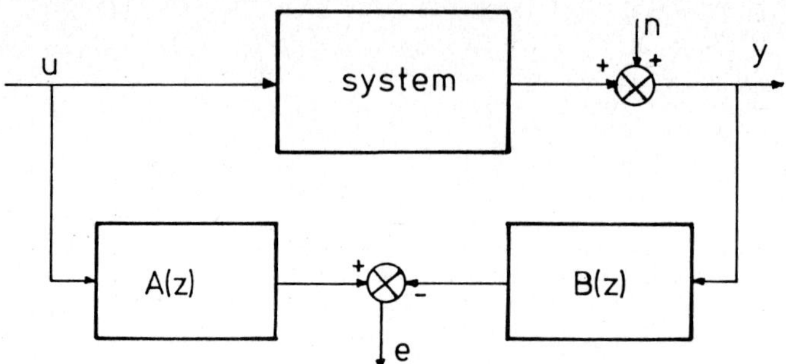

Fig. 1 Least-Squares Method for estimating the discrete transfer function H(z)=A(z)/B(z).

The LS method assumes that the process can be described by the linear model H(z)=A(z)/B(z). The LS method calculates the parameters a_i and b_i ($A(z)= a_0 + a_1 z + \ldots$, $B(z)= b_0 + b_1 z + \ldots$) of the model very quickly when solving a set of n linear equations with n unknown parameters a_i and b_i. The LS method estimates only unbiased parameters a_i and b_i, if the noise n(k) satisfies several conditions, for example, n(k) has to be colored noise arising from white noise filtered by means of $B^{-1}(z)$. If this condition is not met, the noise characteristics have to be estimated too. This can be achieved by, for example, the Extended Matrix Estimator [12]. This extension introduces an iterative solution procedure which increases the calculation time considerably. When a priori information about the structure and/or parameters of the process is available, it is difficult or even impossible to make use of this knowledge.

Another approach to system identification is the use of simulation and optimization. This allows some a priori information of the system to be taken into account, for example, some knowledge about the internal structure, some known parameters, in some cases the shape or even the exact values of a non-linearity, etc. This a priori information can be obtained from additional measurements or from an understanding of the physical laws that describe the system under consideration. This additional a priori information can be very useful in finding an appropriate model of the system.

By means of simulation and optimization we can calculate the "best-fit" model of the system. Linearity assumptions are no longer necessary. Such an approach is illustrated in figure 2. A criterion is defined, based on the error between the output of the system and the output of the model, which uses the input signal of the system. The output of the model is obtained by using simulation techniques. Therefore, the model may be described by continuous parts, discrete parts, non-linear or logical elements or any combination of these. Then an optimization algorithm is able to find optimal model parameters of a (non-)linear model with a user-defined structure and a user-defined criterion. For example, if we know in advance that the system under consideration has two time constants (and thus two real poles) this knowledge can be used in the identification scheme of figure 2, but not in the LS method. This flexibility is achieved at the expense of calculation time. Optimization is inherently a non-linear iterative procedure. Each iteration requires a complete simulation run so that more calculation time is needed than with the LS method. For system analysis this is not a real limitation. However, real-time identification for adaptive control poses very strict limitations on the calculation-time requirements, so that the proposed identification method cannot always be used. For interactive use of this capability, the number of parameters has to be limited to maximally about 5 to 10.

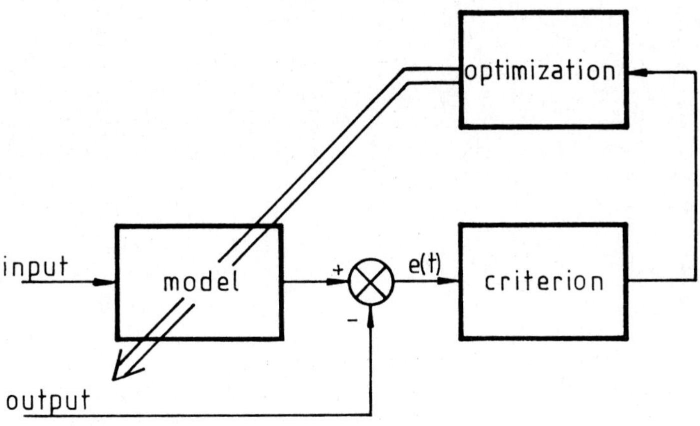

Fig. 2 Identification via simulation and optimization.

It should be stressed that both approaches attempt to find a model whose output is as far as possible equivalent to the output measured. It turns out that there are many models with different orders and completely different sets of parameters that offer about the same time responses. So, if we want to determine the time constants or gains of a system, we may obtain, after identification with the LS method, values that have some deviations or are even completely erroneous, although the time response is quite appropriate. In some cases, especially when a model has to be obtained in order to design a controller for the system, this situation may be acceptable as long as the calculated model and the system have about the same dynamical behavior. When we want to know the accurate values of some parameters, for instance, to find certain physical parameters from some experiments, serious problems may arise. In such a case a priori information may improve the results which favors the simulation and optimization approach over the LS method.

SYSTEM DESIGN

There are many ways to design a system such that it satisfies pre-defined design requirements. In general, some control structure has to be implemented to improve the system behavior. In designing such a controller we can use several graphical representations of the system in order to study its dynamic behavior and to find ways to define controllers such that the system behavior will improve. Linear, single-input single-output systems can be designed by using the Bode or Nyquist diagrams or root loci. Linear multivariable systems can be treated by graphic design methods such as the Inverse Nyquist Array method, the Characteristic Locus Design method, etc. For non-linear systems the describing function method and the circle criterion are available, although they are rather conservative. These graphic design methods offer much qualitative and quantitative information about the system behavior. Nevertheless, if the system is complex much experience and knowledge is required to be able to design an appropriate controller which satisfies the design requirements.

Another approach to designing systems is to formulate the design problem in terms of an optimization problem: formulate a criterion, parameters of a controller that have to be optimized and constraints. The criterion has to satsify two requirements, namely it has to express the design objectives and has to be easy to calculate. In choosing a mathematically oriented criterion the optimization process can be quite fast, but the link with the design objectives, such as overshoot, rise time, damping etc., may be weak or even non-existent. For example, the linear optimal state feedback matrix, according to the quadratic functional J:

$$J = \int_0^\infty (x^TQx + u^TRu)\,dt \qquad (1)$$

taking into account the state x and the input u, can be easily found by solving a Riccati equation. There are also fast algorithms for pole placement, etc.

Output feedback, instead of state feedback, complicates the optimization considerably. Then, mathematical expressions exist to calculate both functional J (1) and its gradient with respect to the coefficients of the feedback matrix. Hirzinger [9] has proposed a usable output-feedback configuration for multivariable systems. His dynamic controller has both feedback and feedforward. The design requirements placed on dynamic behavior and decoupling are expressed in a parallel reference model, which causes an unconstrained optimization problem to arise with functional J (1) as criterion. The dimension of the state now becomes the sum of the dimension of the states of the original system, of the controller and of the parallel model. The value of J and its gradient are calculated by solving Lyapunov equations.

Another way to calculate an arbitrary functional of the output(s) or states of a system is to use the fast numeric inverse Laplace method as proposed by Zakian [13]. The system still has to be linear, but now the criterion can be any linear or non-linear functional of, for example, the output(s).

These approaches deal with unconstrained optimization problems. All design objectives have to be implemented in the criterion. Constraints are incorporated into the criterion via penalty functions. Alternatively, they can be incorporated directly into the problem formulation. Then, Zakian's Method of Inequalities [14] or the more powerful optimization as proposed by Mayne and Polak [10] become attractive. The latter, in particular, can handle infinite-dimensional constraints such as:

$$\underset{t}{\mathrm{MAX}}\ \{y(t)\} < 1.05.y,ref$$

This constraint limits the overshoot to a maximum of 5%. So, Mayne and Polak's design method based on optimization with infinite-dimensional constraints offers increased flexiblity. An important requirement is that both the criterion and the constraints be continuously differentiable and that their gradients be available for the optimization algorithm. These requirements restrict the application of their design procedure.

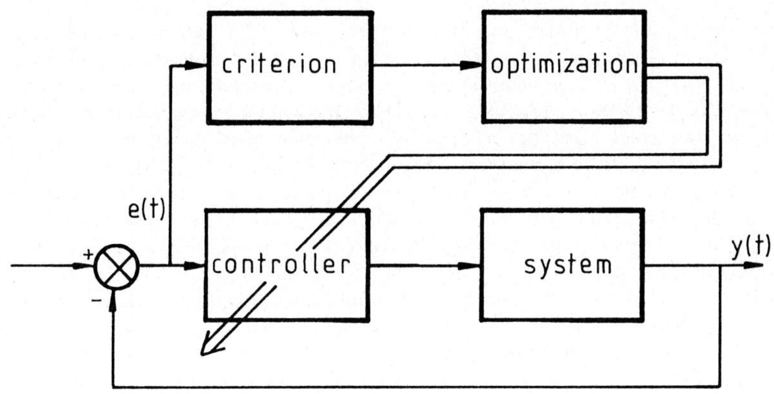

Fig. 3 System design using simulation and optimization.

Even more flexible is the approach based on using simulation and optimization, as illustrated in figure 3. Simulation techniques are used to calculate the error signal e(t) due to the controller and the system. A criterion can be defined, based on this error signal and/or the output, which can be optimized with respect to the parameters of the controller. So, any (non-)linear system and any controller configuration can be used with any criterion. Finite or infinite-dimensional constraints can be included, via penalty functions, in the criterion. Even the combination of a discrete controller which controls a (non-)linear continuous system offers no problems.

Van den Bosch [4] has illustrated that calculation-time requirements of the simulation and optimization approach are comparable with these requirements for solving the linear output feedback problem.

Yet, optimization is just a design tool. It does not directly give the required controller parameters. The solution of an optimization problem offers optimal controller settings, determined by the selection of the criterion, the controller structure and the free controller parameters.

From the point of view of accuracy, simulation suffers less from numerical errors. Especially for high-order systems, numerical solution methods for Riccati or Lyapunov equations may lead to inaccurate or erroneous results.

Therefore, it may be concluded that, even when analysis or design is otherwise possible, it may still be profitable to use simulation and optimization owing to their inherent flexibility.

REQUIREMENTS

General Requirements

User interface:
If a program has attractive and reliable mathematical algorithms, it may still be inferior with respect to control system design when the interaction between the program and the user is not accepted by the user. This interaction is determined by a number of factors, but especially by the communication between the user and the program and the presentation of graphic information [3]. Only an interactive program can support a designer-oriented environment.
Either a command language or a question/answer approach can take care of the communication between the program and the user. We have selected a command language for our CAD programs because it yields the necessary flexibility for most users. However, unexperienced users lack the guidance of a question/answer approach. By means of extensive supporting capabilities, clear error messages and a macro facility some guidance can be incorporated into a command language. A macro facility combines separate commands to obtain a higher-level command language which is easier to use, but, at the same time, is less flexible.

In designing control systems, graphic representations of the system behavior are of paramount importance. Although numbers are much more exact, design considerations mainly deal with graphic representations of a system. For example, linear optimal state feedback, output feedback or pole placement are well-established, mathematically oriented methods for control system design. However, whether or not such a design meets the ultimate design requirements cannot be judged by looking at only the value of the criterion or at the feedback matrix. In general, only time (or freqency) responses offer enough information to make possible a judgement of the ultimate system behavior.
So, a graphics display, which is very fast, or a plotter is almost unavoidable when analyzing or designing systems.

Documentation:
Programs produce large amounts of data. In order to keep track of all this data, it is necessary to have good documentation facilities on all printed output. So, all plots and models printed for later use need some headings that contain information concerning date, time, user and file names.

Compiler:
In spite of the availability of many new programming languages, such as Pascal, Ada and Modula-2, Fortran is still widely used and accepted for technical and scientific work. The majority of subroutine libraries for control system design is Fortran oriented. Consequently, a Fortran compiler has to be available. This compiler has to satisfy at least the Fortran 66 or Fortran 77 standards. Although double-precision complex numbers are not incorporated in the Fortran 77 standard, this data type is very useful and almost necessary in many algorithms.

Calculating speed:
Simulation and optimization severely limit on the calculating speed of computers used in an interactive environment. Other design programs, for example those based on pole placement, require less calculating speed.

Memory requirements:
Design programs can be quite large. Especially when high-order systems are allowed, much memory has to be available, for example 200k to 1M bytes internal memory. Overlay or virtual memory techniques reduce the memory requirements. Owing to their limited memory capacity of 64k bytes and their lack of overlay or virtual memory techniques, the 8-bit microcomputers are not suitable for large CAD programs.

Requirements For Linear Design Programs

For many years the theory of control system analysis and design has been developed with little or no consideration for the reliability of the underlying numerical calculations. In spite of this lack of interest, these algorithms still work satisfactorily for low-order systems, especially if double-precision accuracy has been implemented. However, the growing interest in the application of multivariable system theory to large and complex systems has exposed the computational deficiencies. This emphasizes the need for efficient, reliable and numerical stable algorithms. Recently, much research has been devoted to linear algebra to derive and describe reliable and numerical stable algorithms. Denham et al. [7] indicate that, at least at present, concentration on the use of system models which involve the manipulation of real matrices is more likely to yield numerically robust algorithms than is the case for models involving polynomial manipulations. This favors the use of state space models instead of matrices of tranfer functions. At present there is a class of generally accepted subroutine libraries, such as LINPACK (linear algebra) and EISPACK (eigenvalues) which can be used as a reliable basis for computer-aided design programs. The subroutine package ORACLS [1] supplies the user with Fortran subroutines for designing with the Linear Quadratic Gaussian (LQG) methodology.
Research is going on in several places in the world to implement these subroutine libraries in interactive analysis and design packages.

In conclusion, the main requirement is the availability of efficient, reliable and numerically robust algorithms for the analysis and design of control systems.

Requirements For Simulation Programs

Both digital and hybrid computers can be used for the simulation of continuous systems. Owing to the many advantages of the digital computer over a hybrid one (price, availability, size of the problem, etc.) we shall focus our attention on the requirements for simulation programs for digital computers.

Integration methods:
Simulation programs calculate the solution of sets of linear or non-linear differential and/or difference equations. Digital computers calculate a variable only as a sequence of values at discrete time intervals, determined by the integration interval. Therefore, the continuous integrator has to be approximated. The accuracy with which this approximation can be realized determines the accuracy of the simulation and depends both on the integration method and the integration interval. With a small integration interval and a complex, higher-order integration method more accurate results can be expected than with a larger integration interval and a simpler integration method. But, both a small integration interval and a higher-order integration method increase calculation time. So, a compromise has to be made between calculation time and accuracy.

In using fixed-step integration methods, the second- and fourth-order Runge Kutta integration methods are widely accepted.

Algebraic loops:
A second problem arises in solving a parallel-defined system with a sequentially oriented digital computer. This problem can be solved by using a proper sorting procedure, except when there is an algebraic loop (an equation in which a variable is an algebraic function of its own value), for example, x = sin(x) + y. It is always advisable to avoid algebraic loops. If they cannot be avoided they have to be solved with the aid of time-consuming, iterative algorithms, which can be used not only for the solution of algebraic loops, but also for the solution of any general, non-linear algebraic equation.

Multi-Run facilities:
There is an important distinction between preprocessor-like programs (in general batch-oriented) and interpreter-like programs (in general interactive) for simulation purposes. The former allows statements of a high-level programming language to be included in the simulation-model description. Therefore, these programs can be made as flexible as, for example, a Fortran program. Interpreter-like programs lack this capability, so that special commands have to be implemented to realize, for example, multi-run facilities such as optimization, comparison of variables between different runs, storage of several sets of initial conditions, etc.

EXAMPLES

Interactive TRansformation and Identification Program TRIP

At the Laboratory for Control Engineering an interactive program TRIP has been designed and realized [6] for the analysis and design of linear, single-input single-output systems. TRIP is in use now at many educational and research institutes and private companies in The Netherlands and abroad. The program is based on the assumption that a linear, continuous system can be described by means of a number of different models, namely:

- a transfer function H(s), called SS
- a state-space model, called MA
- a frequency response, called FR and
- a time response, called TY.

In the same way we can distinguish for linear discrete systems the following models:

- a transfer function H(z), called ZZ
- a state-space model, called MZ
- a freqency response, called WW and
- a time response, called TY.

Because these models can describe the same linear system, it is possible to calculate one model from another model. For example, starting with H(s) it is possible to calculate the discrete transfer function H(z), the state-space models and the time and frequency responses. These calculations are called transformations. Not all transformations are straightforward. The time response and the frequency response do not contain order information. Consequently, each transformation from a time or frequency response to a transfer function or to a state-space model requires an order estimation.

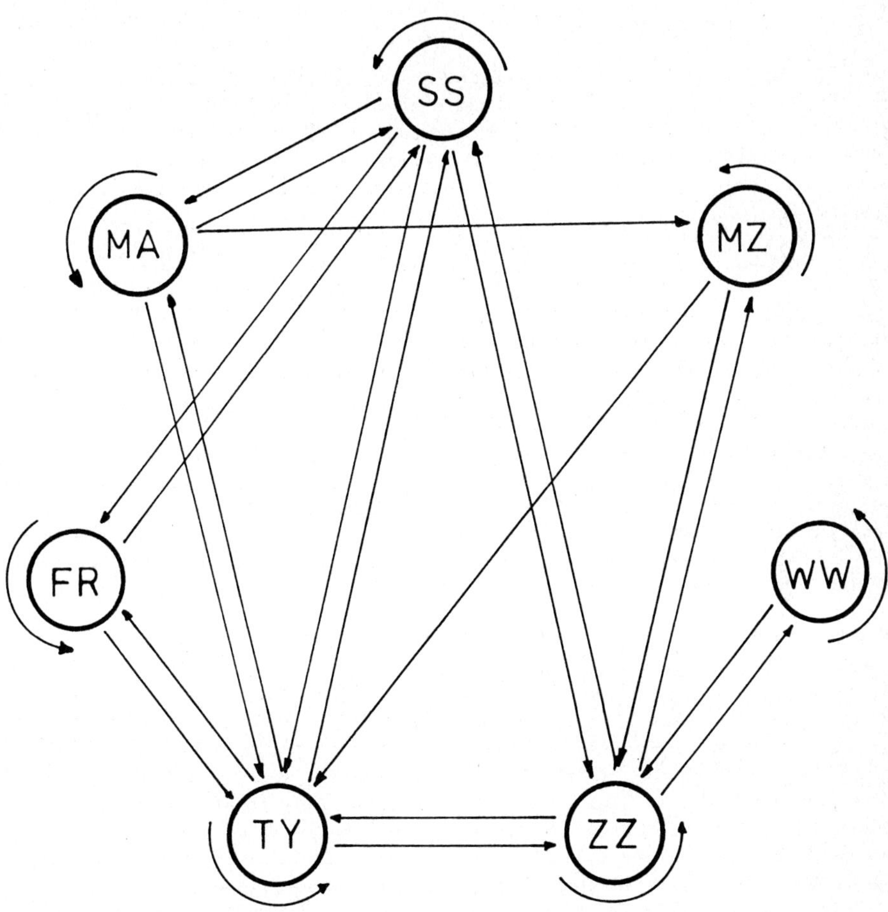

Fig. 4 TRansformation and Identification Program TRIP.

Figure 4 illustrates the possibilities of the program TRIP. Each arrow represents a transformation. Moreover, other capabilities have been implemented in TRIP, such as calculation of the root locus, solution of Lyapunov and Ricatti equations, optimal state feedback, etc.

The program TRIP allows a maximum order of 10. For low-order systems the numerical behavior of the underlying algorithms was quite appropriate. For higher-order systems, this behavior was not sufficient. The development of TRIP was started about 15 years ago and has required about 6 manyears programming activity. Recently, two additional manyears have been used to incorporate state-of-the-art algorithms for the numerical calculations. TRIP occupies about 375k bytes of internal memory.

Interactive Simulation Program PSI

At the Laboratory for Control Engineering an Interactive Simulation Program (PSI) [2,5], which satisfies almost all of the stated requirements has been designed and realized. It is in use now at many educational institutes (universities, technical schools), research institutes and private companies in The Netherlands and abroad. This interpreter-like, block-oriented simulation program offers, for example, the following facilities:

- About 90 commands are available for the user to realize his objectives.
- Five numerical integration methods are available, namely four fixed-step methods (Euler, Adams Bashfort 2, Runge Kutta 2 and Runge Kutta 4) and one variable-step-size method (Runge Kutta 4).
- Solution of algebraic equations is realized by a fast Newton-Raphson algorithm. If this procedure fails, a more reliable, although slower, optimization algorithm is used.
- Optimization with scaling and constraints is supported. In PSI the user can define the output of an arbitrary block as the criterion and up to eight arbitrary parameters of the simulation model as parameters of the optimization. The parameters that offer the smallest value of the criterion will be accepted as the solution of the optimization procedure. Pattern Search [8] has been selected as the minimization procedure, due to its robustness and lack of a line-minimization procedure. Moreover, as soon as the value of the criterion during a simulation run exceeds the present optimal value, this run is terminated prematurely, because the selected set of parameters will never lead to a lower value of the criterion.
Although Pattern Search adjusts its search step size according to the "shape" of the criterion, the speed of convergence can be improved by scaling. Scaling can make each parameter about equally important for the optimization algorithm. Not only is scaling supported by PSI, but constraints are also allowed. Each parameter may have an upper and a lower limit. The optimization algorithm will only search for an optimum in the feasible region of the parameter space.
- Multi-run facilities are available. For example, run-control blocks, comparison of signals between several runs, storage of initial conditions, etc.
- Extensive tests on all user-supplied information is implemented. Each error is indicated by a meaningful error message, of which there are about 75.
- About 55 powerful block types are available, including 5 different types of integrators. Fortran programming in a non-interactive mode is required to define new block types. The user only needs to write a subroutine in which the output is defined as a function of the input(s) and parameter(s) and to compile it, and after a link step, his block is available.
- There are memories to store signals during a simulation run. These signals can be studied after the simulation run, can be saved on and read from disk or can be used as inputs for future runs.
- Variables can be shown on the screen as time responses, as phase trajectories, as numerical values or in a user-defined mode. This last option requires some Fortran programming.

- Symbolic block names can be used. Instead of numbers each block or variable can be assigned a user-selected name of up to eight characters. So blocks can get meaningful names like PRESSURE, SPEED or OUTPUT instead of abstract numbers like block 13, 91 or 512, etc.

The development of PSI was started about 7 years ago and has required about one manyear of programming activity. Depending on the number of blocks PSI occupies about 200k bytes (300 blocks) to 300k bytes (3000 blocks) of internal memory.

This section has described a number of facilities which make programs, such as PSI, highly suited to system analysis and system design. PSI is able to solve (non-linear) differential, difference, algebraic and logical, Boolean equations or any mixture of them. Moreover, attractive and powerful interaction is realized between the user and the program.
A real-time version of PSI can be used to realize a real-time process or a real-time controller which, by means of AD- and DA-converters, is connected with some laboratory equipment or another computer. So, in a flexible and interactive way, quite complex processes and controllers can be easily realized.

Limitations:
Like most other interactive, block-oriented simulation programs, PSI does not support special facilities to solve partial differencal equations, stiff systems and polynomial or matrix equations. These programs deal with single-valued variables, and consequently not with vectors and matrices. The solution of the Ricatti equation of a second-order system is possible, but the solution of this equation of higher-order systems cannot be obtained easily.

Computer Configuration

At the present time many computer configurations can be selected to realize a computer-aided design environment. We can distinguish among superminicomputers, minicomputers and 16- or 32-bit microcomputers. Each installation has its own capabilities, support and price. In this section we will compare these installations for computer-aided design applications. We will discuss a VAX 750 with VMS, the PDP 11/23 with RSX-11M and the Fortran 4 Plus compiler, the TULIP 16-bit microcomputer with MS.DOS and the IBM PC 8/16-bit microcomputer with PC.DOS. The VAX and PDP are manufactured by Digital Equipment Corporation and are equipped with an appropriate floating-point processor. The TULIP has Intel's 8Mhz 8086 processor and the Intel 8087 floating-point processor and is manufactured by the Dutch company Compudata in Den Bosch. The IBM PC has Intel's 4.77 Mhz 8088 and 8087 processor. Both use the Microsoft Fortran V3.2 compiler. All compilers, except the Fortran 4 Plus compiler, support the double precision complex data type.

We have equipped the VAX and the PDP with the VT241 color terminal. A comparison between this terminal and the graphics capabilities of the two microcomputers is shown in table 1, namely the number of different colors, the resolution (in dots) and the possibility to use two separate screens; one alphanumeric screen for the communication between program and user and one color graphics screen for the graphics information.

Table 1. Comparison of graphics capabilities.

Name	number of colors	resolution	separate screens?
VT241	4	800*240	no
TULIP	8	768*288	yes
IBM PC	4	320*200	no

Table 2 illustrates the price per terminal, the maximum size of a program loaded in memory and the calculation speed. The price per terminal is the price of the computer system with terminals, divided by the number of terminals. This price is normalized for the IBM PC. For example, 4 CAD stations can be realized with 4 TULIP computers or one VAX with 4 terminals. The VAX solution is six times more expensive. The calculation speed is normalized for a VAX 750 with one user, running a simulation program. With four users this speed can drop to 0.25.

Table 2. Comparison of computer installations

Name	background memory	number of terminals	price per terminal	maximum program size	calculation speed
VAX 750	60M	4	6	8M+virtual memory	1-0.25
PDP 11/23	10M	2	2	64k+overlay	0.3-0.15
TULIP	2*400k	1	1	896k	0.3
IBM PC	2*360k	1	1	640k	0.14

These tables indicate that a decentralized solution with separate 16-bit microcomputers yields an attractive price-performance ratio. Depending on the number of users of the multi-user VAX, a TULIP can be up to three times slower or a little faster, it has better graphics capabilities and it costs much less than a solution based on a VAX. Although calculating speed of a microcomputer may be less, response times to commands of the user are shorter than in a multi-user environment. The IBM PC offers a less attractive performance.

As soon as a decentralized solution has been selected, the administration becomes more complicated. It takes more time and effort to update all decentralized computers with the newest versions of the software. Moreover, the exchange of data, models and files becomes more complicated, namely via floppy disks instead of via the common large disk. On the other hand, with a decentralized solution each user is responsible for his or her own files and models. If he is willing to work with older versions of the software, there will be fewer conversion problems.

At our laboratory we have decided to decentralize the CAD facilities from a multi-user environment to a number of stand-alone computers. We have selected the TULIP computer for this purpose but fast computers of other manufacturers exist as well, for example the M24 of Olivetti, the Apricot and Wang computers.

CONCLUSIONS

The value of simulation and optimization for system analysis and system design has been discussed. It appears that many systems can only be studied by using simulation techniques. But even when analytical methods are available, simulation and optimization have their own unique merits.

The availability of efficient, reliable and numerically stable algorithms is of paramount importance for programs intended for the analysis and design of control systems. The advent of the 16-bit microcomputer has created possibilities to realize a very flexible and powerful computer-aided design environment which compares favorably with the solutions based on multi-user minicomputers and superminicomputers.

Facilities which allow the use of both simulation and optimization in an interactive way have to be available. It has been illustrated that interactive simulation programs such as PSI are very well suited to use in interactive system analysis and system design.

REFERENCES

[1] Armstrong, E. S., ORACLS, A Design System for Linear Multivariable Control (Marcel Dekker Inc., New York, 1980).

[2] Bosch, P.P.J. van den, PSI-An Extended, Interactive Block-Oriented Simulation Program. Proceedings IFAC Symposium on Computer Aided Design of Control Systems, 459-464 (Pergamon Press, London, 1980).

[3] Bosch, P.P.J. van den and P.M. Bruijn, Requirements and Use of CAD Programs for Control System Design. Proceedings IFAC Symposium on Computer Aided Design of Control Systems, 459-464 (Pergamon Press, London, 1980).

[4] Bosch, P.P.J. van den, Interactive System Analysis and System Design Using Simulation and Optimization. Proceedings IFAC Symposium on Computer Aided Design of Multivariable and Technological Systems, 225-232 (Pergamon Press, London, 1983).

[5] Bosch, P.P.J. van den, Manual of PSI. Laboratory for Control Engineering, Delft University of Technology (96 pages) 1984.

[6] Bosch, P.P.J. van den, Manual of TRIP. Laboratory for Control Engineering, Delft University of Technology (150 pages) 1984.

[7] Denham, M.J., C.J. Benson and T.W.C. Williams, A Robust Computational Approach to Control System Analysis and Design. Proceedings IFAC Symposium on Computer Aided Design of Multivariable and Technological Systems, 667-672 (Pergamon Press, London, 1983).

[8] Hooke, R. and T.A. Jeeves, Direct Seach Solution of Numerical and Statistical Problems, Journal of the ACM, 8 (1961) 212-229.

[9] Hirzinger, G., Decoupling Multivariable Systems by Optimal Control Techniques. Int. J. of Control, 22 (1975) 157-172.

[10] Mayne, D.Q., E. Polak and A. Sangiovanni Vincentelli, Computer Aided Design via Optimization: A Review. Automatica 18 (1982) 147-154.

[11] Sirisena, H.R. and Choi, Minimal Order Compensators for Decoupling and Arbitrary Pole-Placement in Linear Multivariable Systems. Int.J. of Control 25, no 5 (1977).

[12] Talmon, J.L. and Boom, A.J.W. van den, On the Estimation of the Transfer Function of Process and Noise Dynamics Using a Single Estimator. Proc. IFAC Symposium on Identification and Process Parameter Estimation (Pergamon Press, 1974).

[13] Zakian, V., Optimisation of Numerical Inversion of Laplace Transforms. Elec. Letters 6 (1970) 677-679.

[14] Zakian, V., New Formulation for the Method of Inequalities. Proc. IEE 126 (1979) 579-584.

COMPUTER-AIDED CONTROL SYSTEMS ENGINEERING
M. Jamshidi and C.J. Herget (Editors)

L-A-S : A Computer-Aided Control System Design Language

P. J. West
Coordinated Science Laboratory
and Dept. of Electrical
and Computer Engineering
University of Illinois
Urbana, IL. 61801 USA

S. P. Bingulac
Electrical Engineering Dept.
Virginia Polytechnic Institute
and State University
Blacksburg, VA. 24061 USA

On leave from the
University of Belgrade
Dept. of Organizational Sciences
Belgrade, Yugoslavia

W. R. Perkins
Coordinated Science Laboratory
and Dept. of Electrical
and Computer Engineering
University of Illinois
Urbana, IL. 61801 USA

ABSTRACT

This article reviews the basic features of L-A-S, a high-level, interactive, conversational control system language. L-A-S (Linear Algebra and Systems) can be used to construct, test, and evaluate various control algorithms for analysis and design of linear and a wide class of nonlinear systems. L-A-S data structures include matrices, polynomials (continuous or discrete), and matrices of polynomials.

We describe the mechanics of writing L-A-S programs and the theory behind the L-A-S language syntax. Advanced L-A-S features and concepts are briefly discussed. Illustrative examples are also included. The Appendix serves as a quick reference to the L-A-S language.

I. INTRODUCTION

L-A-S (Linear Algebra and Systems) is an interactive, conversational high-level control system programming language. As a Computer-Aided Design (CAD) tool, it can be used to construct, test, and evaluate various control algorithms for linear and some nonlinear systems. L-A-S should be distinguished from the so-called menu-driven CAD packages. L-A-S does not ask you what choice you would like. Rather, you instruct L-A-S as to exactly what you want it to do. Thus, the user is not explicitly limited to the choices on a menu. Instead, the simple fact that L-A-S is a language implies that programs can be written on-line, stored, modified, and re-executed at a later time with the same or a different data set.

The fundamental concept behind an L-A-S program is the notion of an L-A-S *operator*. Essentially, operators should be viewed as generalizations of choices on a menu. That is, consider an arbitrary linear algebra or control system menu of allowable computations. Then, an L-A-S operator simply extracts the function performed by any particular choice and attaches a mnemonic name to that function. In other words, L-A-S operators perform some desired calculation on input data, indicate errors if any, and generate output data. Operators are issued to the L-A-S language interpreter along with the names of the data used and generated. Operators are divided into five groups : Input/Output, Data Handling, Linear Algebra, Control Systems, and L-A-S Program Control. Presently, there are more than 100 L-A-S operators, (see Table 1). Furthermore, the L-A-S user can currently define a maximum of 100 matrices with the total number of matrix elements not to exceed 50,000. The maximum order of any particular matrix is not explicitly limited.

L-A-S programs are written by combining one or more operator statements. In the normal interactive mode, the user types statements directly to the L-A-S language interpreter. L-A-S is a *free-format* language. This means that unless a comment is entered, blanks are removed prior to scanning by the interpreter. Thus, the format in which a program is typed is unrestricted or free. Upon receiving a line of text, the interpreter parses and subsequently analyzes the L-A-S statement. If the syntax and semantics of the statement are correct, then the L-A-S driver library calls upon the necessary FORTRAN subroutines to perform the computations.

The L-A-S computer source code is written in ANSI standard FORTRAN ('66 or '77), but use of L-A-S does not require knowledge of any programming language. Structured programming techniques have yielded a highly modular code. Further, modularity permits simple modifications, updates, and extentions as the need arises. The L-A-S language syntax and semantics are straightforward so that a good working knowledge of the language may be easily and rapidily attained. L-A-S uses the EISPACK and LINPACK mathematical software packages extensively as well as other "state-of-the-art" control system software so that the numerical integrity of the final results is assured.

Thus far, various versions of the L-A-S software have been implemented at the following computer installations : University of Illinois (Vax/UNIX); General Electric, Schenectady, N.Y. (Vax/VMS); Michigan State University (PRIME-750); Washington State University (HP-1000); University of Southern California (Vax/VMS); Ohio State University (PDP-11/70); Drexel University (Vax/VMS); Polytechique University of Valencia, Spain (Univac-1100); and at several locations in Yugoslavia (DEC-20, Vax/VMS, PDP-11/34, Perkin-Elmer 8/32, CDC Cyber).

This article is organized in the following manner. Section II describes the L-A-S operator statement syntax. The basic statement structure is introduced. Then advanced language features are briefly discussed. Finally a formal specification of the L-A-S syntax is presented. Section III is devoted to examples of L-A-S programs. Section IV summarizes and concludes the article. The Appendix provides a brief description of all L-A-S operators to date. The References should be consulted for additional information about the L-A-S language.

Table 1. Summary of L-A-S Operators by Groups

INPUT / OUTPUT OPERATORS

(CDI), (DDM), (DIM), (DIS), (DPM), (DSC), (DSM),
(DVC), (DZM), (INP), (NIK), (OUT), (PLL), (PLT),
(RDF), (RDI), (RGE), (TXT), (WDF), (WGE)

DATA HANDLING OPERATORS

(CTC), (CTI), (CTR), (EMD), (MCP), (RTI), (SHD),
(SHL), (SHR), (SHU), (TVC)

LINEAR ALGEBRA OPERATORS

(+), (-), (*), (-1), and

(ATG), (CHD), (CHE), (COS), (CUR), (EGC), (EGV),
(EXP), (F*), (F/), (GIN), (INV), (JFR), (JFS),
(LOG), (MEA), (NRC), (NRR), (NSP), (P+), (P*),
(PMA), (PMM), (PNR), (RKC), (RKR), (RSP), (S*),
(SIN), (SLE), (SQR), (SVC), (SVD), (T), (TR)

CONTROL SYSTEM OPERATORS

(BOD), (CLS), (COT), (CRE), (CRL), (DRE), (DRL),
(DRO), (EAT), (GBN), (GS), (KFL), (LAP), (LYP),
(MIN), (MTF), (NYQ), (OBS), (OFP), (PPL), (RCS),
(RCT), (RDS), (RDT), (RIC), (RIK), (RLC), (STR),
(TZS)

L-A-S PROGRAM CONTROL OPERATORS

(ELM), (IFJ), (JMP), (LIS), (NLI), (NOP), (NTE),
(NTR), (NTY), (RET), (STO), (TES), (TRA), (TYP)

II. L-A-S OPERATOR STATEMENT SYNTAX

1. Single L-A-S Operator Statement Syntax

The general structure of a single operator statement is as follows:

label : *input-field* (*operator-field*) = *output-field* . (1)

The symbols :, (,) and = act as field delimiters. The *label*, which is optional, is usually used in conjunction with the L-A-S Program Control group of operators (Table 1). The *input-field* consists of zero or more variables and/or constants. The *operator-field* can be decomposed into 2 subfields, the <operator mnemonic> and the <option> subfields. The decomposition has the following form:

<operator-field> ::= <op-mnemonic> [, <option>] (2)

where

<option> ::= E | L | P | T . (3)

In the programming language literature [1,2,3], (2) and (3) represent what are commonly called *language production rules*. The so-called metasymbol "::=" can be interpreted as "is produced by". Thus, (3) can be read as "the symbol <option> is produced by a choice of one of the following letters : E, L, P, or T". In particular,

E means print numbers in an Exponential (E) format,
L means append to the Lineprinter data file,
P means Plot the results on a graphics terminal, and
T means Type the result(s) of this calculation.

Finally, the *output-field* consists of zero or more variables that are defined upon successful completion of the calculation.

At this juncture, it is useful to illustrate the preceeding ideas with a simple example. We emphasize the fact that single L-A-S operator statements resemble, to a high degree, standard linear algebra notation and that the "overhead" notation required by the L-A-S language interpreter is minimal.

Consider the following three linear algebra operations : matrix addition, matrix multiplication, and matrix inversion. Suppose that we are already given two square matrices, A and B. Suppose further that we want to perform the following sequence of calculations:

$$C = A + B \tag{4}$$

$$D = C * A \tag{5}$$

$$E = B^{-1} \tag{6}$$

Notice that we implicitly assume that the assignment statement computes the quantity on the right hand side and subsequently stores it in the array named on the left hand side of the equation, that is, the equal sign acts "right to left". Conversely, L-A-S operators act "left to right". Hence, the L-A-S statements that implement the computations (4), (5), and (6) above are, respectively,

```
A,B (+)   = C        (7)
C,A (*)   = D        (8)
B  (-1)   = E        (9)
```

Because the operator follows the input variables, we say that single L-A-S operator statements are written in a *post fix* notation [3]. Also, the same variable name may be used in both input and output fields.

2. Multiple L-A-S Operator Statements

It is possible to perform more than one operation per L-A-S statement. Let us continue with our simple example of linear matrix algebra. Suppose that the matrix C in (7) and (8) is computed only as an intermediate quantity. That is, the result of adding A and B in (7) is needed implicitly in the ensuing calculation (8), but not explicitly. Thus, we view C as a temporary variable.

Given this situation, it would be desirable to eliminate the explicit computation of the array C. Not only would this save user memory space, but in addition it would lessen the overall quantity of L-A-S computer code. This savings can be accomplished by entering the following *Multiple Operator Statement* (MOS) to the L-A-S interpreter.

```
A,B (+) ,A (*)  = D        (10)
```

To understand the effect of (10), one must understand the L-A-S parsing process. Upon receipt of any statement, the interpreter begins scanning left-to-right. When the first operator is encountered, in this case (+), L-A-S immediately performs the indicated operation. The result is stored in a temporary array, call it T1. Thus after one pass through the MOS, (10) becomes :

$$\text{T1 ,A (*)} \quad = \quad \text{D} \tag{11}$$

and (7) has been performed. The scanning process is repeated until all operators are exhausted. Clearly, (10) is equivalent to (7) followed by (8), where the array C has been replaced by the temporary array T1.

3. User-Defined L-A-S Subroutines

A very useful feature of a programming language is a subroutine capability. The ability to write user-defined L-A-S subroutines permits the construction of structured, compact, and efficient L-A-S programs.

To illustrate this, consider again our simple example. Suppose that the calculations defined by (4), (5), and (6) are to be performed repeatedly throughout some more complex algorithm. Then the definition of a subroutine that performs this task is a rational solution. Assume that, in this subroutine, we wish to use the MOS version of (7) and (8) given by (10). For lack of a more meaningful name, we will call our routine XYZ. Then, the L-A-S subroutine definition for XYZ takes the following form :

$$\text{A,B (XYZ,SUB)} \quad = \quad \text{D,E} \tag{12}$$

$$\text{A,B (+), A (*)} \quad = \quad \text{D} \tag{13}$$

$$\text{B (-1)} \quad = \quad \text{E} \tag{14}$$

$$\text{(RET)} \quad = \tag{15}$$

The subroutine XYZ is declared by (12). There are two array names in the input field, (A & B), and two names in the output field, (D & E). The body of the routine consists of (13) and (14). Finally, a return to the calling routine in the L-A-S language is performed by executing the RETurn operator displayed in (15). Typical calls to this subroutine might look like :

$$\text{A,B (XYZ,SUB)} \quad = \quad \text{D,E} \tag{16}$$

or

$$\text{R,S (XYZ,SUB)} \quad = \quad \text{A1,A2} \tag{17}$$

or even

$$\text{Y,X (XYZ,SUB)} \quad = \quad \text{Z,A .} \tag{18}$$

L-A-S subroutines are "call-by-address". Thus, the *address* of the named array is passed to the subroutine and that array is subsequently operated on directly during the execution of the routine.

4. Syntax, Semantics, and the L-A-S Language Grammar

Syntax is concerned with the *form* of a programming language whereas semantics is concerned with the *meaning* of a particular program written in that language. Obviously, a syntactically correct program need not make any sense, semantically. Semantics is a more complicated topic as compared to syntax, hence it is not surprising that there remain a number of unresolved problems in semantics research. By contrast, the theory of syntactical specifications of programming languages has reached a reasonable level of standardization.

A formal definition of the syntax of a programming language is called a *grammar*. Thus, a grammar is a set of rules that specify the form that a program may take. A *formal grammar* is simply a grammar that is specified by using a strictly defined notation. The BNF (Backus-Normal Form) grammar is by far the most well-known type of formal language grammar [2,3].

Now, the purpose of the preceeding discussion is to point out that a "true" programming language involves some precise mathematical formulation. That is, a purported programming language is strengthened by the fact that it can be defined by a formal grammar. On the other hand, a language which has no formal syntactical specification may still qualify as a bona fide programming language. In that light, we will now refer to the L-A-S language and show how the BNF notation can be used to define its syntax.

The metasymbols for defining a BNF production rule are given by the following [2,3]:

< > enclose nonterminal symbols,
::= means "is produced by",
| means "or",
{ } means "zero or more of the enclosed quantity",
[ ] enclose an optional quantity, and
... means "et cetera in a clearly defined way".

The L-A-S production rules are grouped according to the order in which they are referenced. That is, nonterminai symbols which appear on the right-hand side of "::=" are defined in a subsequent grouping of rules. Thus, the L-A-S language grammar is characterized by the set of BNF production rules listed in Table 2.

Table 2. L-A-S Language Grammar

<table>
<tr><td><L-A-S Program></td><td>::=</td><td>{ <L-A-S statement> }</td><td>(R.1)</td></tr>
<tr><td><L-A-S statement></td><td>::=</td><td><comment> | <L-A-S operation></td><td>(R.2)</td></tr>
<tr><td><comment></td><td>::=</td><td>; <text></td><td>(R.3)</td></tr>
<tr><td><L-A-S operation></td><td>::=</td><td>[<label> :] <generalized-op> <output-field></td><td>(R.4)</td></tr>
<tr><td><text></td><td>::=</td><td><L-A-S character> | <L-A-S character> <text></td><td>(R.5)</td></tr>
<tr><td><generalized-op></td><td>::=</td><td><op-field> | <input-field> <op-field></td><td>(R.6)</td></tr>
<tr><td><output-field></td><td>::=</td><td>= | = <array-list></td><td>(R.7)</td></tr>
<tr><td><L-A-S character></td><td>::=</td><td><label> | <digit> | <other-letter> | <symbol></td><td>(R.8)</td></tr>
<tr><td><op-field></td><td>::=</td><td>(<op-mnemonic> [, <option>])</td><td>(R.9)</td></tr>
<tr><td><input-field></td><td>::=</td><td><generalized-var> | <input-field> , <generalized-var></td><td>(R.10)</td></tr>
<tr><td><array-list></td><td>::=</td><td><array-name> { , <array-name> }</td><td>(R.11)</td></tr>
<tr><td><generalized-var></td><td>::=</td><td><array-name> | <constant> | <generalized-op></td><td>(R.12)</td></tr>
<tr><td><array-name></td><td>::=</td><td><label> | <label> <char> | <label> <char> <char></td><td>(R.13)</td></tr>
<tr><td><char></td><td>::=</td><td><label> | <digit></td><td>(R.14)</td></tr>
<tr><td><constant></td><td>::=</td><td><real> | <integer></td><td>(R.15)</td></tr>
<tr><td><real></td><td>::=</td><td>[<integer>] . [<integer>]</td><td>(R.16)</td></tr>
<tr><td><integer></td><td>::=</td><td><digit> | <integer> <digit></td><td>(R.17)</td></tr>
<tr><td><digit></td><td>::=</td><td>0 | 1 | 2 | ... | 7 | 8 | 9</td><td>(R.18)</td></tr>
<tr><td><label></td><td>::=</td><td>A | B | C | ... | X | Y | Z</td><td>(R.19)</td></tr>
<tr><td><other-letter></td><td>::=</td><td>a | b | c | ... | x | y | z</td><td>(R.20)</td></tr>
<tr><td><option></td><td>::=</td><td>E | L | P | T | SUB</td><td>(R.21)</td></tr>
<tr><td><symbol></td><td>::=</td><td>+ | - | * | / | | ! | & | , | .</td><td>(R.22)</td></tr>
<tr><td><op-mnemonic></td><td>::=</td><td>+ | - | * | -1 | "any other 1, 2, or 3 character mnemonic listed in Table 1"</td><td>(R.23)</td></tr>
</table>

III. EXAMPLES OF L-A-S PROGRAMS

In this section, we present several examples of how the L-A-S language can be used to implement various algorithms. The examples become progressively more complex, both conceptually and algorithmically, as we proceed. Furthermore, note that each example simply illustates the form or structure of an L-A-S program. The text that is printed during an execution of an L-A-S program is not shown.

1. Square Root of a Symmetric Matrix

The square root, X, of a square symmetric matrix A satisfies

$$A = X^T * X. \tag{19}$$

The matrix X can be calculated by the recursive formula

$$X_{i+1} = \frac{(X_i^T + A*(X_i)^{-1})^T}{2}. \tag{20}$$

provided that the iterates, X_i, converge to a limit, X. We must define a tolerance for convergence of this algorithm. Call the tolerance variable EPS. For the initial guess, we take $X_0 = A$.

The corresponding L-A-S program written entirely with single operator statements is :

```
;
;   Square Root of the Square Matrix A
;   Single Operator Statement Version
;
                 (INP)   =  A
                 (DSC)   =  S,EPS
               A (MCP)   =  X
;
Q :            X (-1)    =  T
               X (T)     =  XT
             A,T (*)     =  TT
           TT,XT (+)     =  T
             T,S (S*)    =  TT
              TT (T)     =  TT
            TT,X (-)     =  EM
              EM (NRR)   =  EMS
              TT (MCP)   =  X
         EMS,EPS (IFJ)   =  R,R,Q
R :            X (T)     =  XT
            XT,X (*)     =  T
           X,T,A (OUT)   =
```

An MOS version of this same algorithm would look like :

```
;
;            Square Root of the Square Matrix A
;            Multiple Operator Statement Version
;
                                        (INP)   =  A
                                        (DSC)   =  S,EPS
                                      A (MCP)   =  X
;
 Q : X (T), A, X (-1) (*) (+) ,S (S*) (T)   =  TT
                              TT,X (-) (NRR)   =  EMS
                                    TT (MCP)   =  X
                              EMS,EPS (IFJ)   =  R,R,Q
 R :                          X (T),X (*)     =  T
                                X,T,A (OUT)   =
```

Now, assume that the square root calculation must be performed on several different matrices. Then it would be desirable to make the iterative portion of the computation into an L-A-S user-defined subroutine which could be called as often as needed. Such a routine might take the following form:

```
                           A,S,EPS (SQM,SUB)   =  X
                                      A (MCP)   =  X
;
 Q : X (T), A, X (-1) (*) (+) ,S (S*) (T)   =  TT
                              TT,X (-) (NRR)   =  EMS
                                    TT (MCP)   =  X
                              EMS,EPS (IFJ)   =  R,R,Q
 R :                                  (RET)   =
```

Then the MOS program written earlier would be equivalent to:

```
;
;            Square Root of the Square Matrix A
;            Multiple Operator Statement Version
;            Calls L-A-S subroutine : SQM
;
                                        (INP)   =  A
                                        (DSC)   =  S,EPS
;
                           A,S,EPS (SQM,SUB)   =  X
                                X (T),X (*)     =  T
                                X,T,A (OUT)   =
```

2. Response of an Optimal Linear Quadratic Regulator [4,5]

Suppose that we are given a Linear Time-Invariant (LTI) system described by {A,B} where all states are measurable (e.g., C is an N'th order identity matrix). Moreover, suppose we have an infinite-time Linear Quadratic (LQ) performance index characterized by {Q,R}. That is, given Q, positive semi-definite, and R, positive definite, we want to find the gain matrix, K, that minimizes

$$J(\mathrm{K}) = \frac{1}{2}\int_0^\infty x^T \mathrm{Q} x + u^T \mathrm{R} u \; dt . \tag{21}$$

Then, the problem is to solve for the optimal Riccati gain matrix, K, form the closed-loop system matrix, AC, set the initial conditions, and compute the response. Note that the closed-loop system matrix is defined by

$$\mathrm{AC} = \mathrm{A} - \mathrm{S} * \mathrm{K} \tag{22}$$

where

$$\mathrm{S} = \mathrm{B} * \mathrm{R}^{-1} * \mathrm{B}^{\mathrm{T}} \tag{23}$$

and K is the unique positive definite solution to the Algebraic Riccati Equation (ARE)

$$\mathrm{K} * \mathrm{A} + \mathrm{A}^{\mathrm{T}} * \mathrm{K} - \mathrm{K} * \mathrm{S} * \mathrm{K} + \mathrm{Q} = 0. \tag{24}$$

A single operator statement L-A-S program to do this task consists of the following sequence of statements :

```
;
;   Response of an Optimal LQ Regulator
;   Single Operator Statement Version
;
             (INP)   =  A,B,Q,R
           R (INV)   =  RI
           B (T)     =  BT
       RI,BT (*)     =  RBT
       B,RBT (*)     =  S
;
       A,Q,S (RIC)   =  K
;
         S,K (*)     =  SK
        A,SK (-)     =  AC
;
             (INP)   =  X0
       AC,X0 (RCS)   =  Y
```

An equivalent MOS version of the L-A-S program given immediately above is :

```
;
;   Response of an Optimal LQ Regulator
;   Multiple Operator Statement Version
;
                        (INP)   =   A,B,Q,R
    B, R(INV), B(T) (*) (*)     =   S
   A, S, A,Q,S(RIC) (*) (-)     =   AC
;
                        (INP)   =   X0
              AC,X0 (RCS)       =   Y
```

The savings in the number of L-A-S statements is clear.

3. Narendra-Kudva Discrete-Time Identification Algorithm [6]

We shall conclude this section with a more sophisticated application of the L-A-S programming language. In the interest of brevity, we will not formulate the problem here. Rather, we refer the reader to [6]. The bulk of the computations in this discrete-time algorithm centers around the "update equation" which is equivalently Equation (12) or (13) in [6, p.550]. The L-A-S program that implements the identification algorithm [6] is shown in Table 3.

IV. CONCLUSION

We have described the highlights of the L-A-S programming language. The fundamental concept of an L-A-S operator is introduced. The utility of operators as algorithm building blocks is established. The L-A-S statement syntax and language grammar are explicitly defined via the Backus-Normal Form (BNF) notation. An entire section of this article has been dedicated to examples of L-A-S language use. Furthermore, a short description of each operator is provided in the Appendix. The examples coupled with the Appendix make this article a self-contained tutorial. Finally, several references [7-10] have been included should the reader desire additional information about L-A-S.

Acknowledgement - This work was supported in part by the Joint Services Electronics Program under Contract N00014-84-C-0149, and in part by the University of Belgrade, Department of Organizational Sciences.

Table 3. Narendra-Kudva Discrete-Time Identification Algorithm

```
;
;    Narendra-Kudva Discrete-Time Identification Algorithm
;          Implemented via the L-A-S Language
;
;
                                       1  (DVC)  =  O
                                  AK,BK  (CTI)  =  LK
;
  Z :                                    (RDF)  =  CNK
                                         (RDF)  =  INF
                                  INF,O  (CTR)  =  CK,SW
                                     CK  (MCP)  =  KK
                                         (JMP)  =  A
                                  U,CNK  (CTR)  =  UU1,UU2
                                    UU1  (MCP)  =  U
                                 XP,CNK  (CTR)  =  XX1,XX2
                                    XX1  (MCP)  =  XP
;
  A :                               U,O  (CTR)  =  U1,U2
                                   XP,O  (CTR)  =  XP1,XP2
                                  ALF,O  (CTR)  =  AL1,AL2
                                 XP1,U1  (CTI)  =  YT
                                     YT  (T)    =  Y
                                   YT,Y  (*)    =  YTY
                                    YTY  (-1)   =  DEN
                                AL1,DEN  (*)    =  KON
                                   LK,Y  (*)    =  XMK
                                  XP2,O  (CTR)  =  XPT,XP0
                                    XPT  (T)    =  XPK
                                XMK,XPK  (-)    =  EK
                 LK,EK,YT(*),KON(S*)  (-)    =  LK1
                                   II,O  (-)    =  II1
                                   KK,O  (-)    =  KK1
  U2,XP2,LK1,AL2,II1,KK1  (MCP)  =  U,XP,LK,ALF,II,KK
                                   SW,O  (IFJ)  =  D,E,C
;
  E : KK (OUT,T,Counter equals)  =
  D :                                KK  (IFJ)  =  C,X,F
  F :                              II,O  (IFJ)  =  C,B,A
  X :                                    (NOP)  =
                                         (STO)  =
;
  B :                              LK,2  (CTC)  =  AK,BK
AK,BK (OUT,T,Data has been exhausted)  =
                                         (STO)  =
;
  C :                                    (NOP)  =
                          O (OUT,T,ERROR)  =
```

APPENDIX

SUMMARY AND BRIEF DESCRIPTION OF THE L-A-S OPERATORS

This appendix is provided as a quick reference guide to all of the L-A-S operators. For each group of operators listed in Table 1, there is a corresponding entry here consisting of a 1 or 2 line description of each operator and its mnemonic. Capital letters within the operator description are used to correlate the mnemonic name with its explanation. For example, "Array INPut from terminal keyboard" is the explanation for the INP operator.

Mnemonic Name	*Explanation*
INPUT / OUTPUT GROUP	
CDI	Extract Column DImension of a matrix
DDM	Define Diagonal Matrix
DIM	Definition of an Identity Matrix
DIS	DISplay time response on a Tektronix terminal
DPM	Define Pseudorandom Matrix
DSC	Define SCalars (input from terminal keyboard)
DSM	Define Selector (permutation) Matrix
DVC	Define VeCtor (joins scalars into a vector)
DZM	Definition of a Zero Matrix
INP	Array INPut from terminal keyboard
NIK	Frequency (Nyquist) response on a graphics terminal
OUT	OUTput the results on the terminal screen
PLL	"Bar chart" PLot + disk file for Lineprinter
PLT	"Bar chart" PLoT on the terminal (CRT) screen
RDF	Read a Data File from disk (made by L-A-S)
RDI	Extract Row DImension of a matrix
RGE	Read ABCD matrices in GEneral format from disk
TXT	Displays TeXT on an output device (e.g., Terminal)
WDF	Write Data File; writing onto a disk data file
WGE	Write ABCD matrices onto disk in GEneral format

SUMMARY AND BRIEF DESCRIPTION OF THE L-A-S OPERATORS (Continued)

Mnemonic Name	*Explanation*
	DATA HANDLING GROUP
CTC	Matrix CuT by Columns
CTI	Matrix Column TIe (tie by columns)
CTR	Matrix CuT by Rows
EMD	Extract Main Diagonal
MCP	Matrix CoPy
RTI	Matrix Row TIe (tie by rows)
SHD	Matrix SHift Down one row
SHL	Matrix SHift Left one column
SHR	Matrix SHift Right one column
SHU	Matrix SHift Up one row
TVC	Transform (partition) VeCtor into scalars
	LINEAR ALGEBRA GROUP
+	Matrix addition
-	Matrix subtraction
*	Matrix multiplication
-1	Matrix inversion and determinant calculation
ATG	Calculation of the ArcTanGent function
CHD	CHaracteristic polynomial (Discrete) --> eigenvalues of equivalent continuous system
CHE	CHaracteristic monic polynomial --> Eigenvalues + graphics plotting, if desired
COS	Calculation of the COS function
CUR	Calculation of the CUbic Root
EGC	EiGenvalues --> Characteristic polynomial
EGV	EiGenValues of a square matrix + plotting
EXP	Calculation of the EXP function
F*	Function multiplication
F/	Function division
GIN	Generalized INverse of a rectangular matrix
INV	Matrix/scalar INVersion using singular value decomposition
JFR	Modal matrix and Jordan FoRm of a square cyclic matrix
JFS	Modal matrix and Jordan Form of a Square non-cyclic matrix
LOG	Calculation of the LN function
MEA	Maximum Element in Absolute value (Quick Norm)
NRC	Matrix NoRm and norms of each Column
NRR	Matrix NoRm and norms of each Row
NSP	Calculation of the Null SPace
P+	Polynomial addition (non-monic)
P*	Polynomial multiplication (non-monic)
PMA	Polynomial Matrix Addition.

SUMMARY AND BRIEF DESCRIPTION OF THE L-A-S OPERATORS (Continued)

Mnemonic Name	*Explanation*
PMM	Polynomial Matrix Multiplication
PNR	Polynomial NoRmalization (reduce to monic form)
RKC	RanK calculation and separation of linearly independent and dependent Columns
RKR	RanK calculation and separation of linearly independent and dependent Rows
RSP	Calculation of the Range SPace
S*	Matrix multiplication by a Scalar
SIN	Calculation of the SIN function
SLE	Solution of Linear Equations
SQR	Calculation of the SQuare Root function
SVC	Singular Value decomposition of a Complex-valued matrix
SVD	Matrix Singular Value Decomposition
T	Matrix Transposition
TR	Matrix TRace

CONTROL SYSTEM GROUP

BOD	Calculation of the frequency (BODe) diagram
CLS	Closed Loop System (SISO)
COT	Controllability and Observability Tests
CRE	Continuous-time differential Riccati Equation solver
CRL	Continuous ReaLization of a discrete system
DRE	Discrete-time Riccati Equation solver
DRL	Discrete ReaLization of a continuous system
DRO	Discrete-time Riccati Optimal response
EAT	Calculation of state transition matrix (E**AT)
GBN	Generation of input vector (B - New) for output feedback
GS	Polynomial matrix G(S) evaluated at a particular S = a + jb
KFL	Kalman FiLter estimate solver
LAP	Solution of the linear matrix LyAPunov equation
LYP	Solution of the equation : AX + XB = C (Bartels and Stewart Algorithm)
MIN	Determination of the MINimal realization
MTF	Calculation of the Matrix Transfer Function
NYQ	Calculation of the frequency (NYQuist) diagram (N(s)/D(s) --> g(jw) = Re(w) + j Im(w))
OBS	Design of full order OBServer
OFP	Partial pole placement by Output Feedback
PPL	Pole PLacement by state feedback
RCS	Response of a Continuous system in State space
RCT	Response of a Continuous system given by a Transfer function matrix
RDS	Response of a Discrete system in State space
RDT	Response of a Discrete system given by a Transfer function matrix

SUMMARY AND BRIEF DESCRIPTION OF THE L-A-S OPERATORS (Continued)

Mnemonic Name	*Explanation*
RIC	Solution of the algebraic matrix RICcati equation using eigenvector Hamiltonian approach
RIK	Solution of the algebraic matrix RIccati equation using Newton iterative approach
RLC	Root Locus Calculation
STR	State space TRansfomation
TZS	Transmission ZeroS (generalized eigenproblem)

L-A-S PROGRAM CONTROL GROUP

IFJ	IF Jump (conditional jump)
JMP	Unconditional JuMP
NOP	No OPeration
RET	RETurn from an L-A-S user-defined subroutine
STO	STOp
ELM	ELimination of Matrices
LIS	Enter LISt Mode
NLI	Exit List Mode (No LIst)
TES	Enter TESt Mode
NTE	Exit Test Mode (No TESt)
TRA	Enter TRAce Mode
NTR	Exit Trace Mode (No TRAce)
TYP	Enter TYPe Mode
NTY	Exit Type Mode (No TYpe)

REFERENCES

[1] A. Aho and J. Ullman, *The Theory of Parsing, Translation and Compiling*, Prentice-Hall, Englewood Cliffs, N.J., 1972.

[2] J. Backus, "The syntax and semantics of the Proposed International Algebraic Language of Zurich ACM-GAMM Conference," *Information Processing*, UNESCO, Paris, 1960, pp. 125-132.

[3] T. W. Pratt, *Programming Languages: Design and Implementation*, Prentice-Hall, Englewood Cliffs, N.J., 1975.

[4] K. Martensson, "On the Matrix Riccati Equation," *Information Sciences*, vol. 3, 1971, pp. 17-49.

[5] H. Kwakernaak and R. Sivan, *Linear Optimal Control Systems*, Wiley-Interscience, New York, 1972.

[6] P. Kudva and K. S. Narendra, "An Identification Procedure for Discrete Multivariable Systems," *IEEE Trans. AC*, vol. 19, no. 5, October 1974, pp. 549-552.

[7] S. P. Bingulac and N. Gluhajic, "Computer Aided Design of Control Systems on Mini Computers using the L-A-S Language," Proc. of 1982 IFAC Symposium on Computer Aided Design of Multivariable Technological Systems, Purdue University, Indiana, September 15-17, 1982.

[8] S. P. Bingulac, "Recent Modifications in the L-A-S (Linear Algebra and Systems) Language and Its Use in CAD of Control Systems," Proc. of 1983 Allerton Conference on Communication, Control, and Computing, Monticello, Illinois, October 1983.

[9] S. P. Bingulac and P. J. West, "Computer-Aided Design of Control Systems : An Example of Working Software," Proc. of 1983 Allerton Conference on Communication, Control, and Computing, Monticello, Illinois, October 1983.

[10] S. P. Bingulac, J. H. Chow, S. H. Javid and H. R. Dowse, "User's Manual for Linear Algebra and Systems (L-A-S) Language," Report No. 83-EUE-205, (Internal Report), Electric Utility Systems Engineering Department, General Electric, Schenectady, New York, 12345, November 1983.

Section 2. GRAPHICS

COMPUTER-AIDED CONTROL SYSTEMS ENGINEERING
M. Jamshidi and C.J. Herget (Editors)
Elsevier Science Publishers B.V. (North-Holland), 1985

COMPUTER-AIDED CONTROL SYSTEM ANALYSIS AND DESIGN USING INTERACTIVE COMPUTER GRAPHICS

Dean K. Frederick, Electrical, Computer and Systems Engineering Department
Rensselaer Polytechnic Institute, Troy, NY 12180-3590

Tahm Sadeghi, Analytical Science Department, Fairchild Republic Company
Farmingdale, NY 11735

Russell P. Kraft, Mechanical Technologies, Inc.
Latham, NY 12110

A variety of programs involving interactive computer graphics are available at Rensselaer Polytechnic Institute for the analysis and design of both univariable and multivariable control systems and for related instructional purposes. The principal features of these programs and the ways in which they are used in the curriculum are discussed.

I. INTRODUCTION

Because many of the techniques for the analysis and design of control systems rely on graphical methods, the facilities of the School of Engineering's Center for Interactive Computer Graphics (CICG) have played a key role in control system education and research at Rensselaer since its inception in 1977. These programs will be described briefly and samples of the graphical input and output capabilities will be presented. The early activities in the development of control-system software for the CICG are described in [1], and a more recent survey of the use of this facility for electrical engineering is given in [2]. For most of these programs articles or masters theses exist that can provide further details (see References).

Because Imlac vector refresh terminals with light pens and programmable function keys have been used for this work, the software is not transferrable to most other graphics facilities. However, it is hoped that by describing the capabilities of the software and illustrating some of the graphical output the authors will provide ideas that can be used by others to guide software developments of their own. A television tape has been prepared that shows these programs in operation. Interested readers may contact the first author regarding the availability of this tape.

We begin by describing programs available for the analysis and design of univariable (single-input, single-output) control systems. Then two programs for multivariable control systems and several instructional programs will be described. The article concludes with a description of ongoing and future projects.

II. PROGRAMS FOR UNIVARIABLE CONTROL SYSTEMS

A number of programs are available on the interactive graphics system for analyzing and designing univariable systems with time-response solutions, root-locus plots, and frequency-response plots. Some programs provide only one of these capabilities, while others provide combinations of capabilities. There is a wide range of user interfaces among these programs, although each of them allows a considerable amount of interaction through the light pen or keyboard. For the most part, these programs are adaptations of software obtained from sources outside Rensselaer and modified to take advantage of the interactive and graphical capabilities offered by the Imlac terminals. They are most heavily used in a senior-level control system design course but find application in a number of other courses.

A. Four Programs Derived from COINGRAD

These programs were formed from portions of the COINGRAD package that was developed by Volz at the University of Michigan [3]. These are:

TDS (Time Domain Solutions),
RTLOCUS (Root Locus)
DUFS (Design Using Frequency Response), and
SLAP (Single-Loop Analysis Program).

The TDS program [4] generates time-domain solutions and allows the user to save up to nine response plots and redisplay them simultaneously for comparison.

The graphical portion of RTLOCUS [5] is directed by a light pen with a menu that allows the user to blow-up selected points on the loci, add lines of constant damping ratio for the s-plane, and add the unit circle for z-plant loci. The program has the ability to compute the loci of high-order systems and can handle the intersection of multiple loci.

The third program derived from the COINGRAD package computes the frequency-response for fixed, linear systems. By selecting items from menus with the light pen, the user can obtain the frequency response as Nyquist, Nichols, or Bode plots. In addition, these plots can be saved and redisplayed together in simultaneous form, as indicated in Fig. 1.

Because the three programs described above do not share a common data base, the program SLAP has been developed to allow time responses, root-locus plots, and Bode plots to be developed within a single program and data base [6]. The program uses TDS for specifying and modifying the system model and allows the user to activate RTLOCUS merely by issuing a SLAP command. Before doing a root-locus or Bode plot, the program checks the interconnections to ensure that the model is in a single-loop configuration. The time-responses, root-locus plots, and Bode plots that have been drawn for a particular system can be saved and redisplayed simultaneously in reduced form, as shown in Fig. 2.

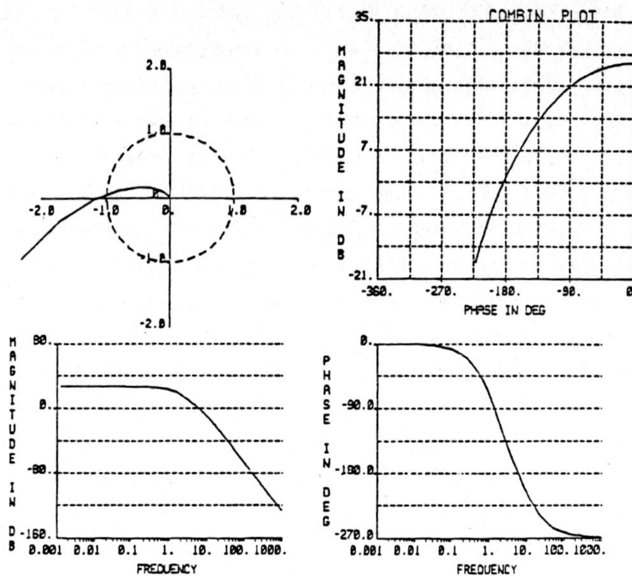

Fig. 1. Frequency-response plots in Bode, Nyquist, and Nichols form (DUFS).

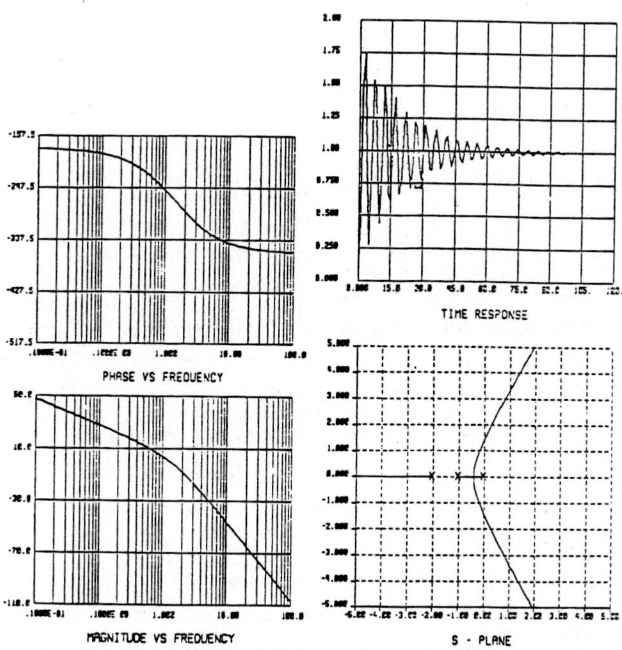

Fig. 2. Impulse response, frequency response, and root locus comparison (SLAP)

B. IGPALS (Interactive Graphics Program for Analysis of Linear Systems)

For this program the light pen is used to draw the blocks, summing junctions, and interconnecting leads of the system's block diagram [7,8]. Once a diagram such as that of Fig. 3 has been drawn the transfer functions of the individual blocks are specified in terms of their poles and zeros. Then pole-zero plots, frequency responses, and impulse responses can be computed and displayed for both open- and closed-loop configurations. Changes can be made in the characteristics of any of the individual blocks such as adding or deleting poles and/or zeros. Also the block diagram can be modified by using the light pen to add and delete leads, summing junctions, and blocks.

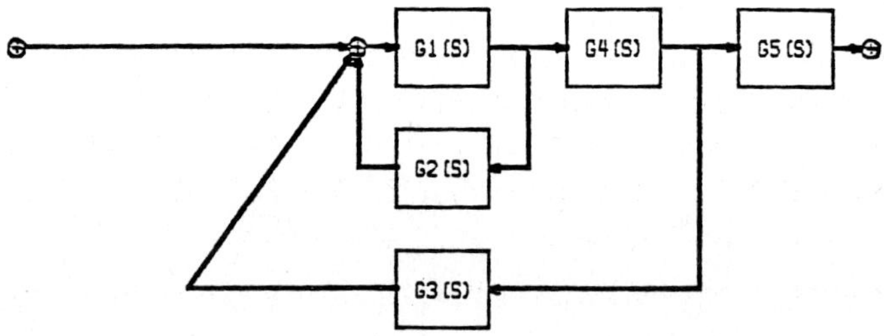

Fig. 3. Block diagram drawn using IGPALS.

C. NDTRAN (Nonlinear Simulation)

The program NDTRAN [9], which has a syntax that is very similar to that of DYNAMO, is used for general nonlinear simulations. The graphical output displays can be controlled by using a light pen to select the variables to be displayed. It is possible to plot several variables simultaneously and to select different scaling options for these combined plots. Plots can also be made with an independent variable other than time, such as in a phase-plane plot.

III. PROGRAMS FOR MULTIVARIABLE CONTROL SYSTEMS

In parallel with the development of the programs described above, a comprehensive program package for Computer-Aided Multivariable Control System Design (CAMCSD) has been developed for the analysis and design of multivariable control systems [10]. The full range of capabilities offered by the interactive graphics facility has been used, including a synchronous control of the software to allow the user to intervene in an optimization process and restart it with modified parameter values. A wide array of comprehensive design algorithms and analysis methods has been included and arrays have been sized to allow up to 30 state variables. The main objective of these efforts by the second author has been to develop a tool that will allow the serious user to tackle complex design problems involving multivariable systems in an efficient manner that would not be possible to attempt without a comprehensive software package. User interaction is possible at all stages of the process and often several alternative algorithms are available to perform a particular task.

The instructions are given to the package by using a light pen to select items from a set of menus that are organized in a hierarchical structure. At the top level is the supervisor, from which the user can enter the menu tree for: input and output, design methods, analysis methods, modification of parameters, or exiting the package. Portions of the software developed at Rensselaer include the following capabilities:

- constant optimal output-feedback design via parameter optimization,
- command-generator tracker design,
- eigenvalue/eigenvector assignment via quadratic weight selection,
- nonoptimal eigenvalue/eigenvector assignment,
- inverse transfer function matrix computation,
- inverse polynomial matrix computation,
- proportional-plus-integral control and tracking,
- transmission and decoupling zero computation,
- transfer-function matrix computation,
- a polynomial matrix library, and
- transient-response plotting.

Also included are the following features from the ORACLS package of subroutines: implicit and explicit model following, Kalman-Bucy filter, and linear quadratic regulator. The EISPACK library is used for the eigenvalue and eigenvector calculations.

IV. INSTRUCTIONAL PROGRAMS

The graphical nature of many important concepts in systems analysis makes them candidates for the development of instructional software using an interactive graphics system. Programs of this type and the topics to which they are directed are:

CONVOL the convolution integral,

PZTR the relationship between the poles and zeros of a transfer function and the system's step and impulse responses,

DTS the relationship between the poles and zeros of a digital filter and the magnitude of its frequency response; also some basic design algorithms,

BASMAT the state transition-matrix, and

STICKBAL a flexible stick being balanced by a motor-driven cart.

In each case the objective of the program is to help the user grasp a particular concept. This is done by generating graphical displays in an interactive fashion.

A. CONVOL (Convolution)

This program [11] leads the user through each step of the convolution process in a graphical fashion. The two functions f(t) and g(t) that are to be convolved are generated by the user and a value of time is specified. Plots of f(λ), g(t-λ), the product f(λ)g(t-λ), and the integral of this product are generated. Finally a smooth curve of the convolution result is displayed.

B. PZTR (Poles, Zeros, and Time Response)

Students in an introductory course covering the modeling and analysis of dynamic systems can use this program to help them relate pole and zero locations to step and impulse responses [12]. This is accomplished by having the user specify the poles and zeros of a transfer function and then request evaluation of the step and impulse responses. After viewing these curves individually, the user can request a combined plot, such as that shown in Fig. 4, that shows the poles and zeros in the s-plane along with the step and impulse responses.

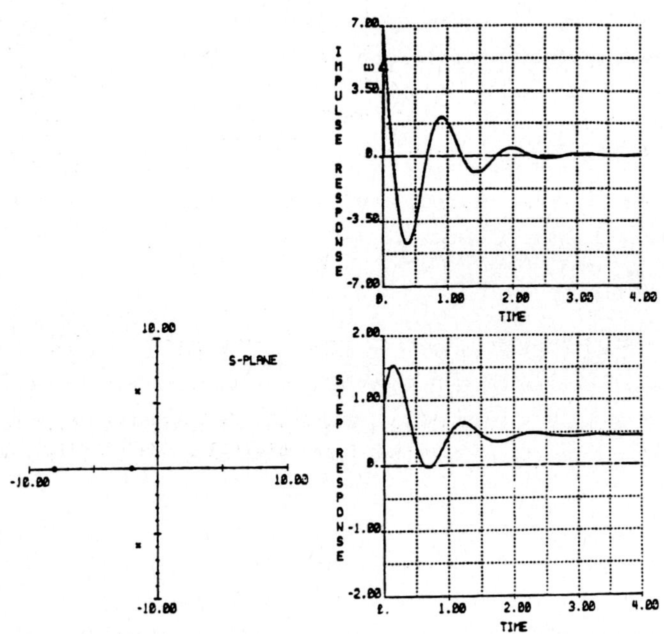

Fig. 4. Pole-zero pattern, step response, and impulse response (PZTR).

C. DTS (Discrete-Time Systems)

This program has three features for helping the user to understand discrete-time systems and digital filters [13]. First, the poles and zeros of a transfer function can be located in the z-plane using a light pen or keyboard entry. Then the magnitude of the frequency response is computed and displayed in a dynamic fashion, along with the pole-zero plots.

In Fig. 5, the small boxes on the unit circle and on the magnitude plot represent the point z=exp (jωt) where ω is the frequency and T is the sampling interval, and they move as ω increases from 0 to ω/T. The vectors in the z-plane from the zeros and poles to the moving box can be used to construct the frequency response.

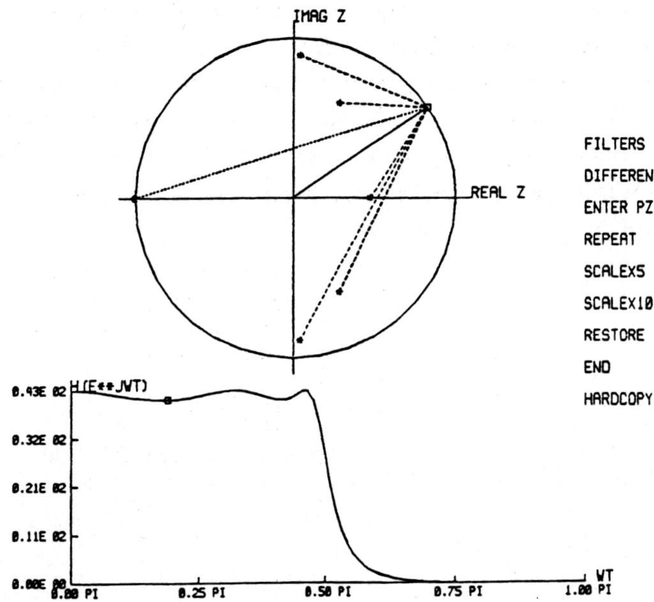

Fig. 5. Snapshot of dynamic pole-zero and frequency-response magnitude plots for a digital filter (DTS).

The second feature of the program is a low-pass filter design section that allows the user to select one of several design algorithms. Once the method has been selected from a light pen menu and the pass- and stop-band frequencies and ripple limits have been set, the filter's transfer function is computed, and the dynamic pole/zero frequency-response plot feature can be executed. The third portion of the program computes the response of a digital filter that has been specified in terms of its difference equation.

D. BASMAT (Basic Matrix Program)

Melsa and Jones [14] developed a program to do basic matrix computations, including the state-transition matrix. As implemented on the Rensselaer interactive graphics systems, this program allows the user to view the time functions that comprise the elements of the state-transition matrix. The light pen is used with a menu to identify those elements of the state-transition matrix that are to be viewed, up to four at a time.

E. STICKBAL (Animated Stick Balancer)

This program [15] employs an animated pictorial representation of a stick balanced on a cart to inform the user of the response of the control system, rather than using the conventional time plots of variables. The stick can be flexible (using a single bending mode in the model), or it can be rigid. The user can select several combinations of mathematical models and control laws, namely a rigid or flexible stick and state-variable or output feedback. Controller gains can be entered by the user to modify the default values. Time plots of variables can be displayed as can a simulation diagram for either the rigid or the flexible case.

V. CONCLUSION

A wide range of programs has been described that are in use at Rensselaer's Center for Interactive Computer Graphics for instruction in control systems and for their simulation, analysis, and design. These programs have proven valuable in the educational process at both the undergraduate and graduate levels and have virtually superceded use of the central computing facility which has at present only a limited graphical display capability and no interactive graphics.

Even with the accomplishments to date, much remains to be done to expand the capabilities of the software and to make improvements in the user/machine interface. For example, efforts are underway in the CICG to develop a common data base that can be shared by programs. This feature would be particularly helpful in the design of univariable control systems using the programs TDS, RTLOCUS, and DUFS.

For the Imlac terminals, the programmable function keys that have been used for the programs PZTR and STICKBAL have helped to provide a more versatile user interface than that obtained by using the light pen or the keyboard. Because the preferences of individual users will vary, it would be desirable to allow the user to select from among these three types of input devices. One advantage of having a strictly keyboard-entry mode is that the individual programs could be made to respond to a command file containing program instructions. Such a capability would allow the user to write command macros. One project that is in progress is directed to adapting the subroutine package INTRAC that has been developed at Lund University, Sweden to the Prime computer. This step will permit the attainment of a high degree of user interaction with only a modest expenditure of programming effort. In another project the graphical input portion of IGPALS is being adapted for use on a VAX computer with the simulation program SIMNON, again from Lund University.

Another advantage of having a command file capability is that instructional command files can be prepared that will allow students to instruct themselves in the use of the particular program in question or to use the program to instruct themselves in some particular topic relating to their course work [16]. For example, with such a capability one could construct a computer-aided instructional unit using TDS, RTLOCUS, and DUFS that explained and demonstrated the related concepts of damping ratio, undamped natural frequency, and gain and phase margins.

APPENDIX

The system on which the programs described above are run consists of two Prime 750 Computers and 36 Imlac vector-refresh terminals. Twenty of these terminals are available 24 hours per day for use by those students at all levels whose courses or projects involve the running or development of interactive graphics software. The remaining terminals are usually reserved for personnel of the CICG for software development and advanced project work.

ACKNOWLEDGEMENT

The authors are indebted to the director, Professor Michael J. Wozny, and the staff members of the Center for Interactive Computer Graphics. The multivariable software described in Section III was developed by the second author (T.S.) with the support of the National Science Foundation under Grant ISP79-2040 and the Industrial Sponsors of the Center for Interactive Computer Graphics. The results reported here do not reflect either the opinions or the approval of the grant sponsors.

We are also indebted to Professor Richard A. Volz of the University of Michigan (TDS, RTLOCUS, and DUFS) and Professors William I. Davisson and John T. Uhran, Jr., of Notre Dame (NDTRAN) who supplied software that was used as the basis for several of these programs. Finally, we wish to acknowledge the contributions of the many students, both undergraduate and graduate, who have been involved in the development and refinement of code over the past seven years.

REFERENCES

1. Frederick, D. K., H. Kaufman, and M. J. Wozny, The Role of Computer Graphics in Control System Studies, Proc. Summer Computer Simulation Conf., Newport Beach, CA, July 1978.

2. Frederick, D. K.,and M. J. Wozny, Computer Graphics in Electrical Eningeering at Rensselaer, Proc. ASEE National Meeting, Los Angeles, CA, July 1981.

3. Volz, R. A., M. Dever, T. J. Johnson, and D. C. Conliff, COINGRAD-Control Oriented Interactive Graphical Analysis and Design, IEEE Trans. Education, Vol. E-17,, Aug. 1974.

4. Buckeley, P.E., TDS-An Interactive Computer Program for Determining the Transient Response of Dynamic Systems, Masters Project Report, Rensselaer Polytechnic Institute, Troy, NY, Aug. 1980.

5. Staudinger, J., Implementation of a Root Locus Program on an Interactive Graphics System, Masters Project Report, Rensselaer Polytechnic Institute, Troy, NY, May 1978.

6. Chin, C. Y., SLAP-An Interactive Computer Program for Analyzing Single-Loop Systems, Masters Project Report, Rensselaer Polytechnic Institute, Troy, NY, Aug. 1980.

7. Lupton, G., Interactive Graphics Program for Analysis of Linear Systems, Masters Project Report, Rensselaer Polytechnic Institute, Troy, NY, Dec. 1974.

8. Killeavy, W., Interactive Computer Graphics Program for Linear Systems Analysis and Optimization, Masters Project Report, Rensselaer Polytechnic Institute, Troy, NY, May 1978.

9. Frederick, D. K., An Implementation of NDTRAN on an Interactive Computer Graphics System, Proc. Pittsburgh Conf. on Modeling and Control, Pittsburgh, PA Apr. 1979.

10. Sadeghi, T. and M. J. Wozny, An Interactive Computer Graphics Package for Linear Multivariable System Design, Proc. IFAC Symp. on Multivariable Technological Systems, West Lafayette, IN, Sept. 1982.

11. Frederick, D. K. and G. L. Waag, An Interactive Graphics Program for Assistance in Learning Convolution, Trans. Computers in Education Div. ASEE, Vol. XII, No. 7/8, July/Aug. 1980.

12. Frederick, D. K. and A. S. Leong, Computer-Graphics Approach to Relating Pole-Zero Plots and Time-Domain Responses, Proc. ASEE National Meeting, College Station, TX, Junne 1982.

13. Frederick, D. K. and L. A. Gerhardt, A Computer Graphics Program for the Analysis and Design of Digital Filters, Proc. Intl. Conf. on Cybernetics and Society, Cambridge, MA, Oct. 1980.

14. Melsa, J. L. and S. K. Jones, Computer Programs for Computational Assistance in the Study of Linear Control Theory, McGraw-Hill Book Co., New York, 1973.

15. Frederick, D. K. and H. T. Nguyen, A Computer Graphics Animation of a Flexible Stick Balancer, Trans. Computer in Education Div. ASEE, Vol. XIII, No. 3/4, March/April 1981.

16. Frederick, D. K., On Using the Computer to Teach the Use of Computer Programs, Proc. IFAC Symp. on Multivariable Technological Systems, West Lafayette, IN, Sept. 1982.

Section 3. ALGORITHMS

COMPUTER-AIDED CONTROL SYSTEMS ENGINEERING
M. Jamshidi and C.J. Herget (Editors)
Elsevier Science Publishers B.V. (North-Holland), 1985

GENERALIZED EIGENPROBLEM ALGORITHMS AND SOFTWARE FOR ALGEBRAIC RICCATI EQUATIONS

William F. Arnold III
Naval Weapons Center
China Lake, California 93555

Alan J. Laub
University of California
Santa Barbara, California 93106

Numerical issues related to the computational solution of the algebraic matrix Riccati equation are discussed. The approach presented uses the generalized eigenproblem formulation for the solution of general forms of algebraic Riccati equations arising in both continuous- and discrete-time applications. These general forms result from control and filtering problems for systems in generalized (or implicit or descriptor) state space form. A Newton-type iterative refinement procedure for the generalized Riccati solution is given. The issue of numerical condition of the Riccati problem is addressed. Balancing to improve numerical condition is discussed. An overview of a software package, RICPACK, coded in portable, reliable FORTRAN is given. Results of numerical experiments are reported.

I. INTRODUCTION

One of the most deeply studied nonlinear matrix equations arising in mathematics and engineering is the Riccati equation. The generic term "Riccati equation" can mean any of a class of matrix "quadratic" algebraic or differential or difference equations of symmetric or nonsymmetric type arising in the study of continuous-time or discrete-time dynamical systems. In this paper we shall discuss algorithms and software for certain classes of algebraic Riccati equations. The resulting software package, called RICPACK, can be used as a module in a larger computer-aided control system design package or environment.

Riccati equations arise naturally in a rich variety of situations, and their role and use in systems and control theory, in particular, has been well-established over the past 25 years. A representative but by no means exhaustive sample of such applications can be found in standard "classical" textbooks on optimal control (see [1]–[5] and the references therein) and filtering and prediction (see [3], [5]–[8] and the references therein). One of the finest mathematical treatises on Riccati equations is the book of Reid [9] that, in addition to control and estimation applications, discusses applications to partial differential equations, multiple uniform and nonuniform transmission lines, the Mycielski-Paszkowski diffusion problem, and neutron transport theory. The transport problem is an illustration of the intimate role played by Riccati equations in the method of invariant embedding (see [10], [11] and the references therein). Further details on the use of Riccati equations in solving two-point boundary value problems are discussed in [12] and various papers in [13], particularly those of Denman, Bramley, and Casti.

One aspect of Riccati equations that has always been significant, and which has received increasing attention over the past 5 years, is effective algorithms for their reliable numerical solution in the finite arithmetic environment of a digital computer. Reliable implementation of a numerical algorithm involves attention to at least the following three concerns:

1. The condition of the underlying problem.
2. The numerical stability of the algorithm being used to solve the problem.
3. The robustness of the actual software implementation.

Each of these will be illustrated further in the sequel in the context of solving algebraic Riccati equations.

Reliable numerical algorithms and software now exist for the solution of linear equations, singular value decomposition, linear least squares, and both standard and generalized eigenvalue problems [14]–[16]. This has not heretofore been the case for the solution of algebraic Riccati equations. This paper will outline, in a tutorial fashion, a class of algorithms that are quite generally reliable and are readily extendable to a variety of related problems. A FORTRAN software package, RICPACK, that implements the preferred algorithmic approach by building on and emulating [14]–[16] is described in detail.

Briefly, the rest of this paper is organized as follows. Section II contains general background information on Schur-type techniques for the solution of various types of algebraic Riccati equations. Section III gives algorithmic details for generalized eigenvalue methods for Riccati equation solution. Iterative refinement techniques are discussed in Section IV. General comments on mathematical software for Riccati equations along with details concerning the software package RICPACK are presented in Section V, and Section VI gives numerical results for example problems. Some concluding remarks are made in Section VII.

II. BACKGROUND

In this section, we provide some background material on Schur-type techniques for the solution of various types of algebraic Riccati equations by means of certain associated generalized eigenvalue/eigenvector problems. The generalized eigenproblem framework provides a unifying methodology that facilitates the reliable numerical solution of very general classes of Riccati equations arising in optimal control or filtering problems (and elsewhere), including those with "nonstandard" features such as singular control weighting (or measurement noise covariance) matrices, cross-weighting (cross-correlation) matrices, and singular transition matrices (discrete-time). Generalized state space models can also be considered and give rise to "generalized" Riccati equations [12].

The basic algebraic Riccati equation (ARE) arising in continuous-time problems takes the form

$$A^TXE + E^TXA - (E^TXB + S)R^{-1}(B^TXE + S^T) + C^TQC = 0 \tag{1a}$$

or

$$\hat{A}^TXE + E^TX\hat{A} - E^TXBR^{-1}B^TXE + C^TQC - SR^{-1}S^T = 0 \tag{1b}$$

where

$$\hat{A} := A - BR^{-1}S^T.$$

This ARE arises in connection with finding a feedback control $u(t) = Kx(t)$ which solves the problem

$$\operatorname{Min}\int_0^{+\infty} \frac{1}{2}[y^TQy + 2x^TSu + u^TRu]\,dt$$

$$\text{subject to: } E\dot{x} = Ax + Bu;\; x\in \mathbb{R}^n,\; u\in \mathbb{R}^m$$

$$y = Cx;\; y\in \mathbb{R}^p$$

Under mild technical conditions on the matrices, the minimizing control which is stabilizing (the generalized eigenvalues of $\lambda E - (A + BK)$ have negative real parts) is given by $K = -R^{-1}(B^TXE + S^T)$ where X is the unique non-negative definite solution of (1). We are assuming E is nonsingular. This situation arises, for example, from first-order formulations of second-order models of the form

$$M\ddot{\xi} + D\dot{\xi} + K\xi = \bar{B}u$$

in which $M = M^T > 0$ but M^{-1} is not to be used—largely for numerical reasons, but possibly also to preserve and possibly exploit the structure.

The more familiar form of (1) occurs when $S = 0$, $E = I$, and $C = I$:

$$A^TX + XA - XBR^{-1}B^TX + Q = 0 \tag{2}$$

(or even $F^TX + XF - XGR^{-1}G^TX + H = 0$, according to taste). Equation (2) can be solved by finding an orthogonal matrix U such that

$$U^TMU = S$$

where

$$U = \begin{pmatrix} U_{11} & U_{12} \\ U_{21} & U_{22} \end{pmatrix};\quad U_{ij} \in \mathbb{R}^{nxn}$$

$$M = \begin{pmatrix} A & -BR^{-1}B^T \\ -Q & -A^T \end{pmatrix};\ \text{a Hamiltonian matrix } \left(M^TJ = -MJ \text{ where } J = \begin{pmatrix} 0 & I \\ -I & 0 \end{pmatrix}\right)$$

and $S = \begin{pmatrix} S_{11} & S_{12} \\ 0 & S_{22} \end{pmatrix}$; a quasi-upper-triangular matrix with all eigenvalues of S_{11} in the strict left-half-plane. The n "Schur vectors" comprising $\begin{pmatrix} U_{11} \\ U_{21} \end{pmatrix}$ span the stable invariant subspace and the solution of (2) is given by $X = U_{21}U_{11}^{-1}$. This so-called Schur method, which is described in detail in [17] and [18], has proved to be a rather reliable, general-purpose method for solving (2) for modest-sized problems (say $n \lesssim 100$) where there is no usefully exploitable structure in the coefficient matrices.

Many other solution techniques exist for (2). Most of these methods fall into one of the following broad categories:

1. Methods based on certain special canonical forms
2. Doubling and other direct integration (of the associated differential equation) techniques
3. Newton's method
4. Parameter embedding methods
5. Chandrasekhar-type algorithms
6. Methods based on use of the matrix sign function
7. Spectral factorization techniques
8. "Square root" formulations

Each of these methods has interesting features but most do not have satisfactory numerical behavior in finite arithmetic. Moreover, many of the methods do not extend easily to the solution of (1) nor to certain related nonsymmetric Riccati equations.

For the solution of (1) we consider the 2n x 2n matrix pencil

$$\lambda \begin{pmatrix} E & 0 \\ 0 & E^T \end{pmatrix} - \begin{pmatrix} A - BR^{-1}S^T & -BR^{-1}B^T \\ SR^{-1}S^T - C^TQC & -(A - BR^{-1}S^T)^T \end{pmatrix} \tag{3}$$

The resulting generalized eigenvalue problem is then transformed by orthogonal matrices V and U to the form

$$V(\lambda L - M)U = \lambda\hat{L} - \hat{M} \tag{4}$$

where $\hat{L}$ is upper-triangular and $\hat{M}$ is quasi-upper-triangular, and the stable generalized eigenvalues are determined by the upper left nxn blocks. The Schur vectors $\begin{pmatrix} U_{11} \\ U_{21} \end{pmatrix}$ span the stable deflating subspace and the feedback K is given by $K = -R^{-1}(B^TU_{21}U_{11}^{-1} + S^T)$. The solution of (1) is given by $X = W_{21}W_{11}^{-1}$ where $W = \begin{pmatrix} E & 0 \\ 0 & I \end{pmatrix} U$; see [12], [19].

Although the matrix R above is often diagonal, or even the identity, which makes R^{-1} quite trivial to determine, it may instead be nondiagonal and ill-conditioned with respect to inversion or possibly even singular. In such cases, the $(2n + m) \times (2n + m)$ extended pencil

$$\lambda \begin{pmatrix} E & 0 & 0 \\ 0 & E^T & 0 \\ 0 & 0 & 0 \end{pmatrix} - \begin{pmatrix} A & 0 & B \\ -C^TQC & -A^T & -S \\ S^T & B^T & R \end{pmatrix} \tag{5}$$

can be used. This pencil can be reduced or compressed to an equivalent $2n \times 2n$ pencil [20] whereupon a procedure similar to that described above is followed.

Of course, all the above has an analogue for the discrete-time counterpart of the linear-quadratic problem described. In this case, the ARE corresponding to (1) takes the form

$$E^TXE = A^TXA - (A^TXB + S)(B^TXB + R)^{-1}(A^TXB + S)^T + C^TQC \tag{6}$$

$$= \hat{A}^TX\hat{A} - \hat{A}^TXB(B^TXB + R)^{-1}B^TX\hat{A} + C^TQC - SR^{-1}S^T \tag{7}$$

where $\hat{A} := A - BR^{-1}S^T$. Equation (6) arises naturally in stochastic realization problems while form (7) arises naturally in LQG problems.

In this case, the analogue of (3) is the pencil

$$\lambda \begin{pmatrix} E & BR^{-1}B^T \\ 0 & (A - BR^{-1}S^T)^T \end{pmatrix} - \begin{pmatrix} A - BR^{-1}S^T & 0 \\ SR^{-1}S^T - C^TQC & E^T \end{pmatrix} \tag{8}$$

$$=: \lambda L - M$$

while if R^{-1} is to be avoided (singular R being not uncommon in discrete-time systems), the analogue of (5) is the extended pencil

$$\lambda \begin{pmatrix} E & 0 & 0 \\ 0 & A^T & 0 \\ 0 & -B^T & 0 \end{pmatrix} - \begin{pmatrix} A & 0 & B \\ -C^TQC & E^T & -S \\ S^T & 0 & R \end{pmatrix} \tag{9}$$

The solution procedure involving (8) or a compressed version of (9) is essentially the same as that outlined above for the continuous-time case. In the special case S = 0, E = I, and C = I, (8) takes the form

$$\lambda \begin{pmatrix} I & BR^{-1}B^T \\ 0 & A^T \end{pmatrix} - \begin{pmatrix} A & 0 \\ -Q & 0 \end{pmatrix}$$

that was suggested in [18] and used in [21]–[24] to avoid the need for A^{-1} in the eigenproblem for the symplectic matrix (if A^{-1} exists):

$$\begin{pmatrix} I & BR^{-1}B^T \\ 0 & A^T \end{pmatrix}^{-1} \begin{pmatrix} A & 0 \\ -Q & I \end{pmatrix} \tag{10}$$

A similar approach was discussed in [25].

III. ALGORITHMIC DETAILS

The main algorithmic issues associated with the Schur-type generalized eigenproblem method are the compression of the matrix pencil (5) or (9), the solution of the generalized eigenproblem (4), and the "ordering" of the generalized eigenvalues so that the stable generalized eigenvalues are contained in the upper left n x n blocks of $\lambda\hat{L} - \hat{M}$. Some details concerning these issues are given in this section. The related issues of numerical condition are also discussed.

Should R in (5) or (9) be singular or near singular (and not easily invertible), then the following compression procedure due to Van Dooren [20] can be employed. Determine an orthogonal matrix $P \in \mathbb{R}^{(2n + m) \times (2n + m)}$ such that

$$\begin{pmatrix} P_{11} & P_{12} \\ & \\ P_{21} & P_{22} \end{pmatrix} \begin{pmatrix} B \\ -S \\ R \end{pmatrix} = \begin{pmatrix} 0 \\ 0 \\ \bar{R} \end{pmatrix} \tag{11}$$

where $\bar{R} \in \mathbb{R}^{m \times m}$ and is nonsingular. P can be formed from a series of Householder transformations. The remainder of the compression technique will be illustrated using the pencil corresponding to the continuous-time problem (5). The discrete-time problem can be handled analogously as illustrated in [23]. Applying P to the pencil (5) we see that the "infinite generalized eigenvalues" are "deflated out" so that we need only work with the 2n x 2n pencil

$$\lambda P_{11} \begin{pmatrix} E & 0 \\ 0 & E^T \end{pmatrix} - \left(P_{11} \begin{pmatrix} A & 0 \\ -C^TQC & -A^T \end{pmatrix} + P_{12} \begin{pmatrix} S^T & B^T \end{pmatrix} \right) \tag{12}$$

It can be shown [20] that the pencils (12) and (5) are equivalent. Therefore, the pencil (12) can be used for the Riccati solution and does not involve the explicit inversion of R. This extended pencil approach is used in [19] and [24] and is implicit in the work of Campbell and others (summarized in [26] and [27]).

One must essentially solve the appropriate generalized eigenproblem to transform the pencil to quasi-upper-triangular form as indicated in (4). Basic references for the numerical solution of the fundamental (unordered) generalized eignevalue problem include [28] and [29]. The algorithm is a generalization of the QR algorithm used for the standard eigenvalue problem and is referred to as the QZ algorithm. Application of the QZ algorithm to the appropriate pencil yields the transformed matrices $\hat{L}$ and $\hat{M}$ as well as the orthogonal transformation U of (4). Unfortunately, the QZ algorithm does not result in $\hat{L}$ and $\hat{M}$ with the desired ordering of the generalized eigenvalues.

However, the eigenvalues can subsequently be reordered to any desired order. Algorithmic and numerical details of the "reordering problem" for the generalized eigenproblem were given in [20]. The reordering is accomplished by orthogonal transformations and the method is proved to be numerically stable.

An important aspect of the analysis of any numerical problem is its condition. That is, if the data of the problem are perturbed slightly, is the resulting change in the solution "large" (an ill-conditioned problem) or "small" (a well-conditioned problem)? Understanding the conditioning of Riccati equations is, of course, crucial to their reliable numerical solution, but Riccati equation condition is apparently an exceedingly complex problem. This is borne out by both analysis and empirical study in [30]. Several proposed "condition numbers" for the Riccati problem are compared and all are shown to have deficiencies for some classes of problems. Among those compared are:

1. The condition of U_{11} or W_{11} with respect to inversion (see Section II) [12], [17] since the final Riccati solution derives from the solution of a linear system of the form $XU_{11} = U_{21}$ or $XW_{11} = W_{21}$. It is known that U_{11} or W_{11} are singular if the underlying model is unstabilizable and/or if E is singular. Thus near-unstabilizability or near-singularity of E can be expected to cause ill-conditioning, which it does. Unfortunately, a Riccati equation can still be ill-conditioned with a well-conditioned U_{11} or W_{11}.

2. A condition number in terms of the singular values of the singular value decomposition of the orthogonal U or W found in the reduction of generalized eigenproblem. This result was developed in [31] and extended in [19], but turns out to be essentially the same as the one discussed in 1 above; see [30].

3. Various condition numbers based on a first-order perturbation analysis of a Riccati equation. One such derivation is given in [32] while still others are developed in [30]. Examples can be shown (see [30]) where such analyses are inadequate in the sense of signaling ill-conditioning falsely or of not detecting ill-conditioning for a known ill-conditioned equation.

Further pertinent discussions, results, and conjectures can be found in [17] and [30]–[32].

The coefficient matrices of many Riccati equations arising in practice contain numbers of widely different magnitudes. It is thus reasonable to expect that some sort of scaling and/or balancing strategy could be employed to improve the accuracy of a numerical solution of a Riccati equation in finite arithmetic. This expectation is, in fact, borne out in numerical experiments reported in [32] and [33]. The main technique employed in those experiments involved the balancing of the generalized eigenvalue problem prior to its solution. Since the Schur-type solution technique for Riccati equations involves solution of the generalized eigenproblem as a fundamental step, this balancing technique seems reasonable.

Ward [34] has developed a balancing algorithm designed especially for QZ numerical solution methods for generalized eigenproblems. The procedure consists of permutations and two-sided diagonal transformations to attempt to scale L and M in (4) so that their elements have magnitudes as close to unity as possible. This balancing strategy is relatively inexpensive, but can improve accuracy substantially and also increase the reliability of the condition of U_{11} or W_{11} with respect to inversion as an indicator of condition of the Riccati problem.

Another potential balancing scheme was investigated in [30]. There, as a preprocessing step before the Riccati solution, the underlying system matrices (A, B, C, etc.) are "balanced" in the system-theoretic sense [35] and [36]. This is an attempt to give equal weight to both controllability and observability, but has the rather severe disadvantages of requiring open-loop stability of $A - \lambda E$, and of being relatively expensive computationally. Other state coordinate systems may offer computational advantages. A convenient algorithm for applying the associated change of coordinate transformation in the Schur-type Riccati solution method is given in [30].

IV. ITERATIVE REFINEMENT TECHNIQUES

Numerical implementation of the Schur-type solution methods of Section II is relatively straightforward. The proven stable algorithms discussed in Section III are coded into reliable FORTRAN software (see Section V). However, the Riccati problem may be ill-conditioned, and the resulting numerical solution may not be as accurate as desired. This section presents an interative refinement procedure utilizing Newton's method. This procedure is presented for both the continuous- and discrete-time problems. The continuous-time method is based on Kleinman [37], and the discrete-time method is based on Hewer [38]. Other iterative solution methods are discussed briefly.

If X_k, $k = 0,1,\ldots$ is the unique non-negative definite solution of the linear algebraic equation

$$0 = (A - BK_k)^T X_k E + E^T X_k (A - BK_k) + C^T QC + K_k{}^T RK_k - SK_k - (SK_k)^T \tag{13}$$

where recursively,

$$K_k = R^{-1}(B^T X_{k-1} E + S^T), \; k = 1,2,\ldots \tag{14}$$

and K_0 is chosen such that $\lambda E - (A - BK_0)$ has generalized eigenvalues with negative real parts (stable), then it can be shown [30] that

1. $0 \leqslant X \leqslant X_{k+1} \leqslant X_k \leqslant \cdots \leqslant X_0$

2. $\lim_{k\to\infty} X_k = X$

3. in the vicinity of X, $\|X_{k+1} - X\| \leqslant C_2 \|X_k - X\|^2$

where X solves (1) and C_2 is a finite constant. This iterative procedure features monotonic convergence to the non-negative definite solution to (1) when it exists, as long as the starting value K_0 is a stabilizing state feedback matrix. The global convergence can be quite slow, however, and result in excessive computation time if a poor K_0 is chosen. Numerical experience (see [30] for examples) has shown that when this iterative method is applied to an initial solution value produced by the direct Schur-type method, the convergence is quadratic or faster. This hybrid solution method is quite useful for improving the Riccati solution accuracy in the presence of certain sources of problem ill-conditioning. For the case $E = I$ and $S = 0$, the above result becomes the familiar procedure established by Kleinman [37].

An analogous result exists for the discrete-time equations (6) or (7) with the analogs of equations (13) and (14) being [30]

$$E^T X_k E = (A - BK_k)^T X_k (A - BK_k) + K_k^T R K_k + C^T Q C - SK_k - (SK_k)^T \tag{15}$$

and

$$K_k = (R + B^T X_{k-1} B)^{-1} (B^T X_{k-1} A + S^T), \; k = 1,2,... \tag{16}$$

with the requirement that K_0 is chosen such that $\lambda E - (A - BK_0)$ has generalized eigenvalues with magnitude less than unity. For the case $E = I$ and $S = 0$, the result is the same as that established by Hewer [38].

Other iterative solution methods such as those based on the matrix sign function are also potential candidates for hybrid-type methods for solving Riccati equations. Whatever iterative method is chosen, two basic advantages make them well worth including in a Riccati equation solution package:

1. Accuracy of computed solutions of ill-conditioned equations can be improved (even without computing residuals in extended precision).

2. For sufficiently small perturbations in the coefficient matrices, a solution of the perturbed equation can be found more efficiently by a few steps of an iterative method rather than redoing a full-ordered generalized eigenproblem.

V. NUMERICAL SOFTWARE FOR RICCATI EQUATIONS

Even for problems which are apparently "simple" from a mathematical point of view, a vast myriad of little details must be attended to in order to enable a robust or reliable algorithmic implementation in finite arithmetic. These details can become so overwhelming that the only effective means of successfully communicating an algorithm is through its embodiment as mathematical software. Mathematical or numerical software simply means an implementation on a computing machine of an algorithm for solving a mathematical problem. Ideally, such software must be reliable, portable, and unaffected by the machine or system environment in which it is used.

There are many characteristics that can be listed to characterize "good" or robust mathematical software. State-of-the-art discussions are to be found in [39]–[51]. Of course, one of the key features is portability. Inevitably, numerical algorithms are strengthened when their mathematical software is made portable since their widespread use is greatly facilitated. Furthermore, reliable and portable software is usually faster than poorer codes. Portability and speed largely account for the preponderance of good mathematical software being coded in FORTRAN although a strong case could now be made for Ada^{TM}*, at least on portability grounds.

A careful examination of the modern mathematical software literature should lead one to the conclusion that serious evaluation of mathematical software is a highly nontrivial task. Clearly the quality of software is largely a function of its operational specifications. It must also reflect the numerical aspects of the algorithm being implemented. The language used and the compiler (e.g., optimizing or not) used for that language will both have an enormous impact on quality, both perceived and real, as will the underlying hardware and arithmetic. Different implementations of an algorithm can have markedly different properties and behavior, even of the same good underlying algorithm.

Many aspects of systems, control, and estimation theory are ready for the research and design that is necessary to produce reliable, portable mathematical software that performs successfully in finite arithmetic. Certainly many of the underlying linear algebra tools (for example, EISPACK [14], [15] and LINPACK [16]) are considered sufficiently reliable as to be used as black—or at least gray—boxes by control engineers. Much of the theory and methodology used in the production of prototypical mathematical software such as EISPACK and LINPACK can and has been carried over to problems such as the solution of Riccati equations. However, much of the work done in engineering, particularly the design and synthesis aspects, is not amenable to nice, "clean" algorithms, and the ultimate software must have the capability to enable a dialogue between the computing machine and the engineer (or scientist), but with the latter probably still making the final engineering decisions. Most applications software will not ultimately look like EISPACK or LINPACK. To even attempt that would be futile. Instead, a better analogy would be to emulate a good ordinary differential equation or partial differential equation package.

Schur-type methods for the generalized algebraic Riccati equations discussed in Section II have been implemented by the authors utilizing the algorithms of Section III with the Newton iterative refinement procedure of Seciton IV in a FORTRAN software package, RICPACK [33]. RICPACK was developed as a research tool to aid in the study of the numerical conditioning of algebraic Riccati equations. The package consists of modularly designed FORTRAN subroutines (approximately 60) together with a FORTRAN driving program for use in an interactive "terminal"-type environment. The driver prompts for all necessary input, and convenient input default options exist not only for ease of data input, but also for exploitation by the subroutines to reduce the number and complexity of the computations. These options are also designed to speed the input of more "standard"-type problems.

Highlights of RICPACK capabilities include:

1. Choice of calculation of the stabilizing (non-negative definite), antistabilitizing (non-positive definite), or just any (possibly) indefinite solution to a generalized ARE (see Section II).

---

*Ada^{TM} is a trademark of the U.S. Department of Defense.

2. Coordinate or system balancing of the system model [35], [36].

3. Ward's balancing [34] of the generalized eigenproblem (prior to the QZ transformation).

4. Direct handling of singular control weighting or singular measurement noise covariance by compression of the expanded pencils (5) or (9).

5. Direct handling of cross-weighting or noise correlation, i.e., $S \neq 0$.

6. Provision for robustness recovery procedure; i.e., to replace the C^TQC term with $Q + \Upsilon C^TC$ and iterate on Υ, the driver program need only modify one block of the matrix pencil at each iteration.

7. Iterative refinement (or new solutions for small-parameter perturbations) by Newton's method and Sylvester equations; i.e., iteratively solving equations of the form

$$\text{(Continuous)} \quad \bar{A}_k^T X_{k+1} E + E^T X_{k+1} \bar{A}_k + \bar{C}_k = 0$$

$$\text{(Discrete)} \quad E^T X_{k+1} E - \bar{A}_k^T X_{k+1} \bar{A}_k + \bar{C}_k = 0$$

for X which is optionally either

a. the new solution at each step, or

b. the required change to the solution at each step.

8. Model unstabilizability detection as indicated by the condition of U_{11} or W_{11} with respect to inversion (see discussion after (4)); U_{11} or W_{11} is singular for an unstabilizable model.

9. Calculation of unique stabilizing solution for stabilizable models with undetectable modes.

10. Residual calculation of the form

$$r = \frac{\|\text{Residual}\|_1}{\|X\|_1} \tag{17}$$

11. Condition estimates for the Riccati problem which provide information on the following sources of ill-conditioning: model unstabilizability; small separation of closed loop spectrum, near singularity of R (continuous-time case).

The authors chose to implement the algorithms as high-quality mathematical software because to do less would be to evade a major responsibility, for to leave software implementation to the algorithm user/reader has the potential to lead to disastrously bad "versions" of a perfectly good algorithm. This software will be usable as a module or "tool" in a wide variety of Computer-Aided Control System Design (CACSD) environments ranging from small packages developed by individuals to large commercial packages such as MATRIX_X (Integrated Systems, Inc.) or CTRL-C (Systems Control Technology, Inc.).

Figure 1 illustrates the hierarchy of the software routines. That is, the routines on the upper levels employ the routines of the lower levels. At the lowest level we have the basic matrix manipulation routines like add, subtract, multiply, etc., and some simple combinations of these basic operations. The next level consists of standard routines for linear equations, eigenvalues, and singular value decomposition (SVD). Most routines in this level are from LINPACK or EISPACK, or are slight modifications to routines from LINPACK and EISPACK. Subroutines that are modified have the modifications noted in the comment documentation included in the subroutine. A list of the Level 0 and 1 routines is given in Table 1. The SVD routine is listed separately with the BLAS routines that it requires from LINPACK, as it is the only routine to require the BLAS and could be modified to eliminate said BLAS.

LEVEL 4	MAIN PROGRAM			
LEVEL 3	COORDINATE BALANCING (BALCOR) (BLCRDC) (BLCRDD)	RICCATI SOLUTION (RICSOL)	NEWTON ITERATION (NEWT)	
LEVEL 2	SCHUR FORM ORDERING (ORDER) (EXCHQZ)	LYAPUNOV SOLUTION (LYPCND, LYPDSD)	SEPARATION ESTIMATION (SEPEST)	FEEDBACK GAIN (FBGAIN)
	COMPRESSED PENCIL (CMPRS)	PENCIL WITH R^{-1} (RINV)	RESIDUAL CALCULATION (RESID)	WARD BALANCING (BALGEN, BALGBK)
LEVEL 1	LINEAR EQUATIONS LINPACK (BLAS-LESS VERSIONS)	EIGENVALUES EISPACK	SINGULAR VALUE DECOMPOSITION (FROM LINPACK, PLUS APPROPRIATE BLAS)	
LEVEL 0	BASIC MATRIX MANIPULATION			

Figure 1
Hierarchy of Subroutines in Software Package RICPACK

Subroutines at Levels 2 and 3 perform more specialized tasks. The task title is given in Figure 1, and the main subroutines performing the task are shown in parenthesis.

Of course, the main driver program is at the highest level (4). This main program would be rewritten, or at least extensively modified for most applications. This driver was written as a research tool and as such performs calculations not relevant to many analysis and design applications. A higher level language would be more appropriate to interface the Level 0 through Level 3 subroutines with a larger CACSD package and perform the necessary input and output.

Table 1
Subroutine List for Levels 0 and 1

Level 0	Level 1		DSVDC
BCORBK	BALANC	MILINEQ	DAXPY
D1NRM	DGECOM	ORTHES	DDOT
MADD	DGEFAM	ORTRAN	DNRM2
MMUL	DGESLM	QZHESW	DROT
MOUT	DSTSLV	QZITW	DROTG
MQF	ELMHES	QZVAL	DSCAL
MQFA	GIV	REBAKB	DSWAP
MQFWO	GRADBK	REDUCE	
MSCALE	GRADEQ	REDUC2	
MSUB	HQR	ROTC	
MULA	HQRORT	ROTR	
MULB	IMTQL2	SCALBK	
MULWOA	LINEQ	SCALEG	
MULWOB		SYMSLV	
PERMUT		TRED2	
SAVE			
SEQUIV			
TRNATA			
TRNATB			

VI. NUMERICAL EXAMPLES

In this section, we give a few examples to illustrate various points discussed previously and to provide some numerical results for comparison with other approaches. We also note here that all of the examples given in [17], [22], and [23] were solved using RICPACK and the solutions obtained were as accurate as the published results.

The following simple continous-time example is used to illustrate the numerical properties of RICPACK when stabilizability is the key factor:

<u>Example 1</u>

$$\dot{x} = \begin{pmatrix} 1 & 0 \\ 0 & -2 \end{pmatrix} x + \begin{pmatrix} \epsilon \\ 0 \end{pmatrix} u$$

$$y = (1 \quad 1)\, x$$

$$\text{minimize} \int_0^{\infty} (y^T y + u^T u)dt$$

This system is stabilizable for $\epsilon \neq 0$ and completely reconstructible.

The applicable ARE is

$$A^TX + XA - XBB^TX + C^TC = 0$$

The "true" solution for X can be hand-calculated for comparison purposes as

$$X = \begin{bmatrix} \dfrac{1 + \sqrt{1 + \epsilon^2}}{\epsilon^2} & \dfrac{1}{2 + \sqrt{1 + \epsilon^2}} \\[2ex] \dfrac{1}{2 + \sqrt{1 + \epsilon^2}} & \dfrac{1}{4} - \dfrac{\epsilon^2}{4(2 + \sqrt{1 + \epsilon^2})^2} \end{bmatrix} > 0.$$

Note that as $\epsilon \to 0$ the system approaches unstabilizability and the (1,1) element of X tends to infinity.

The solution to this problem was numerically computed on a DEC KL-10 (under TOPS-20) in double precision using RICPACK. The machine precision is near 10^{-18} in this case. The results of interest are summarized in Table 2. The only measure of condition included in the table is the condition of U_{11} with respect to inversion because other measures did not give an indication that the solution accuracy was degenerating as $\epsilon \to 0$. We note here that the data in Table 2 and in the succeeding tables that are expressed as a power of 10 are rounded to the nearest power of 10.

Table 2
Numerical Results for Example 1, $\epsilon = 10^{-N}$

N	$\kappa(U_{11})$	r (17)	Acc[a]	Newton iterations	r (17)	Acc[a]
0	10^0	10^{-18}	17	...	...	...
2	10^4	10^{-14}	14	...	...	...
4	10^8	10^{-10}	10	2	10^{-18}	17
6	10^{12}	10^{-8}	6	3	10^{-20}	17
8	10^{16}	10^{-2}	2	4	10^{-34}	17
9	10^{18}	10^{-1}	0	6	10^{-18}	17
10	10^{20}	10^0	0	...	...	...
The following data include Ward balancing effects						
0	10^0	10^{-18}	17	...	...	...
5	10^7	10^{-15}	15	...	...	...
10	10^{12}	10^{-9}	9	2	10^{-18}	17
11	10^{16}	10^{-7}	7	3	0	17
12	10^{17}	10^{-7}	7	3	0	17
13	10^{17}	10^{-6}	7	3	10^{-18}	17
14	singular	10^1	0	...	...	...

[a]Accuracy in correct significant digits.

Some useful observations can be made on this data. One can see that for this example, $\kappa(U_{11})$ and the residual (r) are both good indicators of the numerical accuracy. Since machine precision is near 10^{-18}, one would expect about 17 correct digits for a well-conditioned problem (which is the case for $\epsilon = 1$). The data indicates that one digit of accuracy is lost for each power of 10 change in $\kappa(U_{11})$ and r. This is desirable behavior of a condition estimate. Note that Ward balancing improves the condition of U_{11} and reduces the value of r for the same value of ϵ. Ward balancing enables solution calculation for smaller values of ϵ, but the accuracy is not as smooth a function of $\kappa(U_{11})$. However, the residual is still a good indicator of accuracy. Note that in all cases with a reasonable starting guess a few iterations of Newton's method restores full accuracy. The generalized eigenvalue solution was used as a starting guess and was considered reasonable if $\kappa(U_{11}) < 1./(\text{machine precision})$. When this condition was not satisfied, the Newton iteration failed to converge to the desired solution.

The above example illustrates that stabilizability of the model does indeed influence the numerical accuracy of the Riccati solution. Also, $\kappa(U_{11})$ and r are good indicators of solution accuracy as the model approaches unstabilizability. However, $\kappa(U_{11})$ may not be a good indicator in other situations as the following example will show:

Example 2

$$\dot{x} = \begin{pmatrix} -\epsilon & 1 & 0 & 0 \\ -1 & -\epsilon & 0 & 0 \\ 0 & 0 & \epsilon & 1 \\ 0 & 0 & -1 & \epsilon \end{pmatrix} x + \begin{pmatrix} 1 \\ 1 \\ 1 \\ 1 \end{pmatrix} u$$

$$y = (1 \quad 1 \quad 1 \quad 1)x$$

$$\text{minimize} \int_0^\infty (y^T y + u^T u)dt$$

The model is completely controllable and observable. The open-loop poles are at $\pm\epsilon \pm j$, and the applicable ARE is

$$A^T X + XA - XBB^T X + C^T C = 0$$

The solution to this problem was numerically computed on a UNIVAC 1100/83 in double precision using RICPACK. The machine precision is near 10^{-18} in this case. The results of interest are summarized in Table 3. Although an exact hand solution was not possible in this case, the behavior of the residual can be used to judge the solution accuracy. This example was designed to assess the effect of the separation of the closed-loop spectrum on solution accuracy and the ability of condition estimates to detect degrading accuracy. The column CLP in Table 3 indicates the position (real part) of the closed-loop pole nearest the imaginary axis in the complex plane.

Table 3
Numerical Results for Example 2, $\epsilon = 10^{-N}$

N	CLP	$\kappa(U_{11})$	$\kappa A_c(X)$ (18)	$\kappa B(X)$ (19)	r (17)	Newton iterations	r (17)
0	10^{0}	10^{1}	10^{0}	10^{2}	10^{-16}	1	10^{-16}
3	10^{-6}	10^{0}	10^{6}	10^{7}	10^{-12}	1	10^{-14}
5	10^{-10}	10^{0}	10^{10}	10^{11}	10^{-8}	1	10^{-12}
7	10^{-14}	10^{0}	10^{18}	10^{15}	10^{-8}	2	10^{-16}
8	10^{-16}	10^{0}	10^{16}	10^{17}	10^{-1}	2	10^{-16}
9	10^{-18}	10^{0}	10^{18}	10^{18}	10^{-1}	...	...

One can see from this example that $\kappa(U_{11})$ provides no indication of loss of accuracy in the solution. However, in this example we employ two other measures of condition defined in [30] and [32], respectively as

$$\kappa A_c(X) = \frac{\kappa(E)\kappa(E^T)\ \|C^TQC - SR^{-1}S^T\|}{\|X\|\ \|E\|\ \|E^T\|\ \mathrm{SEP}[A_cE^{-1}, -(A_cE^{-1})^T]} \tag{18}$$

where

$$A_c = (A - BR^{-1}S^T - BR^{-1}B^TXE);$$

and

$$\kappa B(X) = \frac{\|C^TQC\| + 2\|A\|\ \|X\| + \|BR^{-1}B^T\|\ \|X\|^2}{\|X\|\mathrm{SEP}\ [A_c^T, -A_c]} \tag{19}$$

where

$$A_c = A - BR^{-1}B^TX$$

Note that $\mathrm{SEP}(F,G) := \inf_{\|P\| = 1} \|PF - GP\|$, is a measure of the separation of the closed-loop spectrum in the above definitions. For a detailed definition of SEP(F,G) and a discussion of its properties, see Stewart [52] and [53].

The results in Table 3 show that $\kappa A_c(X)$ and $\kappa B(X)$ correlate directly with the behavior of the residual, and thus, the solution accuracy. Ward balancing of the eigenproblem had no noticeable effect on solution accuracy for a given value of ϵ. Newton iterations did significantly improve solution accuracy, as measured by the residual, until the condition $\kappa A_c(X) = \kappa B(X) = 1./$ (machine precision). At this point, the iterations failed to converge to the desired solution, as was the case in Example 1.

This example shows that separation of the closed-loop spectrum does indeed influence the numerical accuracy of the Riccati solution, and that the condition estimates in which separation is a factor provide good indicators of solution degeneracy.

The following example illustrates the effect of ill-conditioning of the R weighting matrix, with respect to inversion, on the numerical solution for the continuous-time case. Recall that ill-conditioning of R is not necessarily a problem in the discrete-time case since its inverse is not explicitly required.

Example 3

$$\dot{x} = \begin{pmatrix} -.1 & 0 \\ 0 & -.02 \end{pmatrix} x + \begin{pmatrix} .1 & 0 \\ .001 & .01 \end{pmatrix} u$$

$$y = (10. \quad 100.)x$$

$$\text{minimize} \int_0^\infty (y^T y + u^T \begin{pmatrix} 1+\epsilon & 1 \\ 1 & 1 \end{pmatrix} u)dt$$

The system is completely controllable and observable. As $\epsilon \to 0$, the R matrix approaches singularity. The applicable ARE is

$$A^T X + XA - XBR^{-1}B^T X + C^T C = 0.$$

The solution to this problem was numerically computed on a UNIVAC 1100/83 in double precision using RICPACK. The machine precision is near 10^{-18} in this case. The results of interest are summarized in Tables 4 and 5. Ward balancing was employed in the calculations for Table 4, and coordinate balancing [35], [36] was employed for Table 5. Results of calculations where no balancing was applied were nearly identical to those of Table 5 for coordinate balancing.

The data in Table 4 indicate that $\kappa(R)$ with respect to inversion accurately reflects the behavior of the residual. Also, $\kappa(U_{11})$ with respect to inversion is too optimistic in its estimation of the problem condition and $\kappa B(X)$ is too pessimistic. $\kappa A_c(X)$ provides no information in this case. A maximum of 10 Newton iterations was allowed, and where 10 appears in the table, convergence did not occur. The value for the residual in that case is the residual associated with the solution at the tenth iteration. One can see that the ill-conditioning of R begins to dominate the numerical accuracy when $\kappa(R) > 10^6$. Since R^{-1} is involved in the Newton iteration calculations, iterative improvement does not improve accuracy when $\kappa(R)$ dominates. The convergence criteria used for the Newton iterations do not recognize this fact and stop the iterations.

The data in Table 5 are essentially the same as that in Table 4 except for one important aspect. $\kappa(U_{11})$ with respect to inversion provides no information on the problem conditioning once $\kappa(R) \geq \kappa(U_{11})$. This may indicate that Ward balancing will cause other sources of problem ill-conditioning, beside unstabilizability of the model and singularity of E, to be reflected in $\kappa(U_{11})$.

Table 4
Numerical Results for Example 3, Ward Balancing, $\epsilon = 10^{-N}$

N	$\kappa(R)$	$\kappa(U_{11})$	$\kappa A_c(X)$ (18)	$\kappa B(X)$ (19)	r (17)	Newton iterations	r (17)
0	10^1	10^1	10^0	10^2	10^{-17}	1	10^{-17}
2	10^2	10^2	10^2	10^6	10^{-17}	1	10^{-17}
4	10^4	10^3	10^3	10^{10}	10^{-14}	2	10^{-14}
6	10^6	10^4	10^3	10^{11}	10^{-12}	10	10^{-12}
8	10^8	10^5	10^3	10^{13}	10^{-10}	10	10^{-10}
10	10^{10}	10^6	10^3	10^{15}	10^{-8}	10	10^{-8}
12	10^{12}	10^7	10^3	10^{17}	10^{-6}	10	10^{-8}
14	10^{14}	10^8	10^3	10^{19}	10^{-5}	10	10^{-5}
16	10^{16}	10^9	10^3	10^{21}	10^{-3}	10	10^{-3}

Table 5
Numerical Results for Example 3, System Balancing, $\epsilon = 10^{-N}$

N	$\kappa(R)$	$\kappa(U_{11})$	$\kappa A_c(X)$ (18)	$\kappa B(X)$ (19)	r (17)	Newton iterations	r (17)
0	10^1	10^3	10^0	10^3	10^{-15}	2	10^{-17}
2	10^2	10^3	10^2	10^6	10^{-15}	2	10^{-17}
4	10^4	10^3	10^3	10^9	10^{-12}	7	10^{-14}
6	10^6	10^3	10^3	10^{11}	10^{-12}	10	10^{-12}
8	10^8	10^3	10^3	10^{13}	10^{-9}	10	10^{-11}
10	10^{10}	10^3	10^3	10^{15}	10^{-6}	10	10^{-9}
12	10^{12}	10^3	10^3	10^{17}	10^{-6}	10	10^{-6}
14	10^{14}	10^3	10^3	10^{19}	10^{-2}	10	10^{-4}
16	10^{16}	10^3	10^3	10^{21}	10^{-1}	10	10^{-2}

The preceding examples illustrate that none of the potential measures of conditioning are reliable indicators by themselves. However, numerical experience to date has shown that in all cases, at least one of the measures will detect the degeneracy of numerical accuracy as it occurs. These examples were of small order because the problem condition is not directly related to size, so simple examples were sufficient to illustrate the numerical effects of an ill-conditioned problem. However, one may wonder if the size of the problem will affect the numerical results, even for well-conditioned problems. The following example shows that size is not a factor.

Example 4

It is difficult to construct examples of large order for which the exact solution is known, or whose condition can be inferred. For this reason, two examples were taken from the literature. The first is a 64th order example involving circulant matrices that appeared in [17] as example 5. The example was solved using RICPACK on a VAX-11/780 (under VMS) in double precision. The machine precision is near 10^{-17} in this case. The resulting solution and closed-loop eignvalues agreed to within the least significant digit published in [17]. RICPACK indicated the condition of the problem was on the order 10^{1} and the r value for the solution was on the order 10^{-14}. The other large example, also solved on the VAX, appeared as example 3 in [22]. In this case, an 80th order discrete-time problem was solved. The closed-loop eigenvalues were accurate to 16 significant digits and the Riccati solution was accurate to 13. RICPACK indicated the condition was on the order 10^{2} and the r value for the solution was on the order of 10^{-15}.

VII. CONCLUDING REMARKS

While significant progress can be documented already on the numerical solution of Riccati equations, substantial numbers of questions remain to be investigated. The ubiquitous nature of these equations in mathematics and engineering, together with the rapid and fundamental changes in computing environments of the 1980s, offer important benefits to be gained in their further study. It is our contention that further progress is necessary but can only be made through an interdisciplinary approach blending systems theory, numerical analysis, computer science, and mathematical software. The research described in this paper is directed towards that goal.

REFERENCES

[1] Anderson, B. D. O., and J. B. Moore, *Linear Optimal Control,* Prentice-Hall, Englewood Cliffs, NJ, 1971.

[2] Athans, M., and P. L. Falb, *Optimal Control,* McGraw-Hill, New York, 1966.

[3] Kwakernaak, H., and R. Sivan, *Linear Optimal Control Systems,* Wiley, New York, 1972.

[4] Wonham, W. M., *Linear Multivariable Control: a Geometric Approach,* Second Edition, Springer-Verlag, New York, 1979.

[5] *IEEE Transactions on Automatic Control,* Special Issue on Linear Quadratic Gaussian Control, Vol. AC-16, December 1971.

[6] Anderson, B. D. O., and J. B. Moore, *Optimal Filtering,* Prentice-Hall, Englewood Cliffs, NJ, 1979.

[7] Jazwinski, A. H., *Stochastic Processes and Filtering Theory,* Academic Press, New York, 1970.

[8] Åström, K., *Introduction to Stochastic Control Theory*, Academic Press, New York, 1970.

[9] Reid, W. T., *Riccati Differential Equations*, Academic Press, New York, 1972.

[10] Scott, M. R., *Invariant Imbedding and its Applications to Ordinary Differential Equations*, Addison-Wesley, Reading, MA, 1973.

[11] Vandevender, W. H., "On the Stability of an Invariant Imbedding Algorithm for the Solution of Two-Point Boundary Value Problems," Sandia Laboratories Report No. SAND77-1107, August 1977.

[12] Laub, A. J., "Schur Techniques in Invariant Imbedding Methods for Solving Two-Point Boundary Value Problems," *Proc. 21st IEEE CDC*, Orlando, FL, December 1982, pp. 55-61.

[13] Childs, B., M. Scott, J. W. Daniel, E. Denman, and P. Nelson (Eds.), *Codes for Boundary-Value Problems in Ordinary Differential Equations*, Lec. Notes in Computer Sci., Vol. 76, Springer-Verlag, New York, 1979.

[14] Smith, B. T., et al, *Matrix Eigensystem Routines - EISPACK Guide*, Second Ed., Lec. Notes in Comp. Sci., Vol. 6, Springer-Verlag, New York, 1976.

[15] Garbow, B. S., et al, *Matrix Eigensystem Routines - EISPACK Guide Extension*, Lec. Notes in Comp. Sci., Vol. 51, Springer-Verlag, New York, 1977.

[16] Dongarra, J., et al, *LINPACK User's Guide*, SIAM, Philadelphia, 1979.

[17] Laub, A. J., "A Schur Method for Solving Algebraic Riccati Equations," *IEEE Trans. Automatic Control*, AC-24, pp. 913-921, December 1979.

[18] Laub, A. J., "A Schur Method for Solving Algebraic Riccati Equations," LIDS Rept. No. LIDS-R-859, LIDS, MIT, October 1978.

[19] Lee, K. H., "Generalized Eigenproblem Structures and Solution Methods for Riccati Equations," Ph.D. Thesis, Dept. of EE-Systems, USC, January 1983.

[20] Van Dooren, P., "A Generalized Eigenvalue Approach for Solving Riccati Equations," *SIAM J. Sci. Stat. Comp.*, 2, pp. 121-135, 1981.

[21] Pappas, T., "Solution of Discrete-Time LQG Problems with Singular Transition Matrix," B.S. Thesis, Dept. of Elec. Engrg., MIT, May 1979.

[22] Pappas, T., A. J. Laub, and N. R. Sandell, "On the Numerical Solution of the Discrete-Time Algebraic Riccati Equation," *IEEE Trans. Automatic Control*, AC-25, pp. 631-641, August 1980.

[23] Emami-Naeini, A., and G. F. Franklin, "Deadbeat Control and Tracking of Discrete-Time Systems," *IEEE Trans. Automatic Control*, AC-27, pp. 176-181, February 1982.

[24] Walker, R. A., A. Emami-Naeini, and P. Van Dooren, "A General Algorithm for Solving the Algebraic Riccati Equation," *Proc. 21st IEEE CDC*, 1982, pp. 68-72.

[25] Emami-Naeini, A., and G. F. Franklin, "Design of Steady State Quadratic Loss Optimal Digital Controls for Systems with a Singular System Matrix," *Proc. 13th Asilomar Conf. on Circ. Sys. Comp.*, November 1979, pp. 370-374.

[26] Campbell, S. L., *Singular Systems of Differential Equations*, Pitman Publishing Ltd., Marshfield, MA, 1980.

[27] Campbell, S. L., *Singular Systems of Differential Equations II*, Pitman Publishing Ltd., Marshfield, MA, 1982.

[28] Moler, C. B., and G. W. Stewart, "An Algorithm for Generalized Matrix Eigenvalue Problems," *SIAM J. Numer. Anal.*, 10, pp. 241-256, 1973.

[29] Ward, R. C., "The Combination Shift QZ Algorithm," *SIAM J. Numer. Anal.*, 12, pp. 835-853, 1975.

[30] Arnold, W. F., "On the Numerical Solution of Algebraic Matrix Riccati Equations," Ph.D. Thesis, Dept. of EE-Systems, USC, December 1983.

[31] Paige, C. C., and C. F. Van Loan, "A Schur Decomposition for Hamiltonian Matrices," *Linear Algebra and its Applications,* 14, pp.11-32, 1981.

[32] Byers, R., "Hamiltonian and Symplectic Algorithms for the Algebraic Riccati Equation," Ph.D Thesis, Dept. of Comp. Sci., Cornell University, 1983.

[33] Arnold, W. F., and A. J. Laub, "A Software Package for the Solution of Generalized Algebraic Riccati Equations," *Proc. 22nd IEEE CDC,* San Antonio, TX, December 1983, pp. 415-417.

[34] Ward, R. C., "Balancing the Generalized Eigenvalue Problem," *SIAM J. Sci. Stat. Comput.,* 2, pp. 141-152, 1981.

[35] Moore, B. C., "Principal Component Analysis in Linear Systems: Controllability, Observability, and Model Reduction," *IEEE Trans. Automatic Control,* AC-26, pp. 17-32, February 1981.

[36] Laub, A. J., "On Computing "Balancing" Transformations," *Proceedings 1980 JACC,* San Francisco, CA, 1980, pp. FA8-E.

[37] Kleinman, D. L., "On an Iterative Technique for Riccati Equation Computations," *IEEE Trans. Automatic Control,* Vol. 13, pp. 114-115, February 1968.

[38] Hewer, G. A., "An Iterative Technique for the Computation of the Steady State Gains for the Discrete Optimal Regulator," *IEEE Trans. Automatic Control,* Vol. AC-16, pp. 382-384, August 1971.

[39] Aird, T. J., *The Fortran Converter User's Guide,* IMSL, 1975.

[40] Boyle, J., and K. Dritz, "An Automated Programming System to Aid the Development of Quality Mathematical Software," *IFIP Proceedings,* North-Holland, pp. 542-546, 1974.

[41] Cowell, W., *Portability of Numerical Software, Oak Brook, 1976,* Lec. Notes in Comp. Sci., Vol. 57, Springer-Verlag, New York, 1977.

[42] Cowell, W., and L. J. Osterweil, "The Toolpack/IST Programming Environment," Argonne Na. Lab., Appl. Math. Div., Rept. No. ANL/MCS-TM7, 1983.

[43] Crowder, H., R. S. Dembo, and J. M. Mulvey, "On Reporting Computational Experiments with Mathematical Software," *ACM Trans. Math. Software,* 5, pp. 193-203, 1979.

[44] Dorrenbacher, J., D. Paddock, D. Wisneski, and L. D. Fosdick, "POLISH, a Fortran Program to Edit Fortran Programs," Dept. of Comp. Sci., Univ. of Colorado (Boulder), Rept. No. CU-CS-050-74, 1974.

[45] Fosdick, L. D. (Ed.), *Performance Evaluation of Numerical Software,* North-Holland, New York, 1979.

[46] Hennel, M. A., and L. M. Delves, (Eds.), *Production and Assessment of Numerical Software,* Academic Press, New York, 1980.

[47] Messina, P., and A. Murli (Eds.), *Problems and Methodologies in Mathematical Software Production,* Lec. Notes in Comp. Sci., No. 142, Springer-Verlag, New York, 1982.

[48] Reid, J. K. (Ed.), *The Relationship Between Numerical Computation and Programming Languages,* North-Holland, New York, 1982.

[49] Rice, J. R., *Matrix Computations and Mathematical Software,* McGraw-Hill, New York, 1981.

[50] Rice, J. R., *Numerical Methods, Software, and Analysis,* McGraw-Hill, New York, 1983.

[51] Ryder, B. G., "The PFORT Verifier: User's Guide," CS. Tech. Rept. 12, Bell Labs, 1975; also *Software Practice and Experience,* 4, pp. 359-377, 1974.

[52] Stewart, G. W. "Error Bounds for Approximate Invariant Subspaces of Closed Linear Operators." *SIAM J. Numer. Anal.,* Vol. 8, pp. 769-808, December 1971.

[53] Stewart, G. W. "Error and Perturbation Bounds for Subspaces Associated with Certain Eigenvalue Problems." *SIAM Review,* Vol. 15, pp. 727-764, October 1973.

COMPUTER-AIDED CONTROL SYSTEMS ENGINEERING
M. Jamshidi and C.J. Herget (Editors)

A SOFTWARE LIBRARY AND INTERACTIVE DESIGN ENVIRONMENT FOR COMPUTER AIDED CONTROL SYSTEM DESIGN

Michael J. Denham
Department of Computing
Kingston Polytechnic
Kingston-upon-Thames, Surrey
U.K.

There is a major requirement in control system design a library of numerically reliable software for analysis of dynamic system models and an interactive computing environment to provide access to the library and other software tools for model creation, simulation and manipulation. The software design issues involved are discussed and progress in providing such software reported on.

INTRODUCTION

We begin on a historical note by tracing briefly the course of events which has led to the programme of software development which will be described here. As is probably well known, control system design software has been under continuous development in the U.K. since the middle 1960's. At that time, two centres were the main participants in this work, UMIST and Imperial College, London. Later Cambridge was to also become a major contributor. Much of this early work was based on the pioneering multivariable frequency domain system analysis and design methods of Rosenbrock, MacFarlane and Mayne, and was carried out using fairly primitive, by modern standards, hardware and software development tools. Many of the basic computational algorithms employed in the software packages which were developed employed dubious procedures from a numerical computation viewpoint and were largely direct interpretations of the mathematical theory in computational form. Little attention was paid to aspects of numerical stability or potential ill-conditioning of the data, but we must remember also that this was also the time at which the early pioneering work in numerical linear algebra by Wilkinson [1], [2] was being reported. It is also a fact that the main aim of the early work on control system CAD was to prove the viability of various design methods, and the use of "good" computer science was of secondary importance. In any case, the problem was that the "good" tools did not exist at that time.

In the last four to five years, a reappraisal of the state of control system CAD has taken place. Most of the large development projects had completed their funded programmes by the start of this period and new tools, hardware and software, were becoming available. Also industrial interest in the use of CAD techniques for multivariate system design was growing, highlighting some of the inherent problems in using current CAD software in an industrial environment. A body of numerical computation tools had been developed, e.g. NAG, EISPACK, LINPACK, and it was becoming apparent how these could be employed in control system analysis and design algorithms. The use of graphical interaction was becoming widespread in other disciplines such as circuit design, and new candidates for implementation languages such as Pascal were becoming available. Powerful computing facilities made interactive use of complex, time consuming algorithms a feasible proposition and the use of networking made remote access to central facilities from intelligent local workstations a common mode of operation. At the same time the power of the local workstations was growing at an enormous rate to make these viable stand-alone design tools, e.g. the Three Rivers PERQ.

In all ways, the time had arrived to reassess the future of control system CAD research and development and by happy coincidence this also coincided in the U.K. with, in 1975, a significant restructuring of support funding for interactive computing in engineering research. Apart from a programme to provide a powerful network of a large number of multi-user minicomputers and a smaller number of super "number crunchers" such as the CRAY machine, a significant aspect of this was the establishment of "Special Interest Groups" (SIG), tasked with the promotion and co-ordination of software development for research in their respective areas.. The SIG in control engineering was set up in 1978 and began its programme of software development in late 1979. Early work included providing nationwide access to the major design packages at UMIST and Cambridge and to good simulation tools. A significant part of the programme was the software development task reported on here. We must emphasise that the level of funding of this programme is not large, namely 3 man-years per year. Nor does it represent by any means the total effort being expended in this area in the U.K. since, for example, the Cambridge and UMIST groups are still highly active. However, what it does represent is a major part of the effort being devoted to providing the tools of control system CAD which are universally required independent of the particular design methodology which is to be used.

The main aim here will be to summarise the current state of this programme of software development, to review its objectives and present the proposals for future work. We strongly welcome comment and criticism of both our current work and proposals and indeed have had wide discussions with colleagues both in the U.K. and throughout the U.S.A. and Europe. We invite further discussion and information on similar projects in progress elsewhere.

THE MAIN AIMS OF THE SOFTWARE DEVELOPMENT PROGRAMME

The following are amongst our principal aims in carrying out this programme:

1) to avoid unnecessary duplication of software development effort by making well documented, maintained and reliable software freely available to research groups,

2) to provide a sound and powerful basis for research into control system design methods, unimpeded by unnecessary numerical difficulties or unfriendly, weak tools for the development of interactive design software,

3) to improve the likelihood of acceptance of design methods and software by providing an efficient, reliable and friendly user environment,

4) to stimulate research into new computational tools for control system design, by exploring and highlighting the deficiencies of the existing tools and attempting to overcome them,

5) to stimulate research into new design methodologies to take maximum advantage of the new software tools to be made available.

It is proposed at present that these aims are realised by the development of two major software tools for control system design:

1) a subroutine library of computational algorithms for control system analysis and design,

2) a universal interactive design environment.

The exact nature of these two substantial pieces of software and the considerations which have entered into their design will now be described.

THE SUBROUTINE LIBRARY IN CONTROL ENGINEERING (SLICE)

SLICE is a library of FORTRAN subroutines which, it is hoped, will eventually incorporate all the known algorithms used in control system CAD which can be implemented according to defined high standards of operation. This is an ambitious objective, and without doubt we are and will be subject to some compromise! However, many of the high standards hoped for are achievable in a straightforward way, albeit with some considerable effort.

A principal aim of the library is that it should incorporate and make use of reliability coded, numerically stable algorithms as far as possible. This naturally implies that full use be made of existing such algorithms for the basic numerical computations and hence the library is substantially based on routines from the EISPACK [3, 4] and LINPACK [5] libraries. Not only does this diminish the effort involved in software development, but it is of course a total nonsense to even consider rewriting the routines incorporated in such libraries, each of which has received several man-years of attention.

To ensure a user friendly interface and the often quoted and highly important properties of portability and maintainability, standards for both documentation and implementation of SLICE routines have been defined [6]. Each routine has been coded in a portable subset of ANSI FORTRAN IV known as PFORT [7] and has to pass through the PFORT verifier [8] for acceptance into the library. An important aspect of user friendliness is that, in so far as it is feasible, the routines should fail if inevitable in a "soft" manner, i.e. by returning full user information on the reason for failure and providing for an orderly return to the calling program.

Adequate levels of test, maintenance and user support are important requirements of any library, which have strong implications for a continuing need for funding in order to supply a satisfactory service to users.

The SLICE library at the time of writing consists of over thirty routines, documented and coded to the defined standards mentioned above. These routines represent less than half, however, of the eventual library as it is currently planned. The routines in the library are broadly classified into three main sections:

1) Transformation routines - for translation between system models,

2) Analysis routines - for computing the properties of system models,

3) Synthesis routines - for computing parameters of specified controllers.

In addition to these sections, there are a number of routines which can be classified as of a supporting or utility type. These include a number of routines for computing various operations in linear algebra. The majority of the latter, as previously mentioned, are provided by the EISPACK and LINPACK libraries. However some operations have been specialised to the requirements of particular SLICE routines, either for efficiency, by eliminating unrequired suboperations, or to improve the clarity of the resulting software and hence its reliability and maintainability.

In the transformation section of the library, the following model representations are currently considered:

1) state space model

2) polynomial matrix fraction (left or right)

3) transfer function matrix (MIMO)

5) time response

6) frequency response

Translation routines are provided between each of these representations, except from 5) and 6) to the others. In many cases, the path between representations is via the state space model, since it is in this form that the algorithms for reducing the model dimension, by eliminating "uncontrollable" or "unobservable" parts, is best understood from a numerical viewpoint. This comment applies in general throughout the library, in which a strong emphasis on state space models for analysis and synthesis is immediately apparent. Unfortunately, the state of knowledge in the numerical analysis field on the properties of numerical rounding errors in algorithms involving polynomial or rational function manipulations in very limited. This forced us at an early stage to concentrate on state space representations, which involve only real or complex valued matrix manipulations.

A major aim of the library development, as mentioned previously, is reliability. One mechanism of achieving this is to employ a small kernel of well tested code in as many routines as possible in the library. In the current set of implemented routines, the most widely used operation is that of numerical rank determination. Two methods are available for this: singular value decomposition and QR (or QU) decomposition, both of which offer reliable, stable methods with further information available readily on the "closeness" of the matrix in question to matrices of lower rank. This latter information is of great potential use in the robust design of both algorithms and controllers. Of these two methods, the singular value decomposition is marginally more reliable in general yet computationally more expensive in time. We have opted currently to concentrate on the use of the QR method as almost as reliable, except in pathological cases, and much cheaper computationally. The overall reliability of the library is therefore enhanced by ensuring that the same reliable, efficient code for the QR decomposition is used in all routines requiring numerical rank determination.

At a higher level, an operation of widespread applicability, e.g. in computing the "controllable and observable" part of a state space model, and in computing the transmission zeros of the model, is the reduction of a matrix to upper block triangular form by a sequence of row or column compression operations on submatrices. An algorithm requiring such an operation is that for computing controllability [9], [10], [11] which carries out this procedure on the rectangular matrix: [B A], derived from the state space model: $x = Ax + Bu$. We have therefore used, or intend to use a common implementation of this operation in all those algorithms which require it.

Finally, a further important feature of the library is the widespread use of "balancing" or "scaling" of data prior to the application of numerical manipulations involving rounding errors. The need for balancing has long been recognised as a means of improving accuracy, e.g. in eigenvalue computation [12], and this need is obvious in many control engineering applications where the model data is in many cases poorly scaled. The effects of this can be easily observed. For example, poor scaling of the F100 engine model as presented in [13] can result in large relative errors in frequency response computation. Balancing the A matrix in the manner proposed in [12] and further scaling of the B, C and D matrices relative to this balancing of A and their own parameter values, reduces this error to the order of 0.01%.

Future work on development of the library will take several forms:

1) the continued improvement in the efficiency and reliability of algorithms,

2) the implementation of new types of algorithms, e.g. sparse problems, "fast" algorithms,

3) the investigation of the role of numerical computation in the engineering design process, e.g. the need for prior conditioning of the system model data and the proper reporting of potential numerical hazards during the design process and mechanisms for dealing with these.

In the first two areas, we hope to enlist the support and co-operation of the algorithm developing community, in return for extensive distribution of the library on a non-commercial basis. Whilst still maintaining an active role in this field ourselves, we recognise the enormity of the software development task as well outside the scope of our resources. In addition, we feel that a free exchange of algorithms is essential to the healthy growth of this important field, backed up by the mechanism for producing reliable code for the algorithms which we believe the SLICE development provides.

The role of numerical computation in the design process requires a proper vehicle for its investigation, i.e. an interactive control system design environment. For this reason, together with many others, we have been naturally led to consider and plan the design and development of such an environment, the fuller description of which is the subject of the remainder of this paper.

AN INTERACTIVE CONTROL SYSTEM ENVIRONMENT

The SLICE development described above is aimed at providing only one of a number of effective tools for CAD of control systems. It is the aim of a programme of software development, planned to start towards the end of 1982, supported again by the U.K. Science and Engineering Research Council, to provide a number of other software tools essential for the effective use of computers in control system design. The integrated set of tools will provide what could be termed a "design environment", analogous to the kind of support environment becoming increasingly common for software engineering, e.g. UNIX, APSE, InterLisp. It is in fact by using this analogy that it is easiest to see what kind of tools should be provided in such a control system design environment, as we shall show later.

Firstly, however, let us specify the main aims of any such environment. These can be listed as:

1) provision of a rich set of software tools for creation, modification, simulation and analysis of dynamic system models,

2) provision of a powerful and effective man-machine interface to provide access to the software tools in an interactive, design orientated environment,

3) universality, i.e. the ability to support any design methodology, in contrast to most present design packages which are method-orientated,

4) portability, i.e. the ability to transport economically the design environment to any user workplace, either by the transference of software alone, or both hardware and software,

5) expandability, i.e. the ability to enhance the environment in the light of continuing advances in techniques and methods of design,

6) reliability and maintainability.

Currently, to this author's knowledge, no CAD system for control design satisfies the aims listed above. In fact, it is my opinion that relative to, say, the designer of software, the control system designer has a very poor set of design aids. Although some very fine design programs exist in isolation, no attempt has been made to provide the kind of universal, integrated environment proposed here.

As mentioned previously, it is useful in order to identify the tools which the design environment should provide, to consider the analogy with the software engineering situation. Tt is natural that the experts in providing software-based design aids should first provide them for their own task, of designing software.

The aim of software design is to produce a program, which when stimulated by externally generated data, performs in the desired manner. Various aids are available to assisting this task, including: formal requirements specification tools, to aid the definition of the function of the program; editors, to create and modify the program description (code or other symbolic description, e.g. flow chart, data flow diagram); translators, to convert from one symbolic program description to another, e.g. RATFOR to FORTRAN processor; compilers, to convert the symbolic program description into an efficient form for machine execution; run time systems, to support program execution; debuggers, to report on and analyse program execution and correctness; documentation systems, to provide program documentation to assist in analysis and understanding of its operation, purpose, etc., and software module libraries, to aid the incorporation of often used routines in a reliable, cost effective, user-transparent way.

The aim of control system design is directly analogous to this task if it is viewed as that of producing a dynamic system model, which when stimulated by externally generated signals, performs on the desired manner. The dynamic system model is directly analogous to the program in software design, and the tools required for control design are therefore analogous to the editors, translators, etc. of software design.

Not wishing to pursue this analogy too strongly, it is probably clear that it will nevertheless provide us with a list of fully integrated tools for the control system design environment:

1) a formal requirement specification aid: to enable exact statements to be made about the required function and performance of the system model,

2) an interactive man-machine interface for model creation and

modification: this could take the form of an interactive high level language for model description (analogous to a general purpose programming language) together with an interactive editor for the language, or a graphical interface using a symbolic description language, e.g. block diagrams, signal flow graphs, (analogous to a flow charting program) together with an interactive graphical editor, using "mouse" or graphics tablet for interaction,

3) model administration aids: to provide back-up of different generations of the same models after editing, concatenate or merge different models into one and vice versa, etc., together with automatic compatibility checking,

4) a set of transformation routines: to translate between the user preferred model representation and that used by the design environment for internal storage, with the ability to translate in both directions,

5) model documentation aids: to automatically provide formal documentation of models and administer documents,

6) model libraries: to provide models of commonly used system components, e.g. sensors, actuators, digital controllers, and mechanisms to administer libraries and the linking of components into larger systems models,

7) "history" and "undo" facilities: to store the operations performed on a model, e.g. in creation or editing and to display, change, reapply or reverse their effect on request,

8) a set of internally stored representations of the system model and translators (compilers) from the model description language to internal form: to store the model in the most suitable forms for analysis, display, simulation, documentation, etc. and translate between forms,

9) a simulator: to provide a set of simulation tools for computing various forms of dynamic response data, e.g. time, frequency and their derivatives,

10) a debugger: to provide storage of model variables with output of values on request, insertion of breakpoints in the simulation and trace points

to flag debugging information, e.g. constraint violation, non-access to sub-models, etc.

11) a set of model performance analysis routines: to provide analysis of simulation data, e.g. correlation analysis, and of model data, e.g. eigenstructure.

A principal difference between the program design and control system design environments is that there exists for control system design many automatic or semi-automatic controller synthesis methods, usually based on an algorithmic optimisation procedure. This implies the need in the design environment for an interactive algorithmic design language. This is in principle distinct from the modelling language mentioned above but the two should clearly use the same constructs, etc. in an integrated system. An essential feature of the algorithmic design language is that it should have efficient, fast access to all the internal representations of the system model, with the ability to modify these representations directly, e.g. to add connections, controller sub-models, etc.

Both the modelling language and the algorithmic design language should be "incrementally compiled", i.e. editing or other change operations to the model should result only in the reconstruction of that part of the internal representation directly affected by the change. This results in faster interactive model creation and simulation than would be the case if complete re-compilation of the model was required for every change, thus leading to a truly interactive environment.

A prime example of an algorithmic design language of the type described above, is provided by the RATTLE language in the DELIGHT system [14] Moreover, this language exists in a support environment, namely DELIGHT, which provides many of those kinds of tools described for the proposed design environment. It is for this and several additional reasons that it is proposed to make the DELIGHT system a major component part of our proposed environment. It is important to note however that to provide the design system we envisage it will be imperative that DELIGHT, together with all other components of the environment, is fully integrated into the system. Indeed, many of the proposed tools, e.g. the "undo" facility must be integrated to work at all.

As a first step towards portability of the environment, it is normally sensible to develop the software tools in a language and under an operating system with a wide user base. The U.K. Science and Engineering Research Council have adopted a common base policy which dictates the use of FORTRAN, Pascal and the UNIX operating system, and this is to be used in the proposed environment. Another component of the common base policy is the use of single user graphics workstations. These powerful, highly interactive machines provide an excellent basis on which to develop the proposed environment and offer the potential capability of "total system portability", i.e. a new user acquires both hardware and software at an acceptable cost. Such a capability, more realistic as hardware costs fall in relation to software, essentially eliminates all the traditional constraints associated with software portability.

CONCLUSION

The programme of software development described here is an on-going programme aimed at providing the kinds of design tools which hitherto have not been available in the control system field, except in a limited way as a set of partially integrated design programs. Recent advances in the fields of numerical analysis, software engineering and interactive computing have provided techniques which were not available when the present set of tools began their development and have made feasible the proposed design environment and its associated set of design tools. We believe that the time is now right to take advantage of this situation to produce a "second generation" of computer aided control system design facilities. (Further,more recent discussion of these matters can be found in [15] and [16].)

The future beyond the present development presents even greater challenges. The use of expert or knowledge based systems for engineering design is now becoming a real possibility and we are actively seeking ways of incorporating this new field within control system design [16]. A further possibility is the development of a "design-to-implementation" system which will take the desired control structure, as produced by the kind of design environment proposed above, and devise a computer control system to implement this. This implies the selection of such components as processors, interfaces, sensors, actuators, communication networks, real-time algorithms, etc. and the determination of factors such as the hardware/software partition of tasks, optimal allocation of control processes to processors, timing constraints, reliability, etc. This is an apparently formidable task involving the consideration of a profusion of engineering implementation constraints. However such a system together with the proposed design environment, would go some way towards producing a total design and manufacturing (CADCAM) system for control engineering of a type only seen to date in such fields as mechanical engineering, chemical engineering and VLSI design.

REFERENCES

1. J.H. Wilkinson, The Algebraic Eigenvalue Problem, Oxford University Press, London, 1965.

2. J.H. Wilkinson, Rounding Errors in Algebraic Processes, Prentice Hall, New Jersey, 1963.

3. B.T. Smith et al., Matrix Eigensystem Routines - EISPACK Guide, Lecture Notes in Computer Science, 6, Springer-Verlag, New York, 1976.

4. B.S. Garbow et al., Matrix Eigensystem Routines - EISPACK Guide Extension, Lecture Notes in Computer Science, 51, Springer-Verlag, New York, 1977.

5. J. Dongarra et al., LINPACK Users' Guide, SIAM, Philadelphia, 1979.

6. M.J. Denham and C.J. Benson, "Implementation and documentation standards for the Subroutine Library in Control Engineering (SLICE)", Internal Report, School of Electronic Engineering and Computer Science, Kingston Polytechnic, 1981.

7. B.G. Ryder, "The PFORT verifier", Software Praçtices and Experience, 4, pp. 359-377, 1981.

8. B.G. Ryder and A.D. Hall, "The PFORT verifier", Computing Science Technical Report 12, Bell Laboratories, New Jersey, 1973.

9. M.M. Konstantinov et al., "Invariants and canonical forms for linear multivariable systems under the action of orthogonal transformation groups", Kybernetika, 17, pp. 413-424, 1981.

10. C.C. Paige, "Properties of numerical algorithms related to computing

controllability", IEEE Trans. Aut. Contr., AC-26, pp.130-138, 1981.

11. P. Van Dooren, "The generalized eigenstructure problem in linear system theory", IEEE Trans. Aut. Contr., AC-26, pp.111-129, 1981.

12. B.N. Parlett and C. Reinsch, "Balancing a matrix for calculation of eigenvalues and eigenvectors", Numerische Mathematik, 13, pp.293-304, 1969.

13. M.K. Sain et al., Alternatives for Linear Multivariable Control, National Engineering Consortium, Chicago, 1978.

14. W.T. Nye, et al., "DELIGHT : An optimization-based computer aided design system", Proc. IEEE ISCAS, Chicago, April 1981.

15. M.J. Denham, "Design environments and the user interface for CAD of control systems", in AGARD Lecture Series No.128, September 1983.

16. M.J. Denham, "Design issues for CACSD systems", Proc. IEEE, December 1984.

COMPUTER-AIDED CONTROL SYSTEMS ENGINEERING
M. Jamshidi and C.J. Herget (Editors)

ALGORITHMS FOR EIGENVALUE ASSIGNMENT IN MULTIVARIABLE SYSTEMS*

R.V. Patel

Department of Electrical Engineering
Concordia University
Montreal, Canada H3G 1M8

This paper describes a computational method for solving the problem of eigenvalue assignment in linear multivariable systems. A given multi-input system is first reduced to an upper block Hessenberg form by means of orthogonal state coordinate transformations. It is then shown how a sequence of state feedback matrices and orthogonal state coordinate transformations can be applied to obtain a block triangular structure for the resulting state matrix, where the matrices on the diagonal are square matrices in upper Hessenberg form and of dimensions equal to the controllability indices of the multi-input system. Furthermore, the structure of the corresponding input matrix is such that the problem of eigenvalue assignment in the multi-input system can be reduced to several single-input eigenvalue assignment problems where the dimensions of the single-input systems are equal to the controllability indices of the multi-input system. For single-input systems, an algorithm based on the QR algorithm (with implicit shifts) is given for carrying out eigenvalue assignment.

1. INTRODUCTION

The problem of eigenvalue (pole) assignment (e.v.a.) in linear multivariable systems by state feedback has been investigated by many researchers and numerous methods have been proposed for achieving e.v.a., e.g. see [1-4]. However, until recently, not much attention had been given to the numerical aspects of the problem. Such considerations become particularly important when dealing with high order systems.

The conceptual simplicity of the e.v.a. problem tends to hide the potential numerical difficulties that can arise when using many of the well-known algorithms. For instance, some techniques, e.g. [3], require the reduction of a given state space system to a canonical form. Such reduction can be a source of numerical instability. Another weak point from the numerical point of view is the requirement in some methods such as [4], to compute the transfer function matrix of the given state space model.

---

* This work was supported by the Natural Sciences and Engineering Research Council of Canada under Grant A1345.

Recently, several algorithms for e.v.a. by state feedback have been proposed [5-8]. The algorithm described in [5] is applicable to single-input systems. It is based on the QR algorithm for computing the eigenvalues of a matrix and uses only numerically stable transformations. The case of multi-input systems has been treated by the same authors in [6]. However, the approach used is not a straightforward extension of that given in [5] for single-input systems. The algorithms described in [7] and [8], although developed independently, are essentially the same. They are based on the reduction of the state matrix to its real Schur form (RSF) by means of orthogonal state coordinate transformations. The RSF of a matrix is a block triangular form with either scalars or 2×2 blocks along the diagonal. The scalars correspond to real eigenvalues of the matrix and the 2×2 blocks to complex-conjugate pairs. Consequently, the reduction of a matrix to its RSF is equivalent to solving an eigenvalue problem. The reduction of an open-loop state matrix to its RSF required in [7] and [8] will therefore be ill-conditioned if the eigenvalue problem for the state matrix is ill-conditioned. This is, therefore, a disadvantage of these algorithms. In fact, this difficulty is common to all methods which require knowledge of the open-loop eigenvalues. It is worth mentioning here that the algorithms in [5,6] and the one proposed in this paper do not have this disadvantage.

In this paper, we present a method for achieving e.v.a. by state feedback in multi-input systems. Our objective is to provide an algorithm which can be used for high-order systems. We accomplish this by reducing the multi-input e.v.a. problem to a number of single-input e.v.a. problems. The orders of the single-input systems are equal to the controllability indices of the multi-input system, which implies that the mechanism of e.v.a. is performed on single-input systems of considerably lower order than that of the given multi-input system. In this context our approach has some similarities with the method proposed by Anderson and Luenberger [3]. However, from the numerical point of view, our approach is far superior since it uses only orthogonal transformations and does not require a reduction to the Luenberger canonical form. For solving the single-input e.v.a. problems, we give an algorithm based on the well-known QR algorithm, e.g. see [9]. Our approach is similar to the one described in [5], but has the advantage of being conceptually and computationally simpler (see the remarks following Algorithm 4).

This paper is organized as follows: In Section 2, we briefly review some relevant results from numerical linear algebra. Section 3 introduces a preliminary step to to our algorithm: reduction of a multi-input system to a block Hessenberg form. In Section 4, we show how the multi-input e.v.a. problem can be solved via a number of lower order single-input e.v.a. problems. An algorithm for e.v.a. in single-input systems is described in Section 5. The numerical aspects of our algorithms are discussed in Section 6. In the rest of this paper, matrices are denoted by upper case letters and vectors by underlined lower case letters. The notation ' (prime) is used to denote matrix and vector transposition.

2. REVIEW OF SOME RESULTS FROM NUMERICAL ANALYSIS

In this section, we review briefly some of the concepts from numerical analysis which are used in the algorithms to be presented in the next section.

2.1 Singular Value Decomposition (SVD)

Let $Z \in \mathbb{R}^{m\times n}$ have rank r. Then there exist m×n and n×n orthogonal matrices U and V such that

$$UZV' = \begin{bmatrix} \Sigma_r & 0 \\ 0 & 0 \end{bmatrix} \tag{2.1}$$

where Σ_r is a diagonal matrix with the non-zero singular values of A along its diagonal. Algorithms for computing the SVD i.e. the matrices U,V and Σ_R have been described extensively in the numerical analysis literature e.g. [9]. It suffices to mention here that numerically stable algorithms exist for computing the SVD and that the use of the SVD greatly alleviates the problem of rank determination, since only orthogonal matrices are used to transform a given matrix whose rank is required, to a diagonal matrix. On a digital computer, the computation of the SVD is, of course, subject to rounding errors, so that the resulting decomposition is exact for some perturbed matrix $\hat{A} = A + \Delta$. The computed singular values will also be perturbed, in general, and therefore different from those of A. The problem then is to choose a "zero threshold" such that all computed singular values less than or equal to this threshold value can be effectively considered to be "zero". This gives us the numerical rank of A. Note that in some cases the choice of a zero threshold value may be critical. For more details concerning the use of the SVD in the problem of rank determination and other applications the reader is referred to [10].

2.2 Row and Column Compressions

Let Z be an m×n matrix having rank r. The SVD can be used to compress the columns (rows) of Z to yield a matrix $[Z_1 \; 0]$ ($\begin{bmatrix} \tilde{Z}_1 \\ 0 \end{bmatrix}$) where $Z_1(\tilde{Z}_1)$ is an m×r (r×n) matrix of full rank r. For example, the column compression is achieved by postmultiplication of Z by V' where V' is as defined in (2.1). Then by SVD we have

$$ZV' = U' \begin{bmatrix} \Sigma_r & 0 \\ 0 & 0 \end{bmatrix} = [Z_1 \; 0] \tag{2.2}$$

where $Z_1 \triangleq U' \begin{bmatrix} \Sigma_r \\ 0 \end{bmatrix}$, and has rank r. For row compression, we have

$$UZ = \begin{bmatrix} \Sigma_r & 0 \\ 0 & 0 \end{bmatrix} V = \begin{bmatrix} \tilde{Z}_1 \\ 0 \end{bmatrix} \qquad (2.3)$$

where $\tilde{Z}_1 \triangleq [\Sigma_r \;\; 0]V$, and has rank r.

2.3 Orthogonal Triangularization

Let $Z \in \mathbb{R}^{m\times n}$ with $m \geq n$. Then there exists an $m\times m$ orthogonal matrix Q such that

$$QZ = \begin{bmatrix} R \\ 0 \end{bmatrix} \qquad (2.4)$$

where $R \in \mathbb{R}^{n\times n}$ is an upper triangular matrix. This transformation can be carried out using Householder transformations, e.g. see [9].

2.4 Orthogonal Transformations, Backward Error and Backward Stability [9,11]

Backward error analysis is simplified considerably when orthogonal transformations are used. The reason for this is that the 2-norm (spectral norm) and Euclidean norm (which are the ones most commonly used in such analysis) are invariant under orthogonal transformations. In other words, if Z is any matrix and T is orthogonal, then

$$\|ZT\|_2 = \|Z\|_2 = \|TZ\|_2$$

and

$$\|ZT\|_E = \|Z\|_E = \|TZ\|_E$$

As an illustration of a backward error analysis, consider a similarity transformation carried out by an orthogonal matrix T on $A \in \mathbb{R}^{n\times n}$

$$F = TAT'$$

Now suppose that as a result of rounding errors there is an error Δ_F in the computed value F^* of F, i.e.

$$F^* = TAT' + \Delta_F$$

Let $\Delta_A = T'\Delta_F T$, so that $\|\Delta_A\|_2 = \|\Delta_F\|_2$ and $\|\Delta_A\|_E = \|\Delta_F\|_E$, and

$$F^* = T(A + \Delta_A)T'$$

We have therefore shown that a perturbation of Δ_F in the result (matrix F) can be accounted for by a perturbation Δ_A of the same 'magnitude' (as measured by the spectral or Euclidean norms) in the data (matrix A). The perturbation Δ_A is therefore called the 'backward error'. The algorithm which implements the transformation is called 'backward stable' if the perturbation Δ_A is bounded as follows

$$\|\Delta_A\|_2 \leq \phi_A \eta \, \|A\|_2$$

where η is the machine precision and ϕ_A is a function of n.

3. REDUCTION TO A BLOCK HESSENBERG FORM

We consider an nth order, m-input, linear, time-invariant system with the state equation

$$\dot{\underline{x}}(t) = A\underline{x}(t) + B\underline{u}(t) \tag{3.1}$$

where $\underline{x}(t) \in \mathbb{R}^n$ and $\underline{u}(t) \in \mathbb{R}^m$. Since in this paper we are only concerned with the design of state feedback we shall not show the output equation of the system. In order to develop our e.v.a. algorithm, we shall need to reduce the above system description to a "condensed" form called the (upper) block Hessenberg form (BHF) [12,13].

Theorem 1

There exists an n×n orthogonal matrix T such that

$$F = TAT' = \begin{bmatrix} F_{11} & F_{12} & F_{13} & \cdots & F_{1k} & F_{1,k+1} \\ F_{21} & F_{22} & F_{23} & \cdots & F_{2k} & F_{2,k+1} \\ 0 & F_{32} & F_{33} & \cdots & F_{3k} & F_{3,k+1} \\ \vdots & \vdots & \vdots & & \vdots & \vdots \\ 0 & 0 & 0 & \cdots & F_{kk} & F_{k,k+1} \\ 0 & 0 & 0 & \cdots & F_{k+1,k} & F_{k+1,k+1} \end{bmatrix} \tag{3.2a}$$

$$G = TB = [G_1' \;\; 0 \;\; 0 \;\; \cdots \;\; 0 \;\; 0]' \tag{3.2b}$$

where F_{ij} are $m_{i-1} \times m_{j-1}$ matrices and G_1 is an $m_o \times m$ matrix.
The integers $m_o, m_1, \ldots, m_k$ are defined by

$$m_o = \text{rank } (B) = \text{rank } (G)$$

$$m_i = \text{rank } (F_{i+1,i}), \; i = 1, \ldots, k \tag{3.3}$$

and k is the smallest integer such that $m_k = 0$ or $m_o + m_1 + \cdots + m_k = n$. Furthermore, the matrices $F_{i+1,i}$, $i = 1, \ldots, k$ and G_1 have the form

$$F_{i+1,i} = [\underbrace{R_{m_i} : 0}_{m_i - 1}], \; i = 1, \ldots, k \tag{3.4}$$

and

$$G_1 = [R_{m_o} \vdots 0] Q_o \tag{3.5}$$

where R_{m_i}, $i = 0, 1, \ldots, k$ are $m_i \times m_i$ lower triangular matrices (of rank m_i) and Q_o is an m×m orthogonal matrix.

Proof of Theorem 1

We shall prove the theorem by giving an algorithm for constructing the orthogonal matrix T and transforming the pair (A,B) to the pair (F,G) in the required form.

Algorithm 1

Step 1 Set $i = 0$ and integers $\mu_1 = \mu_2 = \cdots = \mu_m = 0$

Step 2 (i) Perform row compressions on B by means of an orthogonal matrix U_o. Let rank (B) = m_o (determined during row compression)

(ii) Let $F_o = U_o A U_o'$, $G = U_o B$ and set $U = U_o$.

(iii) Set $\mu_j = \mu_j + 1$, $j = 1,2,\ldots, m_o$; and $\mu = \sum_j \mu_j$ and partition F_o and G_o as

$$F_o = \left[\begin{array}{c|c} F_{11}^{(0)} & F_{12}^{(0)} \\ \hline F_{21}^{(0)} & F_{22}^{(0)} \end{array}\right] \begin{array}{l} m_o \\ n-m_o \end{array}, \qquad G_o = \left[\begin{array}{c} G_1^{(0)} \\ \hline 0 \end{array}\right] \begin{array}{l} m_o \\ n-m_o \end{array}$$

$$\qquad\quad m_o \qquad n-m_o \qquad\qquad\qquad\qquad\qquad m$$

Note: rank $(G_1^{(0)}) = m_o$

Step 3 (i) Set $i = i+1$ and $B_i = F_{i+1,i}^{(i-1)}$. Perform row compression on B_i by means of an orthogonal matrix U_i. Let rank $(B_i) = m_i$ (determined during row compression).

(ii) Let

$$F_i = \begin{bmatrix} I_\mu & 0 \\ 0 & U_i \end{bmatrix} F_{i-1} \begin{bmatrix} I_\mu & 0 \\ 0 & U_i' \end{bmatrix},$$

$$G_i = \begin{bmatrix} I_\mu & 0 \\ 0 & U_i \end{bmatrix} G_{i-1} \quad (= G_{i-1})$$

and set
$$U = \begin{bmatrix} I_\mu & 0 \\ 0 & U_i \end{bmatrix} U$$

(iii) Set $\mu_j = \mu_j + 1$, $j = 1,2,\ldots, m_i$; $\mu = \sum_j \mu_j$ and partition F_i and G_i as follows:

$$F_i = \begin{bmatrix} F_{11}^{(i)} & F_{12}^{(i)} & F_{13}^{(i)} & \cdots & F_{1,i+1}^{(i)} & F_{1,i+2}^{(i)} \\ F_{21}^{(i)} & F_{22}^{(i)} & F_{23}^{(i)} & \cdots & F_{2,i+1}^{(i)} & F_{2,i+2}^{(i)} \\ 0 & F_{32}^{(i)} & F_{33}^{(i)} & \cdots & F_{3,i+1}^{(i)} & F_{3,i+2}^{(i)} \\ \vdots & \vdots & \vdots & & \vdots & \vdots \\ 0 & 0 & 0 & \cdots & F_{i+1,i+1}^{(i)} & F_{i+1,i+2}^{(i)} \\ 0 & 0 & 0 & \cdots & F_{i+2,i+1}^{(i)} & F_{i+2,i+2}^{(i)} \end{bmatrix} \begin{matrix} m_o \\ m_1 \\ m_2 \\ \\ m_i \\ n-\mu \end{matrix}$$

$$\begin{matrix} m_o & m_1 & m_2 & & m_i & n-\mu \end{matrix}$$

$$G_i = \begin{bmatrix} G_1^{(i)} \\ 0 \\ 0 \\ \vdots \\ 0 \\ 0 \end{bmatrix} \begin{matrix} m_o \\ m_1 \\ m_2 \\ \\ m_i \\ n-\mu \end{matrix}$$

Note: $G_1^{(i)} = G_1^{(i-1)} = G_1^{(0)}$

Step 4 Repeat Step 3 until $m_i = 0$ i.e. $B_i = 0$ or $\mu = n$; set $k = i$.

Step 5 (i) Let $V = \text{block diag.} \{Q_i, \ i = 1,\ldots, k+1\}$
where $Q_{k+1} = I_{m_k}$
and Q_i, $i = k, k-1,\ldots, 1$ are $m_{i-1}\times m_{i-1}$ orthogonal matrices such that

$$Q_{i+1} F_{i+1,i} Q_i' = [R_{m_i} : 0]$$

where R_{m_i} is an $m_i\times m_i$ lower triangular matrix.

(ii) Let Q_o be an $m\times m$ orthogonal matrix such that

$$Q_1 \ G_1^{(0)} \ Q_o' = [R_{m_o} : 0]$$

where R_{m_o} is an $m_o\times m_o$ lower triangular matrix.

Step 6 (i) Let $F = VF_kV'$ and $G = VG_k$. Then F,G are in the required form.

(ii) The transformation matrix T is given by

$$T = VU$$

<u>Note:</u> T is an orthogonal matrix since U and V are orthogonal matrices. This completes the proof of the theorem.

<u>Remarks</u>

(i) The reduction to upper (or lower) block Hessenberg form (with some slight variations) has been obtained in recent years in connection with different applications by several researchers [12-15]. Our interest in this form stems from the fact that the reduction can be achieved using only orthogonal transformations. This has certain important advantages from the point of the stability of numerical computations and performing a backward error analysis. The particular structure of the upper BHF that we have chosen (as specified by (3.4) and (3.5)) will prove useful in the e.v.a. problem that we consider in the next section.

(ii) The integers μ_j, $j = 1,\dots, m$ are the controllability indices of system (3.1). They were obtained from the integers m_i, $i = 0,1,\dots,k$ by setting

$$\mu_j = 0,\ j = 1,\dots, m$$

and $$\mu_j = \mu_j + 1,\ j = 1,\dots, m_i,\ i = 0,\dots, k \tag{3.6}$$

Also note that $\mu_1 \geqslant \mu_2 \geqslant \cdots \geqslant \mu_m \geqslant 0$

and

$$\mu = \sum_{j=1}^{m} \mu_j = \sum_{i=0}^{k} m_i \tag{3.7}$$

Similarly, the integers m_i, $i = 0,\dots, k$ can be calculated from μ_j, $j = 1,\dots, m$ by setting

$$m_i = 0,\quad i = 0,\dots, k$$

and

$$m_i = m_i + 1,\quad i = 0,\dots, \mu_j - 1,\ \mu_j = 1,\dots, m \tag{3.8}$$

Therefore, it follows that $k \geqslant \mu_1 - 1$ where μ_1 is the largest controllability index of system (3.1). If $k > \mu_1 - 1$ then $m_{\mu_1} = m_{\mu_1+1} = \cdots = m_k = 0$. i.e. $k = \mu_1 - 1$ is the largest value of k such that $m_k \neq 0$. Also, note that

$$m_o \geqslant m_1 \geqslant \cdots \geqslant m_k \geqslant 0.$$

(iii) From Theorem 1 we note that k is the smallest integer such that $m_k = 0$ or $m_o + m_1 + \cdots + m_k (= \mu) = n$. If $\mu = n$, then by Remark (ii) above, system (3.1) is controllable. If on the other hand, $m_k = 0$ (but $\mu \neq n$), then $F_{k+1,k} = 0$ and from the structure of F and G in (3.2a,b), we see that the eigenvalues of $F_{k+1,k+1}$ denote the uncontrollable modes of system (3.1).

It is also easy to see (from Remark (ii)) that these are the only uncontrollable modes in the system. Therefore, if the given system is uncontrollable, then by applying Algorithm 1, we can separate the controllable part of the system from the uncontrollable part and at the same time obtain an upper BHF of the controllable part. In this case, we can set k = k-1 and consider only the controllable part of the system. Therefore, we shall assume henceforth (without loss of generality) that we have a controllable representation in upper BHF given by (3.2).

4. STATE FEEDBACK EIGENVALUE ASSIGNMENT ALGORITHM

In this section we consider a controllable system described by equations of the form (3.1). The problem of e.v.a. by state feedback is to find a feedback law.

$$\underline{u}(t) = \underline{v}(t) - K_x\underline{x}(t) \tag{4.1}$$

where K_x is an m×n matrix, such that the resulting closed-loop system state matrix has a desired set of eigenvalues. It is well-known [16] that when the given system is controllable, all the eigenvalues of its state matrix can be positioned arbitrarily (with complex poles occurring in conjugate pairs) by means of state feedback.

We consider the system (3.1) with (A,B) a controllable pair and assume that it has been reduced to the upper BHF (F,G) using Algorithm 1.

Denoting the state vector of the upper BHF (F,G) by $\underline{z}(t)$, we have

$$\underline{z}(t) = T\,\underline{x}(t) \tag{4.2}$$

where T is defined in Algorithm 1. Therefore the state feedback relation corresponding to (4.1) for the system (F,G) is

$$\underline{u}(t) = \underline{v}(t) - K_z\underline{z}(t) \tag{4.3a}$$

where

$$K_z = K_x T' \tag{4.3b}$$

We shall now derive an algorithm for determining K_z so as to obtain a desired set of eigenvalues for the closed-loop system state matrix that results from implementing state feedback (4.3a) on the system (F,G). It is easy to see that the closed-loop system state matrix resulting from applying (4.1) on the system (3.1) will also have this set of eigenvalues.

Implementing (4.3a) on the system (F,G) yields the closed-loop state matrix $F-GK_z$. Next we let

$$K_z = Q_o' K_z \tag{4.4}$$

where Q_o is the m×m orthogonal matrix appearing in (3.5). Then the product $G\,K_z$

becomes

$$G K_z = \begin{bmatrix} G_1 \\ 0 \\ \vdots \\ 0 \end{bmatrix} Q_o' \hat{K}_z = \begin{bmatrix} R_{m_o} & 0 \\ 0 & 0 \\ \vdots & \vdots \\ 0 & 0 \end{bmatrix} Q_o Q_o' \hat{K}_z$$

$$= \begin{bmatrix} R_{m_o} & 0 \\ 0 & 0 \\ \vdots & \vdots \\ 0 & 0 \end{bmatrix} \hat{K}_z \triangleq \hat{G} \hat{K}_z \tag{4.5}$$

Also, the required state feedback will be determined in two stages, i.e.

$$\hat{K}_z = \hat{K}_1 + \hat{K}_2 \tag{4.6}$$

Furthermore, let the rows of $\hat{G}$ be partitioned conformably with those of F, i.e.

$$G = \begin{bmatrix} \hat{\underline{g}}_{11} & \hat{\underline{g}}_{12} & \cdots & \hat{\underline{g}}_{1m} \\ \hat{\underline{g}}_{21} & \hat{\underline{g}}_{22} & \cdots & \hat{\underline{g}}_{2m} \\ \vdots & \vdots & & \vdots \\ \hat{\underline{g}}_{k+1,1} & \hat{\underline{g}}_{k+1,2} & \cdots & \hat{\underline{g}}_{k+1,m} \end{bmatrix}$$

Note that $\hat{\underline{g}}_{ij} = \underline{0}$ for all $j > m_o$ and all $i \geq 2$. Also $\hat{\underline{g}}_{1j}$, $j = 1,\ldots, m_o$ are the columns of R_{m_o}. We shall find the above notation convenient in the development that follows.

Next, we perform a partial reduction on F using state feedback and orthogonal state coordinate transformations. Our aim is to reduce each of the non-zero submatrices F_{ij} of F to lower triangular/trapezoidal form. As we shall see later this will enable us to reduce the multi-input e.v.a. problem to a number of lower order single-input e.v.a. problems. The partial reduction is achieved by the following algorithm:

Algorithm 2 (Partial Reduction of F)

Set $i = 1$, $\Phi^{(1)} = F$, $\Psi^{(1)} = \hat{G}$, $\Phi^{(1)}_{pq} = F_{pq}$ and $\psi^{(1)}_{\underline{pr}} = \hat{\underline{g}}_{pr}$, $p = 1\ldots, k+1$, $q = 1,\ldots, k+1$, $r = 1,\ldots, m$; $R = I_n$ and an m×n matrix $\hat{K}_1 = 0$.

Step I

Implement a state feedback matrix $\theta^{(i)}$ to get $\Phi^{(i)}_c \triangleq \Phi^{(i)} - \Psi^{(i)}\theta^{(i)}$ where the submatrices $\Phi^{(i)}_{c_{1q}}$, $q = 1,\ldots, k+1$ are lower triangular/trapezoidal, and the lower triangular/trapezoidal structure achieved for any of the remaining submatrices is not altered.

<u>Comment</u>: This can always be done since the first m_o rows of $\Psi^{(i)}$ are linearly independent (consisting of the lower triangular matrix R_{m_o}). The problem of finding $\theta^{(i)}$ consists of solving the following equation for $\theta^{(i)}$:

$$[\bar{R}_{m_o} : 0]\ \theta^{(i)} = [\bar{\Phi}_{11}^{(i)}\ \bar{\Phi}_{12}^{(i)} \cdots \bar{\Phi}_{1,k+1}^{(i)}] \tag{4.7}$$

where $\bar{R}_{m_o}$ and $\bar{\Phi}_{1,q}^{(i)}$, $q = 1,\ldots, k+1$ denote the matrices obtained respectively from the first m_o-i rows of R_{m_o} and the upper (above diagonal) triangular/ trapezoidal part of the first m_o-i rows of $\Phi_{1,q}^{(i)}$, $q = 1,\ldots, k+1$. It can be shown that the resulting submatrices $\Phi_{c_{1,q}}^{(i)}$, $q = 1,\ldots, k+1$ are lower triangular/trapezoidal and that the implementation $\theta^{(i)}$ does not spoil the lower triangular structure of any of the remaining submatrices of $\Phi^{(i)}$.

<u>Step II</u>

1. Let α be the largest integer such that

$$m_\alpha = m_{\alpha-1} = \cdots = m_i$$

2. For $j = 1,\ldots, m_i-1$, reduce to zero the $(j,m_i)^{th}$ elements of

$$\Phi_{c_{pq}}^{(i)},\ q = \alpha+1, \alpha,\ldots, \rho,\ p = \alpha+1, \alpha,\ldots, 2$$

by means of plane rotations [9] (on columns of $\Phi_c^{(i)}$) with the (non-zero) $(j,j)^{th}$ elements of the lower triangular matrices $\Phi_{c_{p,p-1}}^{(i)}$.

3. Let an orthogonal matrix R_i' denote the plane rotations carried out in Step II.2. Also, let

$$\hat{K}_1 = \hat{K}_1 + \theta^{(i)} R \tag{4.8}$$

$$\Phi^{(i+1)} = R_i \Phi_c R_i' \tag{4.9}$$

$$\Psi^{(i+1)} = R_i \Psi^{(i)} \tag{4.10}$$

$$R = R_i R. \tag{4.11}$$

4. If $i < k$, set $i = i+1$ and go to Step I; if $i = k$:

   (i) implement a state feedback matrix $\theta^{(k+1)}$ to get $\Phi_c^{(k+1)} \triangleq \Phi^{(k+1)} - \Psi^{(k+1)} \theta^{(k+1)}$ where the submatrices $\Phi_{c_{1q}}^{(k+1)}$, $q = 1,\ldots, k+1$ are lower triangular/trapezoidal and the submatrices $\Phi_{c_{pq}}^{(k+1)}$, $p = 2,\ldots, k+1$, $q = 1,\ldots,$ $k+1$ remain lower triangular/trapezoidal. Then set

$$\hat{K}_1 = \hat{K}_1 + \theta^{(k+1)} \tag{4.12}$$

(ii) reduce to zero the $(j,j)^{th}$ entries of $\Phi_{c_{pq}}^{(k+1)}$, $p = 3,\ldots, k+1$, $q = 1,\ldots, k-1$, and $j = 1,\ldots, m_{p-1}$ by means of plane rotations with the $(j,j)^{th}$ entries of $\Phi_{c_{q+1,q}}^{(k+1)}$.

5. Let an orthogonal matrix S denote the plane rotations carried out in Step II.4.(ii). Then set

$$\tilde{F} = S\,\Phi_c^{(k+1)} S' \tag{4.13}$$

$$\tilde{G} = S\,\Psi^{(k+1)} \tag{4.14}$$

Comment: In this step we use plane rotations (equivalent to orthogonal similarity transformations) and state feedback to reduce to zero some of the elements of $\Phi_c^{(i)}$. Our aim is to make the submatrices $\Phi_{c_{pq}}^{(i)}$, $p = 1,\ldots, k+1$, $q = p,\ldots, k+1$, lower triangular/trapezoidal. The order in which the elements of these matrices are reduced to zero is important. Also, it should be noted that the state matrix $\hat{K}_1$ calculated in Steps II.3 and II.4(i) corresponds to state feedback in the co-ordinate system defined by (4.1) i.e. for the upper BHF (F,G). At the end of Step II.4(i), all the submatrices of $\Phi_c^{(k+1)}$ are lower triangular/trapezoidal. The effect of carrying out Step II.4(ii) is to make the submatrices $\Phi_{c_{pq}}^{(k+1)}$, $p = 3,\ldots, k+1$, $q = 1,\ldots, k-1$ strictly lower triangular. As we shall see later, this structure will result, after row/column permutations, in the submatrices along the diagonal having upper Hessenberg form.

Example

In order to illustrate the operation of Algorithm 2, we carry out the partial reduction on the following matrices.

$$F = \left[\begin{array}{cccc|ccc|cc}
x & x & x & x & x & x & x & x & x \\
x & x & x & x & x & x & x & x & x \\
x & x & x & x & x & x & x & x & x \\
x & x & x & x & x & x & x & x & x \\ \hline
x & 0 & 0 & 0 & x & x & x & x & x \\
x & x & 0 & 0 & x & x & x & x & x \\
x & x & x & 0 & x & x & x & x & x \\ \hline
0 & 0 & 0 & 0 & x & 0 & 0 & x & x \\
0 & 0 & 0 & 0 & x & x & 0 & x & x
\end{array}\right], \quad G = \left[\begin{array}{cccc}
x & 0 & 0 & 0 \\
x & x & 0 & 0 \\
x & x & x & 0 \\
x & x & x & x \\ \hline
0 & 0 & 0 & 0 \\
0 & 0 & 0 & 0 \\
0 & 0 & 0 & 0 \\ \hline
0 & 0 & 0 & 0 \\
0 & 0 & 0 & 0
\end{array}\right]$$

$m = m_o = 4,\ m_1 = 3,\ m_2 = 2,\ n = 9,\ k = 2.$

Our aim is to make all the submatrices lower triangular/trapezoidal.

After one pass through the algorithm (with $i = 1$), we get

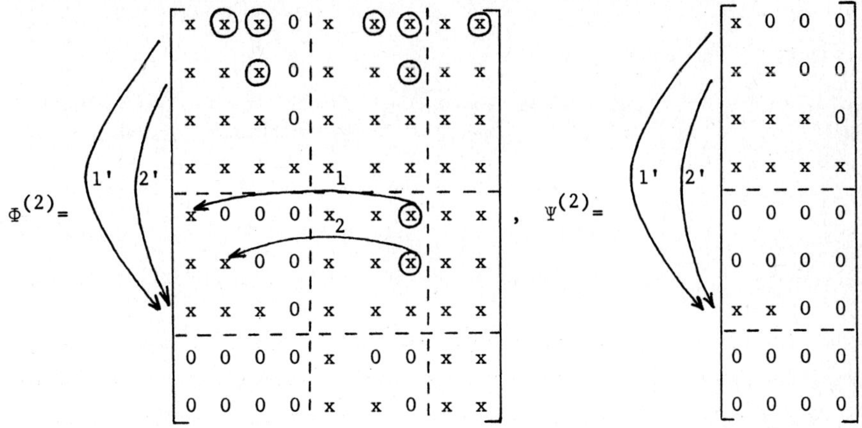

The state feedback matrix $\Theta^{(2)}$ required to make the submatrices $\Phi_{11}^{(2)}, \Phi_{12}^{(2)}$ and $\Phi_{13}^{(2)}$ lower triangular/trapezoidal (i.e. to reduce to zero the entries which have been circled in these matrices), has the following structure:

$$\Theta^{(2)} = \left[\begin{array}{cccc|ccc|cc} 0 & x & x & 0 & 0 & x & x & 0 & x \\ 0 & 0 & x & 0 & 0 & 0 & x & 0 & 0 \\ 0 & 0 & 0 & 0 & 0 & 0 & 0 & 0 & 0 \\ 0 & 0 & 0 & 0 & 0 & 0 & 0 & 0 & 0 \end{array}\right]$$

Therefore, the matrix $\Phi_c^{(2)} = \Phi^{(2)} - \Psi^{(2)}\Theta^{(2)}$ is of the form

$$\Phi_c^{(2)} = \left[\begin{array}{cccc|ccc|cc} x & 0 & 0 & 0 & x & 0 & 0 & x & 0 \\ x & x & 0 & 0 & x & x & 0 & x & x \\ x & x & x & 0 & x & x & x & x & x \\ x & x & x & x & x & x & x & x & x \\ \hline x & 0 & 0 & 0 & x & x & 0 & x & x \\ x & x & 0 & 0 & x & x & 0 & x & x \\ x & x & x & 0 & x & x & x & x & x \\ \hline 0 & 0 & 0 & 0 & x & 0 & 0 & x & x \\ 0 & 0 & 0 & 0 & x & x & 0 & x & x \end{array}\right]$$

At the end of the second pass we get the structure:

$$\Phi^{(k+1)} = \left[\begin{array}{cccc|ccc|cc} x & \textcircled{x} & 0 & 0 & x & \textcircled{x} & 0 & x & \textcircled{x} \\ x & x & 0 & 0 & x & x & 0 & x & x \\ x & x & x & 0 & x & x & x & x & x \\ x & x & x & x & x & x & x & x & x \\ \hline x & 0 & 0 & 0 & x & \textcircled{x} & 0 & x & \textcircled{x} \\ x & x & 0 & 0 & x & x & 0 & x & x \\ x & x & x & 0 & x & x & x & x & x \\ \hline 0 & 0 & 0 & 0 & x & 0 & 0 & x & \textcircled{x} \\ x & 0 & 0 & 0 & x & x & 0 & x & x \end{array}\right], \quad \Psi^{(k+1)} = \left[\begin{array}{cccc} x & 0 & 0 & 0 \\ x & x & 0 & 0 \\ x & x & x & 0 \\ x & x & x & x \\ \hline 0 & 0 & 0 & 0 \\ x & 0 & 0 & 0 \\ x & x & 0 & 0 \\ \hline 0 & 0 & 0 & 0 \\ x & 0 & 0 & 0 \end{array}\right]$$

The state feedback matrix $\theta^{(k+1)}$ is determined so as to reduce to zero the terms which have been circled in $\Phi^{(k+1)}$ above. The matrix $\theta^{(k+1)}$ will have the following structure:

$$\theta^{(k+1)} = \left[\begin{array}{cccc|ccc|cc} 0 & x & 0 & 0 & 0 & x & 0 & 0 & x \\ 0 & 0 & 0 & 0 & 0 & 0 & 0 & 0 & 0 \\ 0 & 0 & 0 & 0 & 0 & 0 & 0 & 0 & 0 \\ 0 & 0 & 0 & 0 & 0 & 0 & 0 & 0 & 0 \end{array}\right]$$

The structure of the resulting matrix $\Phi_c^{(k+1)}$ is

$$\Phi_c^{(k+1)} = \left[\begin{array}{cccc|ccc|cc} x & 0 & 0 & 0 & x & 0 & 0 & x & 0 \\ x & x & 0 & 0 & x & x & 0 & x & x \\ x & x & x & 0 & x & x & x & x & x \\ x & x & x & x & x & x & x & x & x \\ \hline x & 0 & 0 & 0 & x & 0 & 0 & x & 0 \\ x & x & 0 & 0 & x & x & 0 & x & x \\ x & x & x & 0 & x & x & x & x & x \\ \hline 0 & 0 & 0 & 0 & x & 0 & 0 & x & 0 \\ x & \textcircled{x} & 0 & 0 & x & x & 0 & x & x \end{array}\right] \quad \Psi^{(k+1)} = \left[\begin{array}{cccc} x & 0 & 0 & 0 \\ x & x & 0 & 0 \\ x & x & x & 0 \\ x & x & x & x \\ \hline 0 & 0 & 0 & 0 \\ x & 0 & 0 & 0 \\ x & x & 0 & 0 \\ \hline 0 & 0 & 0 & 0 \\ x & 0 & 0 & 0 \end{array}\right]$$

The transformation matrix S is determined to make the submatrix $\Phi_{c_{31}}^{(k+1)}$ strictly lower triangular (by performing the indicated plane rotation). This results in the system $(\tilde{F},\tilde{G})$ given by

$$
\tilde{F} = \left[\begin{array}{cccc|ccc|cc}
x & 0 & 0 & 0 & x & 0 & 0 & x & 0 \\
x & x & 0 & 0 & x & x & 0 & x & x \\
x & x & x & 0 & x & x & x & x & x \\
x & x & x & x & x & x & x & x & x \\
\hline
x & 0 & 0 & 0 & x & 0 & 0 & x & 0 \\
x & x & 0 & 0 & x & x & 0 & x & x \\
x & x & x & 0 & x & x & x & x & x \\
\hline
0 & 0 & 0 & 0 & x & 0 & 0 & x & 0 \\
x & 0 & 0 & 0 & x & x & 0 & x & x
\end{array}\right], \quad
\tilde{G} = \left[\begin{array}{cccc}
x & 0 & 0 & 0 \\
x & x & 0 & 0 \\
x & x & x & 0 \\
x & x & x & x \\
\hline
0 & 0 & 0 & 0 \\
x & 0 & 0 & 0 \\
x & x & 0 & 0 \\
\hline
0 & 0 & 0 & 0 \\
x & 0 & 0 & 0
\end{array}\right]
$$

Next, we carry out a permutation of the rows and columns of $\tilde{F}$ and rows of $\tilde{G}$ so as to obtain a system (F^*, G^*) in which the problem of e.v.a. is reduced to m_o e.v.a. problems for single-input systems of order μ_i, $i = 1, \ldots, m_o$.

Algorithm 3 (Permutation of rows and columns of $\tilde{F}$ and rows of $\tilde{G}$)

Step I

1. Let $\beta_j = m_o + m_1 + \ldots + m_j$, $j = 0, \ldots, k-1$.
2. Let α_ℓ, $\ell = 1, \ldots, \mu$ be given by

$$
\begin{aligned}
\{\alpha_\ell,\ \ell = 1, \ldots, \mu\} = \{1,\ & \beta_o+1,\ \beta_1+1, \ldots,\ \beta_{k-1}+1 \\
2,\ & \beta_o+2,\ \beta_1+2, \ldots,\ \beta_{k-1}+2; \\
& \vdots \\
m_o, & \beta_o+m_o, \beta_1+m_o, \ldots,\ \beta_{k-1}+m_o \,\big|\, \beta_i+j \neq \beta_p+q,\ i \neq p,\ q \neq j \\
& \text{and } \beta_i+j \leq \mu\}
\end{aligned}
\tag{4.15}
$$

Step II

1. Set $P_{i,j} = 0$, $i = 1, \ldots, \mu; j = 1, \ldots, \mu$
2. Set $P_{\ell,\alpha_\ell} = 1$, $\ell = 1, \ldots, \mu$
3. Let $F^* = P\tilde{F}P'$ (4.16)

   and $G^* = P\tilde{G}$ (4.17)

Comments: It can be easily shown that there are altogether $m_o+m_1+\cdots+m_k (=\mu)$ elements in the set $\{\alpha_\ell\}$. This set of integers denotes the required reordering of the rows and columns of $\tilde{F}$ and rows of $\tilde{G}$. The reordering is equivalent to performing state coordinate transformations using the permutation (orthogonal) matrix P.

i.e. $$\underline{z}^*(t) = P\tilde{\underline{z}}(t) \tag{4.18}$$

From the computational point of view, it is of course much more efficient to implement the transformations in Step II.3 by carrying out the appropriate row/column interchanges than to perform the matrix multiplications. The matrices F* and G* resulting from Algorithm 3 have the following form:

$$F^* = \begin{bmatrix} F_{11}^* & 0 & \cdots & 0 \\ F_{21}^* & F_{22}^* & \cdots & 0 \\ \vdots & & & \\ F_{m_o 1}^* & F_{m_o 2}^* & \cdots & F_{m_o m_o}^* \end{bmatrix}, \quad G^* = \begin{bmatrix} \underline{g}_{11}^* & 0 & \cdots & 0 & 0 \\ \underline{g}_{21}^* & \underline{g}_{22}^* & \cdots & 0 & 0 \\ \vdots & \vdots & & \vdots & \vdots \\ \underline{g}_{m_o 1}^* & \underline{g}_{m_o 2}^* & \cdots & \underline{g}_{m_o m_o}^* & 0 \end{bmatrix} \tag{4.19}$$

where F_{ij}^* are $\mu_i \times \mu_j$ matrices and $\underline{g}_{ij}^*$, are $(\mu_i \times 1)$ vectors. Furthermore, the diagonal submatrices F_{ii}^* are in upper Hessenberg form. The diagonal (vector) entries $\underline{g}_{ii}^*$, $i = 1,\ldots, m_o$ are of the form

$$\underline{g}_{ii}^* = \underbrace{[x \quad 0 \quad 0 \quad \ldots \quad 0]'}_{(\mu_i \times 1)} \tag{4.20}$$

The subdiagonal entries F_{ji}^* and $\underline{g}_{ji}^*$, $i = 1,\ldots, m_o-1$, $j = i+1,\ldots, m_o$ do not have any specific structure and will not play any part in the e.v.a. algorithm.

Example

We shall illustrate the effect of performing the row and column permutations using the example introduced earlier.

The integers m_j are given by:

$$m_o = 4, \; m_1 = 3 \text{ and } m_2 = 2.$$

Therefore the integers β_j, $j = 0,1,\ldots, k-1$, are

$$\beta_o = 4, \; \beta_1 = 7 \quad (k = 2)$$

The rows and columns are arranged in the following order:

1,5,8; 2,6,9; 3,7; 4

The matrices F* and G* are given by:

$$F^* = \left[\begin{array}{ccc|ccc|cc|c} x & x & x & 0 & 0 & 0 & 0 & 0 & 0 \\ x & x & x & 0 & 0 & 0 & 0 & 0 & 0 \\ 0 & x & x & 0 & 0 & 0 & 0 & 0 & 0 \\ \hline x & x & x & x & x & x & 0 & 0 & 0 \\ x & x & x & x & x & x & 0 & 0 & 0 \\ x & x & x & 0 & x & x & 0 & 0 & 0 \\ \hline x & x & x & x & x & x & x & x & 0 \\ x & x & x & x & x & x & x & x & 0 \\ \hline x & x & x & x & x & x & x & x & x \end{array}\right] \qquad G^* = \left[\begin{array}{cccc} x & 0 & 0 & 0 \\ 0 & 0 & 0 & 0 \\ 0 & 0 & 0 & 0 \\ \hline x & x & 0 & 0 \\ x & 0 & 0 & 0 \\ x & 0 & 0 & 0 \\ \hline x & x & x & 0 \\ x & x & 0 & 0 \\ \hline x & x & x & x \end{array}\right]$$

Note that (F*,G*) is in the required form.

Incorporating the coordinate transformations R and S (from partial reductions) and P (from permutation of rows and/or columns), we have the following closed-loop state matrix upto this stage:

$$\begin{aligned} A-BK_x &= T'(F-GK_z)\ T \\ &= T'(F-\hat{G}\hat{K}_z)\ T = T'(F-\hat{G}\hat{K}_1)\ T \\ &= T'R'S'\ \tilde{F}\ S\ R\ T \\ &= T'R'S'P'F^*\ P\ S\ R\ T \end{aligned} \tag{4.21}$$

where $\hat{K}_z$ and $\hat{G}$ are defined in (4.4) and (4.5) respectively. Therefore, the closed-loop eigenvalues at this stage are the eigenvalues of F* and from the block triangular structure of F*, these are the eigenvalues of the diagonal submatrices F^*_{ii}. These eigenvalues will not necessarily be at any of the desired locations. So the next step is to find a state feedback matrix K* in order to position all the eigenvalues of F* at desired locations. This can be achieved by solving m_o single-input e.v.a. problems in order to position the eigenvalues of F^*_{ii} by state feedback to the ith input i.e. by finding a $(1\times\mu_i)$ state feedback vector $\underline{k}^{*'}_i$ such that $F^*_{ii}-\underline{g}^*_{ii}\underline{k}^{*'}_i$ has all its μ_i eigenvalues at a subset of the desired locations. The matrix K* is then given by:

$$K^* = \begin{bmatrix} \underline{k}_1^{*\prime} & \underline{0}' & \cdots & \underline{0}' \\ \underline{0}' & \underline{k}_2^{*\prime} & \cdots & \underline{0}' \\ \vdots & \vdots & & \vdots \\ \underline{0}' & \underline{0}' & \cdots & \underline{k}_{m_o}^{*\prime} \\ \underline{0}' & \underline{0}' & \cdots & \underline{0}' \\ \vdots & \vdots & & \vdots \\ \underline{0}' & \underline{0}' & \cdots & \underline{0}' \end{bmatrix} \quad (4.22)$$

where $\underline{k}_i^* \in \mathbb{R}^{\mu_i}$, $i = 1,\ldots, m_o$. The vectors $\underline{k}_i^*$ can be computed using the single-input e.v.a. algorithm which is described in the next section.

To conclude this section, we relate the state feedback matrix K* used to position the eigenvalues of the system (F*,G*) to the required state feedback matrix K_x for the system (A,B). The matrix K* will yield $\hat{K}_2$ in (4.6):

i.e. $$\hat{K}_2 = K^*PSR \quad (4.23)$$

Therefore, using (4.6) and (4.21) we get

$$\begin{aligned} K_x &= K_z T \\ &= Q_o'\hat{K}_z T \\ &= Q_o'(\hat{K}_1 + \hat{K}_2)\ T \\ &= Q_o'(\hat{K}_1 + K^*PSR)\ T \end{aligned} \quad (4.24)$$

5. EIGENVALUE ASSIGNMENT IN SINGLE-INPUT SYSTEMS

In this section we consider single-input systems described by equations of the form

$$\dot{\underline{x}}(t) = F\underline{x}(t) + \underline{g}u(t) \quad (5.1)$$

We shall assume that the pair $(F,\underline{g})$ is controllable. Furthermore, we shall assume without loss of generality that the system description is already in upper Hessenberg form with F an <u>unreduced</u> upper Hessenberg matrix i.e. $f_{i,i-1} \neq 0$, $i = 2,\ldots, n$. Note that Theorem 1 implies that any controllable pair $(A,\underline{b})$ can be reduced to such a form by means of an orthogonal transformation applied to the state of the system.

The problem that we consider is to find a 1×n constant state feedback vector $\underline{k}'$ such that the unreduced upper Hessenberg matrix

$$\hat{F} = F - \underline{g}\ \underline{k}' \quad (5.2)$$

has all its eigenvalues at desired locations in the complex plane. We shall denote these desired locations by the set $\{\lambda_i,\ i = 1,\ldots, n\}$. Also for simplicity of presentation we shall assume that all the λ_i's are real. Later on in this section, we shall show how the algorithm can be modified to handle complex-conjugate pairs of eigenvalues using only real arithmetic.

Our solution to the problem involves finding an orthogonal matrix N and a state feedback vector $\underline{k}'$ such that $N'\hat{F}N$ is a (block) upper triangular matrix with the desired eigenvalues along the diagonal.

<u>Algorithm A</u> (e.v.a. in single-input systems using explicit shifts)

<u>Step 1</u>: Set $\underline{k}' = \underline{k}_o' = \underline{0}'$, $\underline{g}_1 = \underline{g}$, $F_1 = F$, $N = I_n$ and $i = 1$.

<u>Step 2</u>: Perform plane rotations [9] denoted by $M_{i,j}$, $j = 1,\ldots, n-i$, on the columns of $F_i-\lambda_i I_r$ to reduce the matrix to upper triangular form.

<u>Comment</u>: The structure obtained at the end of this step is,

$$
(F_i-\lambda_i I_n)M_{i,1} \cdots M_{i,n-i}
$$

$$
= \left[\begin{array}{cccc|cccccc}
\lambda_1-\lambda_i & x & \cdots & x & x & x & x & \cdots & x & x \\
0 & \lambda_2-\lambda_i & \cdots & x & x & x & x & \cdots & x & x \\
\vdots & \vdots & & \vdots & \vdots & \vdots & \vdots & & \vdots & \vdots \\
0 & 0 & & \lambda_{i-1}-\lambda_i & x & x & x & \cdots & x & x \\
\hline
0 & 0 & \cdots & 0 & x & x & x & \cdots & x & x \\
0 & 0 & \cdots & 0 & \textcircled{x} \xrightarrow{n-i} & x & x & \cdots & x & x \\
0 & 0 & \cdots & 0 & 0 & \textcircled{x} \xrightarrow{n-i-1} & x & \cdots & x & x \\
\vdots & \vdots & & \vdots & \vdots & \vdots & \vdots & & \vdots & \vdots \\
0 & 0 & \cdots & 0 & 0 & 0 & 0 & \cdots & \textcircled{x} \xrightarrow{1} & x
\end{array}\right] \tag{5.3}
$$

where it has been assumed that eigenvalues $\lambda_1,\ldots, \lambda_{i-1}$ have already been assigned. We can also use Householder transformations [9] instead of plane rotations.

<u>Step 3</u>: Determine feedback $\underline{k}_i'$ such that the $(i,i)^{th}$ element of the matrix on the right-hand side of (5.3) is reduced to zero when the feedback is applied, i.e. the $(i,i)^{th}$ term of $(F_i-\lambda_i I_n)M_{i,1}\cdots M_{i,n-i} - \underline{g}_i\underline{k}_i'$ is equal to zero.

<u>Comment</u>: The vector $\underline{k}_i'$ has the form

$$
\underline{k}_i' = [\underbrace{0\ \ 0\ \cdots\ 0}_{i-1 \text{ terms}} \mid x\ \ 0\ \ 0\ \cdots\ 0\ \ 0] \tag{5.4}
$$

<u>Step 4</u>: Let $\quad N_i = M_{i,1} \cdots M_{i,n-i}$ (5.5)

and complete the state coordinate transformation by performing corresponding row operations on $\underline{g}_i$ and $(F_i-\lambda_i I_n)N_i - \underline{g}_i\underline{k}_i'$ to get

$N_i'\underline{g}_i$ and $N_i'\ [(F_i-\lambda_i I_n)N_i - \underline{g}_i k_i']$

<u>Comment</u>: The vector $N_i'\underline{g}_i$ is of the form

$$N_i'\underline{g}_i = [\underbrace{x\ \ x\ \cdots\ x}_{i\ \text{terms}} \mid x\ \ 0\ \cdots\ 0]' \tag{5.6}$$

and the structure of the matrix $N_i'\ [(F_i-\lambda_i I_n)N_i - \underline{g}_i k_i']$ differs from that of the matrix on the right-hand side of (5.3) only in that the subdiagonal elements $(i+2,i+1),(i+3,i+2),\ldots,\ (n,n-1)$ are no longer zero. This is a consequence of the nature of the transformations $M_{i,n-i-1}', M_{i,n-i-2}', \ldots,\ M_{i,1}'$. Note that the elements (i,i) and $(i+1,i)$ remain equal to zero at the end of this step.

<u>Step 5</u>: Set $\underline{g}_{i+1} = N_i'\underline{g}_i$ (5.7)

$$F_{i+1} = N_i'(F_i-\lambda_i I_n)N_i - \underline{g}_i k_i' + \lambda_i I_n$$

$$= N_i'(F_i-\underline{g}_i\underline{k}_i'N_i')N_i \tag{5.8}$$

$$N = N\,N_i \tag{5.9}$$

and $\underline{k}' = \underline{k}' + \underline{k}_i'N'$ (5.10)

If $i = n$ stop; otherwise set $i = i+1$ and go to Step 2.

<u>Comment</u>: The vector $\underline{g}_{i+1}$ has the structure shown in (5.6). The matrix F_{i+1} has the following structure:

$$F_{i+1} = \left[\begin{array}{ccccc|ccccc}
\lambda_1 & x & \cdots & x & x & x & x & \cdots & x & x \\
0 & \lambda_2 & \cdots & x & x & x & x & \cdots & x & x \\
\vdots & \vdots & & \vdots & \vdots & \vdots & \vdots & & \vdots & \vdots \\
0 & 0 & \cdots & \lambda_{i-1} & x & x & x & \cdots & x & x \\
0 & 0 & \cdots & 0 & \lambda_i & x & x & \cdots & x & x \\
\hline
0 & 0 & \cdots & 0 & 0 & x & x & \cdots & x & x \\
0 & 0 & \cdots & 0 & 0 & x & x & \cdots & x & x \\
0 & 0 & \cdots & 0 & 0 & 0 & x & \cdots & x & x \\
\vdots & \vdots & & \vdots & \vdots & \vdots & \vdots & & \vdots & \vdots \\
0 & 0 & \cdots & 0 & 0 & 0 & 0 & \cdots & x & x
\end{array}\right]
\begin{array}{l} \left.\begin{array}{c}\\ \\ \\ \\ \\ \end{array}\right\} i \\ \left.\begin{array}{c}\\ \\ \\ \\ \\ \end{array}\right\} n-i \end{array} \tag{5.11}$$

from where it is clear that an eigenvalue has been assigned at λ_i. Denoting the last $(n-i)$ rows and columns of F_{i+1} by the matrix $\hat{F}_{i+1}$ and the last $(n-i)$ elements of $\underline{g}_{i+1}$ by the vector $\hat{\underline{g}}_{i+1}$, we see that the pair $(\hat{F}_{i+1},\hat{\underline{g}}_{i+1})$ is in upper Hessenberg form. The procedure described by Steps 2-5 can therefore be used repeatedly to assign the remaining eigenvalues at desired locations.

Remarks

It may be noted that Algorithm A is similar to the QR algorithm with explicit shifts. In fact Algorithm A is somewhat simpler than the QR algorithm since in Algorithm A the shifts are the desired eigenvalues and are known a priori, whereas in the QR algorithm, the shifts converge to the eigenvalues to be computed. The other difference is that we compute feedback $\underline{k}_i'$ (Step 3) in order to make the shift correspond to the desired eigenvalue. An algorithm developed recently by Miminis and Paige [5] for e.v.a. in single-input systems is also based on the QR algorithm. Their algorithm differs from Algorithm A in that they start with the closed-loop matrix $F_1 = F-\underline{g}\,\underline{k}'$ where $(F,\underline{g})$ is in upper Hessenberg form and $\underline{k}'$ is to be determined. In other words, F_1 has an unknown first row. They then carry out a forward sweep which uses shifts and orthogonal transformations to place the desired eigenvalues on the diagonal of an upper (block) triangular matrix (real Schur form) with unknown super diagonal terms. A backward sweep is then performed to determine the unknown elements and eventually the unknown first row of F_1 and the feedback vector $\underline{k}'$.

As in the QR algorithm we can perform implicit shifts instead of the explicit shifts described in Algorithm A. We can also carry out two steps with implicit shifts. This enables us to assign complex-conjugate pairs of eigenvalues using real arithmetic. We shall now describe briefly how implicit shifting can be performed to assign a real eigenvalue or a complex- conjugate pair of eigenvalues. It is assumed that the members of a complex- conjugate pair of eigenvalues are arranged consecutively.

Algorithm 4 (e.v.a. in single-input systems using implicit shifts)

Step I: Set $\underline{k}' = \underline{k}_o' = \underline{0}'$, $\underline{g}_1 = \underline{g}$, $F_1 = F$, $N = I_n$ and $i = 1$.

Step II: If λ_i is complex go to Step III; otherwise:

1. (Real eigenvalues): If $i = n$ go to 6; otherwise determine an orthogonal matrix P_i such that

$$\underline{f}_i' P_i = \pm \|\underline{f}_i\|_2\, \underline{e}_n' \tag{5.12}$$

$$\underline{f}_i' = [0,0,\ldots, f_{n,n-1}, f_{nn}-\lambda_i] \tag{5.13}$$

is the last row of $F_i-\lambda_i I_n$, $\|\underline{f}_i\|_2 = \sqrt{\underline{f}_i'\underline{f}_i}$, and

$$\underline{e}_n' = \underbrace{[0\ 0\ \ldots\ 0\ 1]}_{n \text{ elements}}$$

Comment: A matrix P_i can always be found to accomplish this step [9 p. 232]. The aim is to eliminate $f_{n,n-1}$ using $f_{nn}-\lambda_i$.

2. Set $\bar{F}_i = P_i' F_i P_i$ and $\bar{g}_i = P_i' g_i$.

<u>Comment</u>: Note that the transformations represented by P_i are applied to F_i and not to $F_i - \lambda_i I_n$. This is because the shift has been accounted for in P_i. The matrix $\bar{F}_i$ has the structure

$$\bar{F} = \left[\begin{array}{cccc|ccccc} \lambda_1 & x & \cdots & x & x & x & \cdots & x & x & x \\ 0 & \lambda_2 & \cdots & x & x & x & \cdots & x & x & x \\ \vdots & \vdots & & \vdots & \vdots & \vdots & & \vdots & \vdots & \vdots \\ 0 & 0 & \cdots & \lambda_{i-1} & x & x & \cdots & x & x & x \\ \hline 0 & 0 & \cdots & 0 & x & x & \cdots & x & x & x \\ 0 & 0 & \cdots & 0 & x & x & \cdots & x & x & x \\ 0 & 0 & \cdots & 0 & 0 & x & \cdots & x & x & x \\ \vdots & \vdots & & \vdots & \vdots & \vdots & & \vdots & \vdots & \vdots \\ 0 & 0 & \cdots & 0 & 0 & 0 & \cdots & x & x & x \\ 0 & 0 & \cdots & 0 & 0 & 0 & \cdots & \textcircled{x} & x & x \end{array}\right] \triangleq \left[\begin{array}{c|c} \bar{F}_i^{11} & \bar{F}_i^{12} \\ \hline 0 & \bar{F}_i^{22} \end{array}\right] \qquad (5.14)$$

and is not in upper Hessenberg form since its $(n,n-2)^{th}$ element is non-zero (-consequence of performing row operation represented by P_i'). For simplicity, we have assumed in (5.14) that the eigenvalues $\lambda_1, \ldots, \lambda_{i-1}$ are all real.

3. Reduce $\bar{F}_i$ to upper Hessenberg form H_i by means of plane rotations $U_{i,j}$, $j = 1,2,\ldots, n-i-1$ i.e.

$$U_{i,n-i-1}' \cdots U_{i,1}' \bar{F}_i U_{i,1} \cdots U_{i,n-i-1} = H_i \qquad (5.15)$$

Also let $g_{i+1} = U_{i,n-i-1}' \cdots U_{i,1}' \bar{g}_i$ (5.16)

<u>Comment</u>: Note that the $(i-1)\times(i-1)$ upper triangular matrix $\bar{F}_i^{11}$ is not altered by this transformation and only the submatrix $\bar{F}_i^{22}$ is reduced to upper Hessenberg form. The vector g_{i+1} has the following structure:

$$g_{i+1} = [\underbrace{x\ x\ \ldots\ x}_{i-1} \mid \underbrace{x\ x\ 0\ \ldots\ 0\ 0}_{n-i+1}]' \qquad (5.17)$$

4. Determine a feedback vector $\underline{k}_i'$ such that the $(i+1,i)^{th}$ element of $H_i - \underline{g}_{i+1}\underline{k}_i'$ is zero. The vector $\underline{k}_i'$ will have the following structure

$$\underline{k}_i' = [\underbrace{0 \;\; 0 \; \dots \; 0}_{i-1} \,|\, \underbrace{x \;\; 0 \; \dots \; 0 \;\; 0 \;\; 0}_{n-i+1}] \tag{5.18}$$

where the non-zero element is chosen so as to reduce the $(i+1)^{th}$ element of H_i to zero by applying feedback. Then the $(i,i)^{th}$ element of $H_i - \underline{g}_{i+1}\underline{k}_i'$ will be equal to λ_i.

5. Set $N_i = P_i U_{i,1} \dots U_{i,n-i-1}$ (5.19)

$$N = NN_i \tag{5.20}$$

$$\underline{k}' = \underline{k}' + \underline{k}_i'N' \tag{5.21}$$

$$F_{i+1} = H_i - \underline{g}_{i+1}\underline{k}_i' \tag{5.22}$$

Set i = i+1 and go to Step II.

Comment: The vector $\underline{g}_{i+1}$ required for the next iteration is given by (5.16). Equations (5.19) and (5.20) accumulate all the transformations performed upto that stage, and (5.21) gives the state feedback vector in the coordinate system of (5.1).

6. Determine a feedback $\underline{k}_n'$ such that the $(n,n)^{th}$ element of $F_n - \underline{g}_n\underline{k}_n'$ is equal to λ_n. The vector $\underline{k}_n'$ will have the following structure

$$\underline{k}_n' = [\underbrace{0 \;\; 0 \; \dots \; 0}_{n-1} \;\; x]$$

Comment: The non-zero element in $\underline{k}_n'$ is chosen such that

$$(F_n)_{nn} - (\underline{g}_n)_n(\underline{k}_n')_n = \lambda_n$$

where $(F_n)_{nn}$ denotes the $(n,n)^{th}$ element of F_n, and $(\underline{g}_n)_n$ and $(\underline{k}_n')_n$ denote the nth elements of $\underline{g}_n$ and $\underline{k}_n'$ respectively. Note that $(g_n)_n \neq 0$ since the pair $(F,\underline{g})$ is controllable.

7. Set $F_{n+1} = F_n - \underline{g}_n\underline{k}_n'$

$$\underline{g}_{n+1} = \underline{g}_n$$

$$\underline{k}' = \underline{k}' + \underline{k}_n'$$

and stop.

Step III: (Complex-conjugate pair of eigenvalues):

1. If $i = n-1$ go to 6; otherwise determine an orthogonal matrix P_i such that

$$\hat{\underline{f}}_i' P_i = \pm \|\hat{\underline{f}}_i\|_2 \, \underline{e}_n' \tag{5.23}$$

where $\hat{\underline{f}}_i' = [0,0,\ldots,\hat{f}_{n,n-2},\hat{f}_{n,n-1},\hat{f}_{nn}]$ (5.24)

is the last row of $(F_i-\lambda_i I_n)(F_i-\lambda_i^* I_n)$, λ_i^* being the complex-conjugate of λ_i.

The elements $\hat{f}_{n,n-2},\hat{f}_{n,n-1}$ and $\hat{f}_{nn}$ are given by

$$\hat{f}_{n,n-2} = f_{n-1,n-2}\, f_{n,n-1} \tag{5.25}$$

$$\hat{f}_{n,n-1} = f_{n,n-1}\, [f_{n-1,n-1} + f_{nn} - (\lambda_i+\lambda_i^*)] \tag{5.26}$$

$$\hat{f}_{nn} = f_{nn}^2 + f_{n,n-1}\, f_{n-1,n} - f_{nn}(\lambda_i+\lambda_i^*) + \lambda_i\lambda_i^* \tag{5.27}$$

where f_{ij} denotes the $(i,j)^{th}$ element of F_i.

Comment: Note that $\hat{f}_{n,n-2},\hat{f}_{n,n-1}$ and $\hat{f}_{nn}$ are all real since $\lambda_i+\lambda_i^*$ and $\lambda_i\lambda_i^*$ are real. Therefore P_i is real (orthogonal) and can be determined using real arithmetic.

2. Set $\bar{F}_i = P_i' F_i P_i$ and $\bar{\underline{g}}_i = P_i' \underline{g}_i$

Comment: The matrix $\bar{F}_i$ has the structure shown below where for simplicity we have assumed that $\lambda_1,\ldots,\lambda_{i-1}$ are real eigenvalues. For a complex conjugate pair of eigenvalues we would have a 2×2 block on the diagonal.

$$\bar{F}_i = \left[\begin{array}{cccc|ccccccc}
\lambda_1 & x & \cdots & x & x & x & \cdots & x & x & x & x \\
0 & \lambda_2 & \cdots & x & x & x & \cdots & x & x & x & x \\
\vdots & \vdots & & \vdots & \vdots & \vdots & & \vdots & \vdots & \vdots & \vdots \\
0 & 0 & \cdots & \lambda_{i-1} & x & x & \cdots & x & x & x & x \\
\hline
0 & 0 & \cdots & 0 & x & x & \cdots & x & x & x & x \\
0 & 0 & \cdots & 0 & x & x & \cdots & x & x & x & x \\
0 & 0 & \cdots & 0 & 0 & x & \cdots & x & x & x & x \\
\vdots & \vdots & & \vdots & \vdots & \vdots & & \vdots & \vdots & \vdots & \vdots \\
0 & 0 & \cdots & 0 & 0 & 0 & \cdots & \textcircled{x}\,3 & x & x & x \\
0 & 0 & \cdots & 0 & 0 & 0 & \cdots & \textcircled{x}\,2 & \textcircled{x}\,1 & x & x
\end{array}\right] \triangleq \left[\begin{array}{c|c} \bar{F}_i^{11} & \bar{F}_i^{12} \\ \hline 0 & \bar{F}_i^{22} \end{array}\right] \tag{5.28}$$

The matrix $\bar{F}_i$ is not in upper Hessenberg form since its (n-1, n-3), (n,n-3) and (n,n-2) elements are non-zero.

3. Apply plane rotations $U_{i,j}$, j = 1,2,..., 3(n-i-1) in order to make the matrix $\bar{F}_i^{22}$ as close to upper Hessenberg as possible i.e.

$$U'_{i,3(n-i-1)} \cdots U'_{i,1} \bar{F}_i U_{i,1} \cdots U_{i,3(n-i-1)} = H_i \tag{5.29}$$

Also let $\underline{g}_{i+2} = U'_{i,3(n-i-1)} \cdots U'_{i,1}\bar{\underline{g}}_i$ (5.30)

<u>Comment</u>: It may be noted that $\bar{F}_i^{22}$ cannot be reduced to upper Hessenberg form by the transformations specified in this step. The closest we can come to it to get a matrix with the following structure.

$$\begin{bmatrix} x & x & x & \cdots & x & x & x \\ x & x & x & \cdots & x & x & x \\ * & x & x & \cdots & x & x & x \\ 0 & 0 & x & \cdots & x & x & x \\ \vdots & \vdots & \vdots & & \vdots & \vdots & \vdots \\ 0 & 0 & 0 & \cdots & x & x & x \\ 0 & 0 & 0 & \cdots & 0 & x & x \end{bmatrix}$$

where the element denoted by the * cannot be eliminated by the transformations. The upper triangular matrix $\bar{F}_i^{11}$ is unaltered by the transformations. Note that each $U_{i,j}$ performs one plane rotation so that each $U_{i,j}$ eliminates only one non-zero element. This is why we have a total of 3(n-i-1) plane rotations. The vector $\underline{g}_{i+2}$ has the structure

$$\underline{g}_{i+2} = [\underbrace{x \;\; x \;\ldots\; x}_{i-1} \,\vdots\, \underbrace{x \;\; x \;\; x \;\; 0 \;\ldots\; 0}_{n-i+1}]' \tag{5.31}$$

4. Determine a feedback vector $\underline{k}'_i$ such that the $(i+2,i)^{th}$ and $(i+2,i+1)^{th}$ elements of $H_i - \underline{g}_{i+2}\underline{k}'_i$ are zero. The required vector $\underline{k}'_i$ will have the following structure

$$\underline{k}'_i = [\underbrace{0 \;\; 0 \;\ldots\; 0}_{i-1} \,\vdots\, \underbrace{x \;\; x \;\; 0 \;\ldots\; 0}_{n-i+1}] \tag{5.32}$$

where the two non-zero elements are chosen so as to reduce the $(i+2,i)^{th}$ and $(i+2,i+1)^{th}$ elements of H_i to zero by applying feedback. This will result in a 2×2 matrix in the ith and $(i+1)^{th}$ rows and columns of $H_i - \underline{g}_{i+2}\underline{k}_i'$ with eigenvalues λ_i and λ_i^*.

<u>Comment</u>: The matrix $H_i - \underline{g}_{i+2}\underline{k}_i'$ has the following structure

$$\left[\begin{array}{cccc|cc|cccccc} \lambda_1 & x & \cdots & x & x & x & x & x & \cdots & x & x \\ 0 & \lambda_2 & \cdots & x & x & x & x & x & \cdots & x & x \\ \vdots & \vdots & & \vdots & \vdots & \vdots & \vdots & \vdots & & \vdots & \vdots \\ 0 & 0 & \cdots & \lambda_{i-1} & x & x & x & x & \cdots & x & x \\ \hline 0 & 0 & \cdots & 0 & \alpha & \beta & x & x & \cdots & x & x \\ 0 & 0 & \cdots & 0 & \gamma & \delta & x & x & \cdots & x & x \\ \hline 0 & 0 & \cdots & 0 & 0 & 0 & x & x & \cdots & x & x \\ 0 & 0 & \cdots & 0 & 0 & 0 & x & x & \cdots & x & x \\ 0 & 0 & \cdots & 0 & 0 & 0 & 0 & x & \cdots & x & x \\ \vdots & \vdots & & \vdots & \vdots & \vdots & \vdots & \vdots & & \vdots & \vdots \\ 0 & 0 & \cdots & 0 & 0 & 0 & 0 & 0 & \cdots & x & x \end{array}\right]$$

which implies that we have assigned two eigenvalues at the eigenvalues of the matrix $\begin{bmatrix} \alpha & \beta \\ \gamma & \delta \end{bmatrix}$ i.e. at λ_i and λ_i^*.

5. Set $N_i = P_i U_{i,1} \cdots U_{i,3(n-i-1)}$ (5.33)

$N = NN_i$ (5.34)

$\underline{k}' = \underline{k}' + \underline{k}_i'N'$ (5.35)

$F_{i+2} = H_i - \underline{g}_{i+2}\underline{k}_i'$ (5.36)

Set i = i+2 and go to Step II.

<u>Comment</u>: In this step we increment i by 2 because we have assigned two eigenvalues.

6. Determine a feedback vector $\underline{k}_{n-1}'$ such that the 2×2 matrix in the last two columns and rows of $F_{n-1} - \underline{g}_{n-1}\underline{k}_{n-1}'$ has eigenvalues λ_{n-1}, λ_{n-1}^*. The vector $\underline{k}_{n-1}'$ will have the following structure

$$\underline{k}_{n-1}' = [\underbrace{0 \ \ 0 \ \ldots \ 0}_{n-2} \ \ x \ \ x]$$

<u>Comment</u>: The structure of the 2×2 matrix in the last two columns and rows of $F_{n-1} - \underline{g}_{n-1}\underline{k}_{n-1}'$ is as follows:

$$\begin{bmatrix} x & x \\ x & x \end{bmatrix} - \begin{bmatrix} x \\ 0 \end{bmatrix} [x \quad x] \qquad (5.37)$$

The effect of applying the feedback $\underline{k}'_{n-1}$ is to change the first row of the 2×2 matrix, so that by appropriate choice of the two non-zero elements in $\underline{k}'_{n-1}$ we can ensure that the 2×2 matrix in (5.37) has eigenvalues at λ_{n-1} and λ^*_{n-1}.

7. Set $F_{n+1} = F_{n-1} - \underline{g}_{n-1}\underline{k}'_{n-1}$

$\underline{g}_{n+1} = \underline{g}_{n-1}$

$\underline{k}' = \underline{k}' + \underline{k}'_{n-1}$

and stop.

Remarks

1. Algorithm 4 assigns real and complex-conjugate pairs of eigenvalues using real arithmetic. One element of $\underline{k}'_i$ is determined for assigning each real eigenvalue and two elements for assigning a complex-conjugate pair. The results of the algorithm consist of matrices F_{n+1} and N and vectors $\underline{g}_{n+1}$ and $\underline{k}'$ where F_{n+1} is an upper (block) triangular matrix (real Schur form) with each (2×2) matrix on the diagonal having a desired pair of complex-conjugate eigenvalues and each scalar on the diagonal being a desired real eigenvalue. The matrix N is an orthogonal matrix which transforms the closed-loop upper Hessenberg matrix $(F-\underline{g}\ \underline{k}')$ to its Schur form i.e.

$$N(F-\underline{g}\ \underline{k}')N = F_{n+1} \qquad (5.38)$$

The vector $\underline{k}'$ is the desired state feedback vector in the coordinate system of the upper Hessenberg form $(F,\underline{g})$. Also, it suffices to mention that Algorithm 4 can be justified along similar lines as the implicitly shifted QR algorithm, e.g. see [9].

2. It should be noted that we cannot use implicit shifts to assign the last real eigenvalue or the last complex-conjugate pair of eigenvalues. These eigenvalues have to be assigned separately. For the case of real λ_n, the required feedback can be obtained directly. For the case of a complex-conjugate pair λ_{n-1}, λ^*_{n-1} we can proceed as follows:

For the sake of illustration, we denote the matrix in (5.37) as

$$\begin{bmatrix} \alpha_{11} & \alpha_{12} \\ \alpha_{21} & \alpha_{22} \end{bmatrix} - \begin{bmatrix} \beta_1 \\ 0 \end{bmatrix} [\theta_1 \quad \theta_2] \qquad (5.39)$$

where θ_1 and θ_2 are to be determined so that the resulting matrix has eigenvalues at λ_{n-1} and λ^*_{n-1}. The elements $\alpha_{11}, \alpha_{12}, \alpha_{21}, \alpha_{22}$ and β_1 are known. It can be easily verified that θ_1 and θ_2 are given by

$$\theta_1 = \frac{1}{\beta_1}[\alpha_{11} + \alpha_{22} - (\lambda_{n-1} + \lambda^*_{n-1})] \tag{5.40}$$

and

$$\theta_2 = \frac{1}{\alpha_{21}\beta_1}[\lambda_{n-1}\lambda^*_{n-1} + \alpha_{21}\alpha_{12} + \alpha^2_{22} - \alpha_{22}(\lambda_{n-1} + \lambda^*_{n-1})] \tag{5.41}$$

where β_1 and α_{21} are non-zero if the given single-input system is controllable. Note that θ_1 and θ_2 can be computed using real arithmetic.

6. DISCUSSION OF THE ALGORITHMS

Our algorithms for e.v.a. in multi-input systems perform four functions:

Algorithm 1: Reduction to upper BHF (F,G)

Algorithm 2: Partial reduction of F to get $(\tilde{F},\tilde{G})$

Algorithm 3: Permutation of rows and columns to get (F*,G*)

Algorithm 4: Single-input eigenvalue assignment.

The reduction to upper (and lower) BHF has been considered by several researchers in different contexts. Here we have essentially used the reduction described in [12] with a further transformation applied to obtain the lower triangular structure for the non-zero subdiagonal matrices. This structure is required in order to carry out the partial reduction in Algorithm 2. All the transformations used in Algorithm 1 are performed using orthogonal matrices and the algorithm can therefore be shown to be numerically backward stable.

In Algorithm 2, we use state feedback in conjunction with orthogonal state coordinate transformations to reduce each submatrix of F to lower triangular/trapezoidal form. The state feedback matrices are determined by solving sets of linear equations while the orthogonal transformation matrices result from performing plane rotations. The problem of solving sets of linear equations has been treated extensively in the numerical analysis literature e.g. see [9,11] and it suffices to say that it can be solved in a numerically stable manner. In fact, the solutions to the sets of linear equations can be obtained, in our case, directly by "back substitution" because the coefficient matrices (which are the submatrices of $\Psi^{(i)}$) are all lower triangular. As a consequence of this and the fact that only orthogonal transformations are used in Algorithm 2, it is clear that the algorithm will have the property of being numerically backward stable.

Algorithm 3 simply carries out row and column permutations on $\tilde{F}$ and row permutations on $\tilde{G}$. Therefore, there is no question about the numerical stability of this algorithm. Algorithm 4 is based on the implicitly shifted QR algorithm [9] for computing the eigenvalues of a matrix. However, unlike the QR algorithm, our algorithm is not iterative in nature since the shifts, being the desired closed-loop eigenvalues, are known a priori. Essentially, the only significant difference between an iteration of Algorithm 4 and that of the QR algorithm is the step that computes the feedback vector $\underline{k}_i'$ (Step II.4 or III.4). This step requires one division for each eigenvalue to be assigned. Also, the algorithm requires that the single-input system be in upper Hessenberg form which is precisely the structure that we have for $(F_{ii}^*,\underline{g}_{ii}^*)$, $i = 1,\ldots, m_o$. Another algorithm for single-input e.v.a., which is based on the QR algorithm, is the one proposed by Miminis and Paige [5]. The difference between their approach and ours has been discussed in the remarks following Algorithm A. It suffices to mention here that our approach appears to be conceptually and computationally less involved than theirs. In this respect, it is worth mentioning that a program for Algorithm 4 can be obtained by making a few minor changes to a program for the QR algorithm.

It should be noted that the single-input e.v.a. problems resulting from a multi-input e.v.a. problem will, in general, involve systems of considerably lower order than the order of the given multi-input system. In fact, from the generic property of controllability indices [17], it follows that generically, the largest dimension of F_{ii}^* is the smallest integer $\geq n/m_o$. This gives our algorithm the advantage that the actual mechanism of e.v.a. is performed on significantly lower order systems thereby reducing computation time and numerical (rounding) errors. Furthermore, there is the advantage, from the design point of view, that the task of e.v.a. is shared between all the independent inputs since e.v.a. is carried out for each of the systems $(F_{ii}^*,\underline{g}_{ii}^*)$, $i = 1,\ldots, m_o$. Also, it should be noted that Algorithms 1-4, do not require computations of the eigenvalues of the open-loop state matrix. Thus any ill-conditioning in the eigen-system of this matrix will not, in general, affect the computations performed in these algorithms. It must also be pointed out that in the e.v.a. approach described here, certain combinations of complex-conjugate pairs of eigenvalues may not be achievable e.g. if all the controllability indices are odd, then it is not possible to have only complex-conjugate pairs as the desired eigenvalues. However, these cases can be treated by using two inputs to assign a complex-conjugate pair of eigenvalues.

It may be observed that our algorithm has some similarities with the method proposed by Anderson and Luenberger [3]. Their algorithm is based on the Luenberger canonical form for multi-input systems and as such is numerically unstable. It can be shown [13] that the reduction to the Luenberger canonical form can be carried out in two parts - a numerically stable reduction to a BHF and a potentially numerically unstable reduction to a block Frobenius form (BFF) from which the canonical form can be obtained by row and column permutations of the type described in Algorithm 3. In our Algorithm, we have avoided the numerical instability by avoiding the numerically unstable reduction to a BFF and by using only orthogonal matrices for any subsequent reduction/transformations.

7. CONCLUSIONS

A method has been described for carrying out eigenvalue assignment in multi- input systems by means of state feedback. The method consists of four algorithms: Algorithm 1 reduces the given multi-input. system to a condensed form - an upper block Hessenberg form, using orthogonal state coordinate transformations, Algorithm 2 performs partial reduction of the state matrix of the upper block Hessenberg form by means of state feedback and orthogonal state coordinate transformations, Algorithm 3 carries out a permutation of the rows/columns of the resulting system matrices, and Algorithm 4 performs eigenvalue assignment in single-input systems. We have shown that Algorithms 1-3 reduce the multi-input eigenvalue assignment problem to a number (= the number of independent control inputs) of single-input eigenvalue assignment problems for systems of orders equal to the controllability indices of the multi-input system. Assignment of the desired eigenvalues is performed using Algorithm 4 which is based on the well-known QR algorithm. The numerical properties of our approach have been discussed. In particular, we have used only orthogonal transformations in Algorithms 1-4. Our approach should be particularly advantageous when used for high order multi-input systems because of the fact that the actual mechanism of eigenvalue assignment is performed on single-input systems of considerably lower order. In fact, it can be easily shown that the greater the number of independent control inputs (= rank of input matrix B), the more significant is the advantage in using our approach.

8. REFERENCES

1. F. Fallside, Ed., "Control System Design by Pole-Zero Assignment", London: Academic Press, 1977.
2. R.V. Patel and N. Munro, "Multivariable System Theory and Design", Chapter 6, Oxford: Pergamon Press, 1982.
3. B.D.O. Anderson and D.G. Luenberger, "Design of multivariable feedback systems", Proc. Inst. Elec. Engrs., Vol. 114, pp. 395-399, 1967.
4. F. Fallside and H. Seraji, "Direct design procedure for multivariable feedback systems", Proc. Inst. Elec. Engrs., Vol. 118, pp. 797-801, 1971.
5. G.S. Miminis and C.C. Paige, "An algorithm for pole assignment of time-invariant linear systems", Int. J. Control, 35, pp. 341-354, 1982.
6. G.S. Miminis and C.C. Paige, "An algorithm for pole assignment of time invariant multi-input linear systems", Proc. 21st IEEE Conf. on Decision and Control, Orlando, pp. 62-67, 1982.
7. A. Varga, "A Schur method for pole assignment", IEEE Trans. Autom. Contr., AC-26, pp. 517-519, 1981.

8. M.M. Konstantinov, P.H. Petkov and N.D. Christov, "A Schur approach to the pole assignment problem", Proc. IFAC 8th World Congress, Kyoto, Japan, paper 51.1, Vol. XI, 1981.

9. G.W. Stewart, "Introduction to Matrix Computations", New York: Academic Press, 1973.

10. V.C. Klema and A.J. Laub, "The singular value decomposition: its computation and some applications", IEEE Trans. Autom. Contr., AC-25, pp. 164-176, 1980.

11. J.H. Wilkinson, "The Algebraic Eigenvalue Problem", London: Oxford University Press, 1965.

12. R.V. Patel, "Computation of minimal order state-space realizations and observability indices using orthogonal transformations", Int. J. Contr., Vol. 33, pp. 227-246, 1981.

13. R.V. Patel, "Computation of matrix fraction descriptions of linear time-invariant systems", IEEE Trans. Automat. Contr., Vol. AC-26, pp. 148-161.

14. P. Van Dooren, "The generalized eigenstructure problem in linear system theory", IEEE Trans. Autom. Contr., AC-26, pp. 111-129, 1981.

15. M. Konstantinov, P. Petkov and N. Christov, "Invariants and canonical forms for linear multivariable systems under the action of orthogonal transformation groups", Kybernetika, Vol. 17, pp. 413-424, 1981.

16. W.M. Wonham, "On pole assignment in multi-input controllable linear systems", IEEE Trans. Automat. Contr., Vol. AC-12, pp. 660-665, 1967.

17. W.M. Wonham, "Linear Multivariable Control: A Geometric Approach", Second Edition, New York: Springer-Verlag, 1979.

Section 4. SOFTWARE SUMMARIES

COMPUTER-AIDED CONTROL SYSTEMS ENGINEERING
M. Jamshidi and C.J. Herget (Editors)

SOFTWARE SUMMARIES

This section contains brief summaries of computer-aided control system design software packages, as compiled by Professor Dean K. Frederick of Rensselaer Polytechnic Institute and Dr. Charles J. Herget and Fran McFarland of Lawrence Livermore National Laboratory. Updated versions of these summaries will be made available in the future. New material or revisions for inclusion in a future release should be submitted to: **Prof. Dean K. Frederick**, Electrical, Computer, and Systems Engineering Department, Rensselaer Polytechnic Institute, Troy, NY 12180-3590, or to Fran McFarland, Lawrence Livermore National Laboratory, P. O. Box 808 L-156, Livermore, California 94550.

1

Package or program name: EASY5 Dynamic Analysis and Design System

Principal developer: Dr. John D. Burroughs

Software capabilities: Nonlinear simulation/time history generation, steady-state analysis, optimal control synthesis, full state feedback/Kalman state estimator/reduced order controller, linear model generation, linear control system analysis, eigenvalues/stability margins/root locus/frequency response, linear simulation/time history generation, modular modeling language is used, continuous and discrete time models (single and multirate sampling), nonlinear simulations and linear control analysis in one integrated package, system of order 150 has been handled satisfactorily, selective analysis options may handle 500th order system.

Interactive capabilities: Interactive batch (immediate batch execution)

Programming language used: FORTRAN

Computers and terminals on which available: MAINSTREAM-EKS (BCS nationwide network), VAX 11/780, CYBER (Large Scale), IBM (Large Scale), FPS 164, PRIME, and APOLLO.

Documentation: EASY5 User's Guide, BCS Document #20491-0503.

Memory and disk requirements: 100K–300K Octal 60 bit words, (Depends on size of model), on CDC, comparable on other hosts

State of development: Production code

Availability of code: Object Code may be leased or EASY5 may be accessed *via* MAINSTREAM-EKS.

Person to contact for details: Robert J. McRae, 565 Andover Park West, M/S 9C-01, Tukwila, WA 98188; (206)575-5072.

2

Package or program name: SUBOPT

Principal developer: Dr. P. J. Fleming

Software capabilities: Design of controllers for linear optimal regulators and suboptimal linear regulators for linear time-invariant continuous systems and sampled-data systems. Will handle state feedback, output feedback and dynamic compensator controller structures together with the possibility of augmenting the performance index to include sensitivity and model-following terms. Employs a robust gradient minimization technique. State-space input specification. Typically will handle 15-20 varying gain parameters and overall system sizes of at least order 30 satisfactorily. Numerical and graphical evaluation tools for controller designs including comprehensive graphics display facilities (GHOST).

Interactive capabilities: Question and answer; help commands

Programming language used: FORTRAN

Computers and terminals on which available: Any computer which supports FORTRAN and for graphics version, any terminal which GHOST graphics package supports.

Documentation: User's Manual, implementation document, commented source

Memory and disk requirements: Memory requirement is dependent on user specification. Typically an overall system of order 30 will require 190K bytes of executable code (excluding graphics).

State of development: Complete version for continuous-time systems currently available along with option for sampled-data systems.

Availability of code: By license for source code. Software to interface to user graphics packages, other than GHOST, may be supplied, subject to negotiation.

Person to contact for details: Dr. P. J. Fleming, School of Electronic Engineering Science or Managing Director, Industrial Development (Bangor) Ltd., University College of North Wales, Bangor, Gwynedd, LL57 1UT, U.K.

References:

Fleming, P. J., "A CAD Program for Suboptimal Linear Regulators," *Proceedings of the IFAC Symposium on Computer-Aided Design of Control Systems*, Zurich, 259–266, (August 1979).

Fleming, P. J., "A CAD Program for Suboptimal Regulators–Design Applications," *Proceedings of the IASTED Symposium on Modeling, Identification, and Control*, Davos, Switzerland, 136–140, (February 1981).

3

Package or program name: SSPACK^{tm} *S*tate *S*pace Systems–Software *PACK*age

Principal developer: Technical Software Systems (Techni-Soft)

Software capabilities: Linear and nonlinear simulation, linear and (approximate) nonlinear Gauss-Markov simulation with associated statistics and confidence limits, linear state estimation (U-D Kalman filter), nonlinear state estimation (U-D Extended Kalman filter), linear identification (least squares), linear and nonlinear identification (U-D Extended Kalman filter) with associated statistical analysis and display (e.g., whiteness testing).

Modular design including pre-processor to easily enter complex models and a graphics/statistical post-processor to display, statistically test, provide on-line help, and interface to all package algorithms. Dual modes of operation: (1) menu-driven supervisor for casual users and (2) command mode for experienced user.

Interactive capabilities: Interactive graphics capability, in-package file editing and compile/link for nonlinear models.

Programming language used: FORTRAN, pre-processed FORTRAN; all sources available.

Computers and terminals on which available: VMS (DEC VAX), UNIX, TOPS-10/20 (DEC-10/20) with terminals: DEC VT100 with Retrographics board, VT-125, VT-240, VT-241, SUN Workstation, IBM-PC, Tektronix 4010, 4014, 4025, 4105, RAMTEK 9400, HP2647 and 2648, LEXIDATA, VERSATEC printer/plotter, LASERGRAPHIX printer/plotter, METHEUS, ANADEX. Other terminals available on request.

Documentation: SSPACK User's Manual including sample problems (300 pages) from Tech-ni-Soft. Telephone support available.

Memory and disk requirements: Depend on problem size.

State of development: Production code

Availability of code: By license from Techni-Soft.

Person to contact for details: Technical Software Systems, P. O. Box 2525, Livermore, CA 94550. (415) 443-7213.

4

Package or program name: DIGICON/APL

Principal developer: G. F. Franklin and A. Emami-Naeini

Software capabilities: DIGICON design digital (sampled data) controls by pole-placement.
DOPTICON: design digital (sampled data) controls by quadratic loss, including noise.
OPTICON: design continuous controls by quadratic loss (LQG).
CONCON: design continuous controls by pole-placement.

Interactive capabilities: Based on APL, the package is highly interactive.

Programming language used: APL

Computers and terminals on which available: Any APL terminal

Documentation: Listings, description of problems solved and example problems available.

Memory and disk requirements: Problem dependent. Runs on IBM 5100 with 32K bytes. Also available on VMS APL(IBM 370).

State of development: In development since 1974.

Availability of code: Listings only

Person to contact for details: Office of Technology Licensing, 105 Encina Hall, Stanford University, Stanford, CA 94305.

References:

Emami-Naeini, A and Franklin, G. F., "Interactive Computer-Aided Design of Control Systems," *IEEE Control Systems Magazine, 1*, 31-36, (1981).

5

Package or program name: DPACS-F

Principal developer: Furuta Laboratory

Software capabilities: Software for the analysis of multivariable control systems based on state space and frequency-domain approaches. Pole-assignment and LQG method is used for design. Program for identification of multivariable systems based on ML and GLS are also included. 40 x 40 Matrix and Time Series 20 x 2100 data can also be handled. (Dimension will be increased on MV version.)

Interactive capabilities: Command line is employed to perform the program for data. In a command "question and answer" is used.

Programming language used: FORTRAN IV (will be converted into FORTRAN 77).

Computers and terminals on which available: NOVA 3, CRT, PRINTER, PLOTTER, Tektronix 4014 (will be available on MV 4000).

Documentation: User's Manual, Programmer's Guide and Specification of subroutines (in Japanese)

Memory and disk requirements: 96KB cpu, 5MB disk (without source program)

State of development: Now in revision 3 which has been used for three years.

Availability of code: Only available to those pursuing a joint research project with Furuta Laboratory.

Person to contact for details: K. Furuta, Department of Control Engineering, Tokyo Institute of Technology, Oh-Okayama, Meguro-ku, Tokyo, Japan.

References:

Furuta, K., et al., "Computer Aided Design Program for Linear Control Systems," *IFAC CAD Symposium*, Zurich, 267-272, (1980).

Futura, K., et al., "Computer System Design for Furnace by CAD," *IFAC Symposium*, Delhi, 31-38, (1982).

6

Package or program name: HONEYX

Principal developer: Steve Pratt, Honeywell

Software capabilities: Linear time-invariant models, uses LQG compensator structure for design, singular value analysis of multivariable systems, uses LINPACK, EISPACK, PLOT10 routines, have handled systems with 80 states and 10 inputs and outputs, supports both frequency and time domain analysis.

Interactive capabilities: Command driven with arguments provided on command line. Any missing arguments are solicited from terminal input.

Programming language used: FORTRAN, C, and UNIX command environment.

Computers and terminals on which available: Any computer that hosts UNIX, graphics available on Tektronix 4010 series or equivalent.

Documentation: On-line documentation which may be printed as a reference manual (similar to UNIX documentation).

Memory and disk requirements: Limited memory requirements because of program modularity; extensive use of disk I/O, since very little data are actually stored in program.

State of development: Experimental/research

Availability of code: Not available

Person to contact for details: Steve Pratt, Systems and Research Center, Honeywell, Inc., 2600 Ridgway Parkway, Minneapolis, MN 55413

7

Package or program name: (1) Linear Systems Analysis, (2) Digital Filter Design, (3) Linear Multivariable Systems Analysis

Principal developer: H. Elliott, G. K. F. Lee, and L. L. Scharf at Colorado State University for Hewlett-Packard Corporation

Software capabilities: (1) Enter block diagrams, Calculate Step, Impulse, Bode, Nyquist and Root Locus Plots, (2) Design and Analyze FIR and IIR Digital Filters using a variety of automated techniques. Outputs are filter coefficients, impulse, and frequency response, and (3) Controllability, Observability, Stability Analysis of Multivariable Systems using State Space, Transfer Matrix, and Differential Operator Representations, Step, Impulse and Bode Plots. Pole Placement, Observer, Steady-State Regulator, and Steady-State Kalman Filter designs. Matrix utilities.

Interactive capabilities: All commands initiated using programmable function keys. Highly interactive prompting scheme for data entry.

Programming language used: Extended BASIC

Computers and terminals on which available: (1) Hewlett-Packard HP9845 B/C, HP9835, HP85, HP9826, (2) Hewlett-Packard HP9845 B/C, HP85, and (3) Hewlett-Packard HP9845 B/C

Documentation: User's Manual furnished with program.

State of development: (1) Complete, (2) Complete, (3) Software Complete, Documentation Incomplete

Person to contact for details: Local Hewlett-Packard Sales Office

---

8

Package or program name: UMIST Computer Aided Control System Design Package (CONCENTRIC).

Principal developer: Prof. Neil Munro and Mrs. B. J. Bowland

Software capabilities: On-line design, analysis, and simulation of both SISO and MIMO systems. In addition to the design methods, a very flexible data I/O and manipulation facility is provided. SISO analysis and design methods provided: Nyquist, Bode, Nichols, Root Locus. MIMO analysis and design methods provided: Inverse Nyquist array method (continuous and sampled data versions). System models in state space form, transfer function descriptions, and measured frequency responses. Numerical Algorithms used: QR & QZ algorithms, complex matrix eigenvalues, inversion Fadeev and generalized Fadeev, and minimal realiazation algorithm. Highest order systems: 50 states or 50th order polynomials, can be customized.

Interactive capabilities: Question and answer with extensive help facilities

Programming language used: FORTRAN 77.

Computers and terminals on which available: Runs under PRIMOS and RSX-11M; machine dependent features have been isolated to special code section, Tektronix 4010 and 4110 series, DEC VT125, and Sigma 5600 series.

Documentation: User's Manual, Programmer's Manual, commented source

Memory and disk requirements: Disk storage 1.5M bytes. Run-time requirements: Largest module 64K bytes, if overlayed.

State of development: Production code commercially available

Availability of code: For commercial use need UMIST license. For teaching/ research sign confidentiality agreement with UMIST.

Person to contact for details: Prof. Neil Munro, Control Systems Centre, UMIST, Sackville St., Manchester

M60 1QD, England

Additional comments: Synthesis procedures for obtaining and improving dominance automatically in the Inverse Nyquist Array method are currently being implemented, and a new facility for the graphical input of system descriptions has been developed. Also, fully portable versions for small computer systems, DEC PDP-11/34 and ACT SIRIUS desk-top machine (UCSD *p*-system) are currently being developed.

9

Package or program name: I.S.E.R.–C.S.D. (*I*nteractive *S*ystem for *E*ducation and *R*esearch in *C*ontrol *S*ystem *D*esign)

Principal developer: Laboratoire D'Automatique, E.N.S.I.E.G., B.P. 46

Software capabilities: Simulation of dynamic systems in time or frequency-domains, single/multivariable systems, continuous or discrete models, models in transfer function form or state space form. Identification of systems in multi-input/single-output case (four methods), optimal control design based on the Linear Quadratic Gaussian methodology.

Interactive capabilities: High-level graphical output facilities, interactive scheme generation, and block-diagram introductions.

Programming language used: PASCAL and FORTRAN

Computers and terminals on which available: NORD 10/S and 100 computers, Tektronix graphic terminals

Documentation: Documented source (in French), User's Guide and Reference Manual

Memory and disk requirements: 64K words memory, 10M byte disk

State of development: Going on for more flexible data structuring, nonlinear simulation, graphical input facilities, and study of complex systems.

Availability of code: Available

Person to contact for details: Dr. Cevdet Suleyman, Institut Privé Control Data, 59 rue Nationale 75013, PARIS, France.

Additional comments: I.S.E.R.–C.S.D. has the following main characteristics: perform datastructure and dynamic memory management, modular structure for further modifications.

10

Package or program name: Linear Systems Analysis program (LSAP)

Principal developer: Charles J. Herget, Diane M. Tilly, and Thomas P. Weis

Software capabilities: An interactive program with graphics for the classical analysis and design of linear control systems. Allows for the definition of rational transfer functions, either Laplace or

Z-transforms. Manipulation of transfer functions by addition, subtraction, multiplication, or division. Conversion from Laplace to *Z*-transforms is provided. Plots of root loci, time responses, or frequency responses can be obtained for both continuous time and discrete time systems.

Interactive capabilities: Fully interactive, prompts the user for necessary inputs.

Programming language used: Primarily in PASCAL, some subroutines in FORTRAN

Computers and terminals on which available: VAX/VMS with the following terminals: DEC VT100 with retrographics; Tektronix 4010 series, 4025, 4027, and 4105; and HP 2648A. On DEC PDP-11 or LSI-11 with RSX-11M operating system on the Tektronix terminals only.

Documentation: User's Manual, Programmer's Manual, and documented source

Memory and disk requirements: Requires 71K words on the VAX. Runs on 32K words under RSX using an overlay.

State of development: Completed

Availability of code: Code is available from the National Energy Software Center, Argonne National Laboratory, 9700 South Cass Avenue, Argonne, IL 60439.

Person to contact for details: Dr. Charles J. Herget, Lawrence Livermore National Laboratory, P. O. Box 808 L-156, Livermore, CA 94550

References:

Herget, C. J. and Tilly, D. M., "Linear Systems Analysis Program," this volume.

11

Package or program name: Computer Aided Multivariable Control System Design (CAMCSD)

Principal developer: Tahm Sadeghi

Software capabilities: Programs developed: constant suboptimal output feedback design *via* parameter optimization, pole variant zero placement algorithm, command generator tracker, eigenvalue/eigenvector assignment *via* quadratic weight selection, eigenvalue/eigenvector assignment, inverse transfer function matrix computation, polynomial matrix library, inverse polynomial matrix computation, proportional-plus-integral control, transmission and decoupling zeroes computations, transfer function matrix computation: Kaufman, Faddeev, Patel, and Sadeghi, transient response plot. Interactive graphics drivers for the ORACLS routines: implicit and explicit model following, Kalman Bucy filter, and linear quadratic regulator. Interactive drivers for the EISPACK routines: eigenvalue/eigenvector computations. Libraries: NASA's ORACLS and NAL's EISPACK.

Interactive capabilities: Light pen, control devices, and keyboard

Programming language used: FORTRAN IV

Computers and terminals on which available: PRIME 750/IMLAC DYNAGRAPHICS and IBM 4341/Tektronix 4014-15 computer systems

Documentation: On-line help option

Memory and disk requirements: Virtual memory (14M machine) and 700 records of disk space (256 bytes per record)

State of development: In-operation with demonstrative examples

Availability of code: Available (without ORACLS and EISPACK) on exchange basis.

Person to contact for details: Tahm Sadeghi, Fairchild Republic Company, Farmingdale, NY 11735

References:

Frederick, D. K., Kraft, R. P., and Sadeghi, T., "Computer-Aided Control System Analysis and Design Using Interactive Graphics," *IEEE Control Systems Magazine*, 19-23, (December 1982).

12

Package or program name: Cambridge Linear Analysis and Design Program

Principal developer: J. M. Maciejowski (Cambridge University) and J. Edmunds (now UMIST)

Software capabilities: The CLADP package contains a range of techniques for both analysis and design of multivariable control systems. For analysis, the package uses generalized frequency response methods. Included are generalized Nyquist diagrams, principal loci (for the system, its closed loop, and its sensitivity transfer function matrices), and the Nyquist array (including inverse), multivariable root loci, time simulation for a variety of inputs, pole-zero computation, matrix manipulation, and command macro facilities. State space design methods include LQG, Kalman filter evaluation, and a pole placement facility. Both discrete and continuous time systems can be handled, with a multiple parameter facility, scheduled control schemes to be designed over several operating points. Some irrational forms, including time delays and some distributed systems are also handled. Facilities are included to transform between Laplace transfer function, *Z*-transfer function, and state space descriptions.

Interactive capabilities: The package is fully interactive, in question and answer format.

Programming language used: FORTRAN IV

Computers and terminals on which available: Currently on VAX 11/780, Prime 750, and GEC 4090. Older versions on PDP-11 and H6000, using

Tektronix 4010, 4014, and Sigma 5600 series terminals, both monochrome and color.

Documentation: User's Manual and documented source

Memory and disk requirements: For compiling and linking, requires 7M of disk. To run, requires 1.2M bytes of memory (on GEC machine).

State of development: Mixture, some research code, mostly complete.

Availability of code: By license.

Person to contact for details: Cambridge Control, Ltd., Madingley Road, Cambridge, England, CB2 HOP.

References

MacFarlane, A.G.J. and Kouvaritakis, B., "A Design Technique for Linear Multivariable Feedback Systems," *International Journal of Control, 25*, 837-879, (1977).

Edmunds, J.M., "Control System Design and Analysis Using Closed Loop Nyquist and Bode Arrays," *International Journal of Control, 30*, 773-802, (1979).

Postlethwaite, I., Edmunds, J.M., and MacFarlane, A.G.J., "Principal Gains and Principal Phases in the Analysis of Linear Multivariable Feedback Systems," *IEEE Transactions on Automatic Control, AC-26*, 32-46, (1981).

13

Package or program name: KEDDC

Principal developer: Dr.-Ing. Chr. Schmid

Software capabilities: Analysis and synthesis of SISO and MIMO systems: linear, nonlinear, state space, frequency-domain, description using polynomial matrices, continuous, time-discrete. Process identification: deterministic and stochastic methods. Controller design: classical approaches, LQ techniques, zeroes/pole assignment, inverse-Nyquist-array technique, parameter optimization, compensator controllers, observers, Kalman filters. Adaptive control: classical approaches, model reference methods, self-tuning regulators. Utilities: root-locus, transformation into all representation forms, frequency-domain characteristics, simulation.

Interactive capabilities: Global command driven dialog, local question-and-answer dialog. Help commands in case of errors. Menu-driven graphics output according ACM/SIGGRAPH core standard. Dialog is German or English.

Programming language used: Core system: FORTRAN ANSI X3.9-1966. Graphics: Standard PASCAL according to Jensen and Wirth. Interface to operating system: depends on implementation.

Computers and terminals on which available: Core system is computer independent. Interface for: HP2100 (RTEII), HP1000 (RTE), PDP-11/23, 34

(RT-11, RSX-11M), PM800 (MAS), IBM 370 (VMS), DPS8/52 (CP6), VAX (VMS).

Documentation: User's Manual (1200 pages), Programmer's Manual (1400 pages), 40 percent of the source is comments.

Memory and disk requirements: Can run in a 56K byte partition. Source code approximately 10M byte. System dimensions (orders, etc) can be specified using a PASCAL-written FORTRAN preprocessor. Only one percent of the total size is prerequisite for installing small subsystems of any size.

State of development: Complete, revisions and additions made as need arises. Standard updating is done twice a year.

Availability of code: Contact Dr. Schmid for details

Person to contact for details: Dr.-Ing. Chr. Schmid, Lehrstuhl fuer Elektrische Steuerung und Regelung, Ruhr University Bochum, P.O. Box 102148, D-4630 Bochum 1, Federal Republic of Germany

Additional comments: Detailed materials can be requested from Dr. Schmid. You can send benchmark examples to him. KEDDC is accessible by an international computer communication network.

14

Package or program name: TRIP Transformation and Identification Program

Principal developer: P. P. J. van den Bosch

Software capabilities: This interactive program is intended for the analysis and design of linear, single-input single-output systems.

This program distinguishes seven different models for describing linear systems. For a continuous system we have the transfer function $H(s)$, the state-space model, and the frequency response. For a discrete system we have the transfer function $H(z)$, the state-space model, and the frequency response. A time-response model is used for both a continuous and a discrete system. TRIP supports the transformation of one model to another model. For example, starting with $H(s)$ we can calculate the discrete transfer function $H(z)$, the state-space models and the frequency and time responses. These calculations are called transformations. TRIP supports about 35 transformations including optimal state feedback, eigenvalue calculation, calculation of the root locus, fast Fourier transformation, least-squares method, filtering, complex-curve fitting, solution of Lyapunov and Riccati equations, calculation of Bode and Nyquist plots, deviation ratio, etc.

Interactive capabilities: A command language is used. A Help command is available. All user-supplied information is tested for correctness and about 100 error messages can be generated to inform the user of errors in his input data. Responses are presented in numerical and in graphical form. A cursor can be used to ask for numerical values belongining to a graph.

Programming language used: FORTRAN 77

Computers and terminals on which available: PDP-11 running RSX-11M using (color) raster scan displays. Meanwhile, TRIP is implemented on many different installations, among which mainframes, VAX and 16-bit microcomputers. Version for MS.DOS and PC.DOS are available.

Documentation: User's Manual and documentation for the implementation are available.

Memory and disk requirements: The PDP-11 implementation uses overlay techniques and requires 40K bytes memory. The MS.DOS and PC.DOS implementations require about 375K bytes of internal memory. A disk is necessary.

State of development: In 1984 Version 3 was completed and released both for RSX-11M, MS.DOS, and PC.DOS.

Availability of code: Both a load module or the FORTRAN code can be bought.

Person to contact for details: Dr. P. P. J. van den Bosch, Delft University of Technology, P. O. Box 5031, 2600 GA Delft, The Netherlands.

References

van den Bosch, P. P. J., "Interactive Computer-Aided Control System Analysis and Design," in Jamshidi, M. and Herget, C. J., (Eds.), *Advances in Computer-Aided Control System Engineering*, North Holland, Amsterdam, (1985).

15

Package or program name: Relay systems: The software suite SUNS contains several computer programs which can be divided into three areas. Relay systems: SYMRLY, FOROSC, COMPRLY, QTZR, and MVREL. Nonlinear discrete systems: SAMDF, ZSAM and ZASAM. Describing function methods: DF, MV.

Principal developer: U. M. Rao, M. D. Wadey, and O. P. McNamara and A. Goucem.

Software capabilities: The relay system programs involve the exact determination of limit cycles, assessment of their stability and plots of the solution waveforms. Three programs exist for single loop relay systems: SYMRLY is for autonomous systems and caters for the possibility of sliding limit cycles, FOROSC is for systems forced with a sinusoidal input signal, and COMPRLY is a compensator design package. QTZR is for more general single loop nonlinear systems where the nonlinearity is approximated by a quantizer, this can be constructed from several relays in parallel. There is one program available for MIMO systems, MVREL, this is for limit cycle analysis in nonlinear systems which contain either relay or saturation elements.

SAMDF, ZSAM, and ZASAM are programs for the prediction of limit cycles in SISO nonlinear sampled data and digital systems. These also display the solution waveforms and the latter two programs allow for calculations of gain and phase margins.

DF is a describing function based program for the prediction of limit cycles in SISO systems, which uses the cursor and Newton Raphson techniques to find limit cycles. Other methods are also available to improve solution accuracy and to accurately verify any limit cycle predictions.

MV uses the methods implemented in DF for the prediction of limit cycles in nonlinear systems with a general structure. The system can be represented as a number of blocks, each of which may contain a mixture of linear and nonlinear elements.

Interactive capabilities: Describing function package menu driven, other programs question and answer format, limited options. All use GINO graphics.

Programming language used: FORTRAN 77

Computers and terminals on which available: PRIME 550 or VAX 11/780 with Tektronix 4010 or others with Tektronix emulator (e.g., Sigma T5670 or Televideo 920C)

Documentation: Reports of the School of Engineering and Applied Sciences, University of Sussex. Report Nos.: SYMRLY-CE /S /2 and CE /S /14; FOROSC-CE/S/18; COMPRLY-CE/S/3; QTZR-CE /S /21; MVREL-CE /S /20; SAMDF-CE/S/13; ZSAM and ZASAM CE/S/17; DF-CE/S/22; MV not yet documented, has a help system.

Memory and disk requirements: SYMRLY–238 KB FOROSC–206 KB; COMPRLY–138 KB; MVREL–252 KB; SAMDF–256 KB; ZSAM–230 KB; ZASAM–230 KB; QTZR–216 KB; MV–428 KB; DF–344 KB. (These figures are for the executable code with all additional libraries loaded, these can include NAG, SLICE, and GINO).

State of development: Complete apart from minor revisions.

Availability of code: Contact Professor D. P. Atherton for details

Person to contact for details: Professor D. P. Atherton, School of Engineering and Applied Sciences, University of Sussex, Falmer, Brighton, BN1 9QT, U.K.

References

Atherton, D.P., "Oscillations in Relay Systems," *Trans. Inst. M. C.*, *3*(4), 171-184, (1981).

Atherton, D. P. and Wadey, M. D., "Computer Aided analysis and Design of Relay Systems," *IFAC Symposium on CAD of Multivariable Technological Systems*, Purdue, 355-360, (September 1980).

Atherton, D. P., Wadey, M. D., McNamara, O. P., and Goucem, A., "An Overview of the Sussex University Nonlinear Control Systems Software," *IMC/ SERC CACSD Workshop*, 33-38, September, 1984.

16

Package or program name: Linear Control Systems–(SISO)-LINCON; State variable routines–STATE-VARCON; Multivariable nonlinear–MVDFCON; Multivariable Systems–MVCON.

Principal developer: Graduate and undergraduate students working under the supervision of Professors D. P. Atherton and R. Balasubramanian.

Software capabilities: LINCON-Primarily for linear continuous systems. Transfer function input, Bode, Nyquist, Nichols, root locus, transient response, etc. Nonlinear stability, i.e., Popov and describing function (DF) routines can also be used; STATE-VARCON–transformation to controllable companion form of a single and multivariable system. Solution of Riccati equation; MVDFCON–stability of multivariable system using graphical absolute stability and DF criteria; MVCON–inverse Nyquist locus, Gershgorin bands, characteristic loci.

Interactive capabilities: Interactive features are those of APL. One selects a routine, i.e., Nyquist, and input is question and answer. Plot routines call data allowing plots of several loci. If familar with APL can add simple curves.

Programming language used: APL

Computers and terminals on which available: IBM 370/3032; Tektronix 4015 Princeton.

Documentation: Manuals describe routines in packages and how to use them. For LINCON a manual illustrating use for certain examples is also available.

Memory and disk requirements: Routines used in assigned APL work space (128K)

State of development: Continuing. LINCON complete and routinely used in undergraduate laboratory/tutorial.

Availability of code: Contact Dr. Balasubramanian for details.

Person to contact for details: Dr. R. Balasubramanian (or Mr. W. Mersereau–Computing Centre), Electrical Engineering Department, University of New Brunswick, Fredericton, N. B. E3B 5A3 Canada.

17

Package or program name: PSI Interactive Simulation Program

Principal developer: P. P. J. van den Bosch

Software capabilities: This interactive, blockoriented simulation program can be used for the solution of a combination of (nonlinear) differential, difference and algebraic equations, as they arise in the identification and design of (control) systems.

There are 55 different block types, among which are a mode-controlled integrator, a limited integrator, unit delay, time delay, etc. Symbolic block names can be used. Five integration methods are available: four methods with a fixed step size (Euler, Adams-Bashfort-2, Runge-Kutta-2, and Runge-Kutta-4) and one with a variable step size (RungeKutta-4). Three types of function generators are supported with linear and quadratic interpolation from 11 to 201 points. Algebraic loops can be solved iteratively, therefore, nonlinear algebraic equations can be solved. Memories are available to store signals during a simulation run. These signals can be studied afterwards, can be saved on disk or can be used as inputs for future runs. Many multi-run facilities are available. Optimization is directly available. The user can define the output of one block as a criterion and up to nine arbitrary parameters of the simulation model as parameters of the optimization. Each parameter may have an upper and lower limit. To improve the speed of convergence each parameter can be scaled. Combination of several simulation models into one new model. Calculated responses and parameters can be passed from one run to a future run. User-defined block types can be added easily. This requires a new link step.

The program is not suited for the solution of partial differential equations or polynomial and matrix equations.

Interactive capabilities: Mostly a command language is used, switching to a question/answer mode in case a partly wrong command is given. A help command is available. All user-supplied information is tested for correctness and about 60 extensive error messages can be generated to inform the user of errors in his input data.

Responses are presented in numerical and in graphical form. A cursor can be used to ask for numerical values belonging to a graph, such as peak time or overshoot.

Programming language used: FORTRAN 77

Computers and terminals on which available: PDP-11 running RT-11 and RSX-11M using (color) raster-scan displays. Meanwhile, buyers of PSI run the program on many different installations, among which mainframes and 8-bit microcomputers (CP/M) using many different displays. Versions for 16-bit microcomputers (MS.DOS or PC.DOS) are available.

Documentation: A User's Manual (100 pages) and documentation for the implementation are available.

Memory and disk requirements: The PDP-11 implementations use overlay techniques and need 40K bytes memory and 90K bytes disk space. Then 150 blocks are available, each block having up to three inputs and up to three parameters. So, a floppy disk is sufficient. The MS.DOS and PC.DOS version require 256-bytes memory, offering 750 blocks.

State of development: In 1984, Version 6 was completed and released, both for RSX-11M, MS.DOS, and PC.DOS.

Availability of code: The FORTRAN code can be bought.

Person to contact for details: Dr. P. P. J. van den Bosch, Delft University of Technology, P. O. Box 5031, 2600 GA Delft, The Netherlands

References

van den Bosch, P. P. J., "PSI-an Extended, Interactive Block-Oriented Simulation Program," *Proceedings IFAC Symposium on Computer Aided Design of Control Systems*, Zürich, 223-228, (1979).

van den Bosch, P. P. J., "Interactive Computer-Aided Control System Analysis and Design," in Jamshidi, M. and Herget C. J., (Eds.), *Advances in Computer-Aided Control System Engineering*, North-Holland, Amsterdam, (1985).

18

Package or program name: *T*otal *S*ynthesis *P*roblem *S*oftware *P*ackage (TSPSP)

Principal developer: R. M. Schafer and M. K. Sain

Software capabilities: The software package consists of a set of programs that are used in designing multivariable control systems in the Total Synthesis Problem (TSP) framework. The programs include software for calculating a minimal basis from either a transfer function or state space description; a basis manipulation program that allows manipulation of the basis to design the system response and the control necessary to achieve that response; a transfer function algebra program which is used to calculate the configuration dependent controller transfer functions; a realization program which determines a state space description of each of the controllers; and a simulation package. The various programs communicate *via* common data set formats.

Interactive capabilities: All of the design sections of the software are fully interactive, including the generation of time responses *via* partial fraction expansion and singular value plots for robustness considerations during basis manipulation. Most program inputs are read alpha-numerically and internally decoded to prevent input conversion errors, and a question mark will provide more information after a prompt.

Programming language used: FORTRAN 77

Computers and terminals on which available: VAX 11/750 under UNIX of VMS. Graphics on Tektronix 4014, HP plotter or through Tektronix PLOT-10 interface.

Documentation: Commented source code; a User's Manual will be available soon.

Memory and disk requirements: Software can be modified to run on PDP 11/70 with 256K using overlays and virtual arrays.

State of development: Research code

Availability of code: One proprietary part of the package is currently being replaced. The package will be available in the summer of 1985.

Person to contact for details: Dr. R. Michael Schafer, Computer Engineering and Control Laboratory, Department of Electrical Engineering, University of Virginia, Charlottesville, VA 22901.

19

Package or program name: FREDOM/PC, TIMDOM/PC, and LSSPAK/PC

Principal developer: M. Jamshidi, E. Hirt, P. Cunes, and R. Banning, Laboratory for CAD of Systems/Networks, EECE Department, University of New Mexico, Albuquerque, NM 87131.

Software capabilities: (i) FREDOM: a classical control (FREquency DOMain) CAD package capable of analyzing and designing a SISO control system. The tools available are simulation, stability tests, graphical techniques such as Bode, Nyquist and root locus, design *via* compensation, model reduction and optimization, linear algebra, and transform theories. (ii) TIMDOM: A modern control (TIMe DOMain) CAD package capable of analyzing, designing and modeling (model reduction) of multivariable systems. Both multivariable design (Linear state regulator, etc.), estimation, linear algebra, and transform theories. (iii) LSSPAK: A package for modeling and control of large-scale systems. The tools available are frequency-domain and time-domain model reduction techniques for both SISO and MIMO systems, goal coordination and interaction prediction algorithms of hierarchical control, decentralized stabilization and robust controllers, linear algebra, transform theories, etc.

Interactive capabilities: Fully interactive

Programming language used: Microsoft BASIC

Computers and terminals on which available: IBM/PC with one or two floppy disk drives and all IBM Compatibles, also available on IBM PC/XT.

Documentation: Manuals are either available or are in preparation (see References).

Memory and disk requirements: 64K minimum.

State of development: All programs have been written, some housekeeping routines are still being written.

Availability of code: All three are available, please contact Professor Jamshidi for details.

Person to contact for details: Professor M. Jamshidi, Director Laboratory for CAD of Systems/Networks, EECE Department, University of New Mexico, Albuquerque, NM 87131 USA, tele: (505) 277-5538, or (505) 277-0300.

References:

Jamshidi, M. and Malek-Zavarei, M., *Linear Control System–A Computer-Aided Approach*, Pergamon Press, Ltd., Oxford, England, (1985).

Jamshidi, M., *Large-Scale Systems Modeling and Control*, Elseviers, North-Holland, New York, (1983).

20

Package or program name: MATRIX$_X$, SYSTEM_BUILD

Principal developer: Integrated Systems, Inc.

Software capabilities: MATRIX$_X$ is a software system for control systems design and analysis, system identification, data analysis, and simulation. Classical frequency-domain, modern state-space, and other control design techniques are available as single commands. New design approaches can easily be incorporated into the system by the user. System identification tools include nonparametric frequency-domain methods, maximum likelihood algorithms in batch and recursive forms, and adaptive algorithms. Simulation of differential algebraic systems and fast propagation of sparse systems are also available. A powerful two and three dimensional graphics package is included.

The SYSTEM_BUILD capability in MATRIX$_X$ allows graphic building of nonlinear differential/algebraic system models. Each sub-system may be declared to be continuous or discrete with a given sampling interval. Multi-rate time simulations can be performed on sample data systems.

Interactive capabilities: All commands are executed interactively.

Programming language used: ANSI-77 FORTRAN, and assembly language routines.

Computers and terminals on which available: IBM MVS/TSO and VM/CMS, VAX VMS and UNIX APOLLO workstation. DEC VT-100 and 200 series graphics terminals, Tektronix, IBM 327x and others.

Documentation: MATRIX$_X$ User's Guide, MATRIX$_X$ Reference Guide, MATRIX$_X$ Training Guide, Command Summary, and on-line Help.

Memory and disk requirements: System and problem dependent. Executable code requires less than one megabyte.

State of development: Installed and used at many industrial and academic institutions. MATRIX$_X$ has been significantly modified and capabilities are being enhanced continuously. Version 4.0 will be released October '84.

Availability of code: By license, fully supported in US, Japan, and Western Europe.

Person to contact for details: Eleanor Vade Bon Coeur, Integrated Systems, Inc., 101 University Avenue, Palo Alto, CA 94301-1695, (415)853-8400.

References

Walker, R., Shah, S., Gregory, C. Z., and Varvell, D., "MATRIX$_X$: A Model Building, Nonlinear Simulation and Control Design Program." *Advances in Com-*

puter-Aided Control Systems Engineering, Herget, C. J. and Jamshidi, M. (Editors), North Holland, 1985.

Additional comments: Training: A three day class is available either on-site or at ISI.

21

Package or program name: AESOP (Algorithms for EStimator and OPtimal regulator design)

Principal developer: Bruce Lehtinen and Lucille C. Geyser

Software capabilities: The program is mainly geared toward control and estimator design for linear, continuous time-invariant systems described in state space form. Computations performed are: linear quadratic regulator, Kalman filter, transfer functions, frequency responses, transient responses, support functions (error checks, eigenvalues/vectors, controllability and observability, covariance matrices, etc.). Algorithms used (particularly for Riccati equation solution) are moderately sophisticated, and problems up to 41st order have been run with acceptable accuracy.

Interactive capabilities: The basic mode of operation is interactive, using a question-and-answer format. Can also be run batch. User selects desired computations *via* entry of a sequence of function numbers. AESOP aids the user by checking the validity of the requested sequences.

Programming language used: ANSI-66 FORTRAN

Computers and terminals on which available: IBM 370/3033 TSS system. Terminals generally used are Tektronix 4010 or 4014.

Documentation: A User's Manual has been published as NASA TP-2221.

Memory and disk requirements: Source code takes approximately 300K bytes. Present program, which is dimensioned for 50th order problems, requires an additional 1.7M bytes for matrix storage.

State of development: Code is complete. Additions are envisioned as the need arises.

Availability of code: The program (FORTRAN code) may be obtained from COSMIC (University of Georgia, Athens, GA 30602).

Person to contact for details: Bruce Lehtinen, NASA Lewis Research Center, 21000 Brookpark Road, Cleveland, OH 44135.

Additional comments: AESOP uses a proprietary graphics package, thus graphics source code is not available. However, COSMIC does provide functional descriptions of all graphic subroutines to aid in adapting AESOP to a particular graphics environment.

22

Package or program name: SLICE (Subroutine Library in Control Engineering)

Principal developer: Control System Research Group, Department of Computing, Kingston Polytechnic, England.

Software capabilities: A library of FORTRAN subroutines implementing a wide range of algorithms for control system analysis and design.

Interactive capabilities: None

Programming language used: FORTRAN IV (PFORT subset)

Computers and terminals on which available: Any on which suitable FORTRAN compiler is available.

Documentation: User documentation for each subroutine included in source code as initial comment section. User Manual to be produced in near future.

State of development: Approximately 40 subroutines completed in first version out of anticipated 60-70 to be available.

Availability of code: Completed subroutines available on request, at nominal charge.

Person to contact for details: Dr. M. J. Denham, Department of Computing, Kingston Polytechnic, Kingston-upon-Thames, Surrey, KT1 2EE, England

---

23

Package or program name: Waterloo Control Design Suite

Principal developer: J. D. Aplevich

Software capabilities: Classical, optimal, multivariable, frequency-domain, and algebraic design of linear time-invariant, finite-dimensional systems containing single linear operator. File management is performed by the IBM VM/CMS operating environment, under which the programs run. At present linear systems to order about 100 can be handled, with potential for simple extension to multi-dimensional large, sparse systems.

Interactive capabilities: Commands are entered either from the terminal or from an EXEC file or both. Neither a light pen nor a graphical menu is used.

Programming language used: File management and elementary command parsing in EXEC files, numerical operations in FORTRAN and RATFOR.

Computers and terminals on which available: Available *via* computer communications, networks (DATAPAC, TELENET, TYMNET). Compatible with minor changes with any IBM VM/CMS installation. Graphics output on any Tektronix compatible terminal.

Documentation: On-line documentation and archive-retrievable User's Manual, plus a 230-page report describing the theory and design applications of singular pencils (not necessary for routine work).

Memory and disk requirements: Default 1 magabyte virtual machine is used.

State of development: Currently implemented for linear, single-dimensional nonsparse systems.

Availability of code: May be used *via* communication network by arrangement with University of Waterloo. Otherwise by special arrangements.

Person to contact for details: J. D. Aplevich, Electrical Engineering Department, University of Waterloo, Waterloo, Ontario N2L 3G1 Canada.

24

Package or program name: Optimal Regulator Algorithms for the Control of Linear Systems (ORACLS)

Principal developer: Ernest S. Armstrong

Software capabilities: ORACLS is a package for the design of linear time-invariant multivariable observer-based compensators primarily through Linear-Quadratic-Gaussian (LGQ) methodology. A user-provided executive program drives codes selected from a library of 62 special-purpose subroutines classified as input/output, vector/matrix operations, analysis of constant linear systems, and control law design. The library includes subroutines for eigensystem computation, Cholesky and singular value decomposition, matrix exponentials, Sylvester and Lyapunov equation solution, stabilizability/detectability examination, steady-state covariance calculation, and transfer matrix construction. For systems modeled in either continuous or discrete state variable form, the design algorithms generate transient/steady-state LQG, implicit/explicit model following, single-input pole placement, and sampled-data Linear Quadratic Regulator control laws.

Interactive capabilities: None. Developed for batch execution.

Programming language used: FORTRAN IV

Computers and terminals on which available: CDC 6000 series, Tektronix 4051

Documentation: "ORACLS–A Design System for Linear Multivariable Control," Volume *10 Control and System Theory Series*, Marcel Dekker, Inc., New York, (1980).

Memory and disk requirements: 60K Octal of 60 bit words

State of development: Operational and developmental versions

Availability of code: Operational version available from COSMIC, Suite 112, Barrow Hall, University of Georgia, Athens, GA 30602 (Program LAR 12313.)

Person to contact for details: Ernest S. Armstrong, NASA Langley Research Center, Hampton, VA 23665, (804) 865-4591

Additional comments: Preliminary version of documentation found in NASA TP 1106.

25

Package or program name: Control System Design and Analysis Programs

Principal developer: Systems Technology

Software capabilities: AFTF computes longitudinal and/or lateral-directional transfer functions (for control and gust inputs) from aircraft trim and stability and control derivative data. DIGIT converts a system of differential equations in operational form to a system of difference equations in operational form. RESP computes and plots the time response of user-specified transfer function. RMS computes the root mean square response of a user-specified transfer function to a random input or the integral mean-square response to a deterministic input. SYNOPSYS computes the solutions for the filter-observer, the regulator, and the combination of those two into the linear optimal stochastic controller. Also computes a variety of metrics for the assessment of the performance and sensitivity of the synthesized optimal control system. TRFN computes transfer functions (i.e., denomintors, numerators, and coupling numerators) from ordinary differential or difference equations of motion specified in matrix form and operational notation up to 72nd order with six independent forcing functions. MATLIS prints a report-quality TRFN data file in matrix equation format. USAM computes and plots $j\omega$- and σ-Bode and/or root locus diagrams of a user-specified transfer function. Includes modules for computing closed-loop and error transfer functions and opening a closed-loop transfer function and modules for editing and computing lower-order equivalent systems *via* pole-zero cancellation and low-frequency approximations.

Interactive capabilities: Provide for user oriented on-line design. All menu-driven modules are operated from a time-sharing terminal with alphanumeric and graphic capabilities as described above.

Programming language used: ANSI Standard FORTRAN IV

Computers and terminals on which available: All programs are available on a TYMSHARE* PDP-10; some programs are also available on IBM 360/50, CDC 6600, Cyber 176, Apple II CP/M, and/or IBM PC.

Documentation: *User's Manual*, Systems Technology, Inc., Technical Report No. 407-1, July 1970, Revised August 1976 and *Practical Optimal Control for Flight Control Application, Vol. II: Software User's Guide*, NASA CR-152306, March 1979.

Memory and disk requirements: PDP-10: 460K bytes (all programs)

---

* TYMSHARE, Inc. is a timesharing organization with nationwide and international networks of computer access.

State of development: Complete and highly refined.

Availability of code: License or access through TYMSHARE*; purchase for Apple II CP/M, IBM PC or PDP-11 (RSX-11, RT-11). Conversion to VAX is also available.

Person to contact for details: Ray Magdaleno, *Home Office:* 13766 South Hawthorne Blvd., Hawthorne, CA 90250 (213) 679-2281 or Wayne Jewell, *Branch Office:* 2672 Bayshore-Frontage Road, Suite 505, Mountain View, CA 94043 (415)961-4674.

26

Package or program name: (1) Linear Systems Analysis and Feedback Compensator Design and (2) Analysis and Design of Digital Filters and Discrete Time Systems

Principal developer: Parametrics, Inc.,

Software capabilities: (1) This is an interactive program for analyzing linear systems consisting of interconnections of feedforward and feedback blocks, and for designing feedback compensators. The user enters up to 15 transfer function blocks in feedforward, feedback, series, or parallel configurations. From this configuration on overall transfer function is generated whose order must be less than 16. The user may generate and plot (i) impulse and step responses, (ii) Bode plots, (iii) Nyquist plots, and (iv) root-locus plots. Characteristic polynomials may be rooted and gain and phase margins may be computed.

(2) This is an interactive program for designing and analyzing all-zero, all-pole, and pole-zero digital filters and discrete time systems. The user enters filter specifications. FIR filters are designed using Kaiser windows and using weighted least squares. Butterworth and Tchebyshev IIR filters are designed using the bilinear-z transform. Frequency sampling designs are available, as are analog to digital transformations using impulse invariance, covariance invariance, and bilinear-z. The user may compute and plot impulse responses and frequency responses.

Interactive capabilities: All of the software proceeds in an interactive, menu-oriented mode. Graphics may be displayed on a CRT monitor and dumped to a graphics printer or pen plotter.

Programming language used: BASIC

Computers and terminals on which available: IBM PC, HP series 200, HP-85, APPLE II and APPLE II plus.

Documentation: One hundred page User's Manual consisting of theory, practice, instructions, and examples.

Memory and disk requirements: 128K for IBM PC, 256K for HP series 200, 16K MEM Module for HP-85, 48K for APPLE II and APPLE II plus.

State of development: Fully developed production code in use by 500 users worldwide.

Availability of code: Available for immediate shipping

Person to contact for details: Parametrics, Inc., P. O. Box 1576, Fort Collins, CO 80521.

27

Package or program name: SIRENA *Système Informatique pour la Recherche et l'Enseignement en Automatique* (Computer System for Research and Education in Automatic Control)

Principal developer: Laboratoire d'Automatique, INSA de Rennes–France and Société RSI, Meylan, France

Software capabilities: An interactive program with graphics for the description, simulation, and design of dynamical systems. Systems can be continuous, sampled, discrete, linear, and/or nonlinear. SIRENA performs the following points: formal and numerical description of systems, formal and numerical calculation, time response using a variety of inputs, frequency response, root locus plot, computation of digital filters, digital signal processing, and identification. User's definable programs can be easily included in SIRENA.

Interactive capabilities: The package is fully interactive in question-and-answer format with extensive help facilities. High-level interactive dialogue is used to generate graphical displays.

Programming language used: FORTRAN 77 (ANSI X-3-9-1978)

Computers and terminals on which available: Fully portable. Currently on NORSK DATA ND100 (16 bits) and ND500 (32 bits), DEC VAX, Hewlett Packard HP1000 and HP9000, UNIVAC Série 1100, CDC Cyber 170, and Micromega/Fortune (UNIX). Graphical facilities using Tektronix 4010, 4014, and Tektronix compatible SECAPA 741.

Documentation: User's Guide and Reference Manual

State of development: Complete for "Industrial Version," marketed by R.S.I. Going on for the research-code version (INSA).

Availability of code: Source code not available. Direct-executable code available upon request.

Person to contact for details: Mr. Yem, Société R.S.I., ZIRST, Chemin du Pré Carré, 38240 Meylan, France Tél (76) 90 17 52.

28

Package or program name: CYPROS

Principal developer: The staff at the Division of Engineering Cybernetics, The Norwegian Institute of Technology and SINTEF (The Foundation for Scientific and Industrial Research at the Norwegian Institute of Technology), Automatic Control Division.

Software capabilities: The program packages cover a wide range of control engineering problems seen as analysis and design of control systems, system identification, parameter estimation, and simulation. Different model representation forms can be handled. For instance: multivariable state space models in continuous or descrete form, multivariable transfer function in continuous or discrete form or matrix polynomial forms. Transformation programs are available. The control package in CYPROS consists of programs based on multivariable time-domain and frequency-domain methods for analysis and design. An adaptive program based on a Pole Placement (PP) technique and a generalized minimum variance (GMV) control algorithm is available. Classical control methods for single input/single output systems are also included. The identification and parameter estimation package contains programs for single input/single output and multivariable linear and nonlinear models. Methods such as Maximum Likelihood method and extended Kalman filter are implemented. In the simulation package different methods are available for solving differential and difference equations. The user specifies the models by means of FORTRAN subroutines, and the model parameters are stored on a data file and are easily changeable. Packages for general matrix handling and time-series analysis are included.

Interactive capabilities: Output from the CYPROS program is rapidly examined by use of video color graphics or alphanumeric terminals. Hard copies are easily obtained. Command dialog as well as question-answer dialog with the help facility is used. Macros may be used.

Programming language used: FORTRAN 77

Computers and terminals on which available: Nord-10, Nord-100, VAX 11/750. Parts of the system will be available on different micro-computers during 1985. Alphanumeric terminals: Marconi DTI/Genius, Tandberg TDV 2215, Digital VT100, INFOTON 200, Lear Sigler ADM 3A, Data Terminal Incorporation, Tektronix

Documentation: User's Guide and routine documentation

Memory and disk requirements: 128 KB, exclusive operation system. Disk: 3 MB (16 bits)

State of development: Parts of the system are in a complete form, while other programs are continuously revised.

Availability of code: Can be purchased as a total system or as packages.

Person to contact for details: Arne Tysso, CAMO, Computer-Aided Modeling A/S, P. O. Box 2893 Elgesaeter, N-7001, Trondheim, Norway.

Additional comments: Most of the program uses interactive graphics for displaying recordings and results. We have software for displaying on: Sincolar, Tektronix, Versatec Matrix plotter, HP-plotter, and alphanumeric terminals of different types (see above).

References:

Tysso, A., "CYPROS–Cybernetic Program Packages," *MIC, Vol. 1*, No. 4, (October 1980).

29

Package or program name: CTRL-C (pronounced "control-see")

Principal developer: Systems Control Technology

Software capabilities: CTRL-C is an interactive computer language for the analysis and design of multivariable control systems. A powerful matrix·environment provides a workbench for system simulation, signal generation, matrix analysis, and graphics.

Systems may be described in state space, transfer function or discrete time forms. Transformations between representations are available. Design primitives are provided for eigenstructure assignment, LQG regulators, LQG filters, and model following. Time analysis primitives include impulse response, step response, ramp response, and arbitrary input response. Frequency analysis primitives include Bode, Nyquist, Nichols, singular values, and root-locus. Many other matrix analysis and digital signal processing primitives are available.

Interactive capabilities: Fully interactive command driven interpreter. Graphics windows, process spawning, and integration with a screen editor.

Programming language used: ANSI FORTRAN 77

Computers and terminals on which available: VAX/VMS, most common graphics terminals

Documentation: User's Manual, Reference Manual, on-line help

Memory and disk requirements: May be varied by adjusting the stack size.

State of development: Production code

Availability of code: By license, University discount

Person to contact for details: John Little, Systems Control Technology, 1801 Page Mill Road, Palo Alto, CA 94303 (415) 494-2233

30

Package or program name: Multivariable ADaptive control PACKage (MADPAC)

Principal developer: G. Bartolini, G. Casalino, F. Davoli, and R. Minciardi

Software capabilities: Implementation of a wide class of certainty equivalence adaptive control schemes, and simulation of the overall controlled system. Identification routines: Recursive Least/Squares (RLS) and Extended Least Squares (ELS) for multivariable ARMAX models, with Givens orthogonal factorization. Control strategies: LQG type, with or without control weighting and with control horizon ranging from one to several steps (Minimum Variance, Extended Minimum Variance, Multi-step, Receding Horizon). The program allows to simulate adaptive control algorithms based on the indentification of an "implicit" (i.e., closed loop) model by means of simple RLS identification (even in the presence of colored noise). Highest dimensions currently handled: a maximum order of 10 for the system state in the innovations representation.

Interactive capabilities: The input of data by question and answer. A set of graphic routines allows on-line monitoring of the system outputs and inputs, the estimated model parameters, the controller parameters, and the estimated process cost. The dynamic evolution of these quantities over a time window can be interactively followed.

Programming language used: FORTRAN.

Computers and terminals on which available: PDP-11/34 (under RT-11 version 3) plus a graphic terminal (currently in use VDS-501); being developed on VAX 750 under UNIX.

Documentation: No specific one at present, but the references contain detailed descriptions of the algorithms used.

Memory and disk requirements: About 100K bytes of core memory required at run time. Size of executable code about 50K bytes.

State of development: More research code being added. Standard algorithms complete.

Availability of code: Free to nonprofit organizations.

Person to contact for details: Franco Davoli, DIST, University of Genoa, Viale F. Causa, 13, 16145 Genoa, Italy

References:

Bartolini, G., Casalino, G., Davoli, F., Minciardi, R., "A Package for Multivariable Adaptive Control," *Proc. 3rd IFAC Symposium on Software for Computer Control*, Madrid, Spain, Pergamon Press, 229-335, (1982).

31

Package or program name: LISPACK

Principal developer: P. Hr. Petkov

Software capabilities: LISPACK is a collection of more than 20 subroutines for solving the basic problems of the analysis and design of linear multivariable systems described in the state space. Only algorithms that are proved as numerically stable are used. The program implementation is strongly influenced by the style of the EISPACK and LINPACK subroutines. The algorithms are tested for various examples of order at least 50. Except for a few machine dependent parameters the package is portable.

Interactive capabilities: Most of the subroutines are included in the local installation of MATLAB–the interactive matrix laboratory of Professor C. B. Moler.

Programming language used: ANSI FORTRAN IV

Computers and terminals on which available: PDP-11, VAX-11.

Documentation: User's Guide (in preparation). All subroutines have detailed comments.

Memory and disk requirements: 64K word storage, one disk at least 2Mb

State of development: Ready for use.

Availability of code: Available on request, free of charge

Person to contact for details: P. Hr. Petkov, Department of Automatics, Higher Institute of Mechanical and Electrical Engineering, 1156 Sofia, Bulgaria

References:

Konstantinov, M. M., Petkov, P. Hr., and Christov, N. D., "Orthogonal Invariants and Canonical Forms for Linear Controllable Systems," *Proc. 8th IFAC World Congress, Vol. 1*, 49-54, Pergamon Press, (1982).

Petkov, P. Hr., Christov, N. D., Konstantinov, M. M., "A Program Package for Computer-Aided Design of Digital Computer Control Systems," Preprint SOCOCO82, 217-220, Madrid, Spain (1982).

Petkov, P. Hr., Christov, N. D., and Konstantinov, M. M., "A Computational Algorithm for Pole Assignment of Linear Single-Input Systems," *IEEE Trans. Automat. Control* (to be published).

32

Package or program name: The Hull Control System Design Suite (Linear SISO, INA, time and data manipulation facilities based on UMIST CACSD package)

Principal developer: Dr. P. M. Taylor

Software capabilities: SISO linear system analysis and design using Nyquist, Bode, Nichols, root locus, and time response. MIMO linear system analysis and design using inverse Nyquist array, characteristic locus, interaction vectors and time responses. SISO nonlinear system analysis and design using phase plane, describing functions and time responses. MIMO nonlinear system analysis and design using circle criteria, describing function methods (including sequential loop balance method) and time responses. Extensive data manipulation facilites available.

Interactive capabilities: Question and answer with help facilities.

Programming language used: FORTRAN

Computers and terminals on which available: PDP-11/34 or LSI-11/23 under RSX-11M V3.2 (an old version of the linear SISO package also runs under RT-11) with Tektronix compatible terminals (Tektronix 4010, 4014, Tele-video 920C and a BBC microcomputer for color graphics have been used). Hard discs are recommended but the RT-11 version will run slowly with floppy discs.

Documentation: User's Manuals, documented source

Memory and disk requirements: Disc storage 2.0M byte, run time requirements less than 64K bytes including graphics. Each module is separately compiled and linked.

State of development: SISO linear and nonlinear packages are well established. The MIMO packages are being further developed.

Availability of code: Some available, contact Dr. P. M. Taylor for details.

Person to contact for details: Dr. P. M. Taylor, Department of Electronic Engineering, University of Hull, Hull HU6 7RX, U.K.

References:

Gray, J. O. and Taylor, P. M., "Computer Aided Design of Multivariable Nonlinear Control Systems Using Frequency-Domain Techniques," *Automatica Vol. 15*(3), 281-297, (1979).

Taylor, P. M. and Hayton, G. E., "The Manipulation of Interaction Effects in Multivariable Feedback Systems," *Proc. IFAC 8th World Congress*, Kyoto, (1981).

Taylor, P. M., "The Hull Control System Design Suite," *Proc. IMC Workshop on Computer-Aided Control System Design*, Brighton, U.K., 73-78, (1984).

33

Package or program name: AUTOCON (Automated Optimal Control System Synthesis Program)

Principal developer: Control System Automation

Software capabilities: Computer-aided design tool for synthesis and optimal design of linear multivariable control systems with varying plant dynamics. Synthesizes control systems by optimizing, simultaneously, system stability *and* performance subject to classical design objectives and constraints, e.g., stability margins and system performance parameters such as bandwidth, rise-time, overshoot, damping, etc., as required by mil-specs. For user-specified architecture, compensation parameters such as gains, filter coefficients, time constants and parameter scheduling for the case of multiple plant variations, are obtained as a solution to the optimization problem. Optimum nonlinear programming algorithms are employed in the search for local constrained solutions in which violations in stability and performance either vanish or are minimized for a proper selection of the control variables. Output includes printout and neutral plot files of open-loop Nichols responses for each plant variation and closed-loop Bode and transient responses for all output transfer functions and plant variations. Plot device driving routines can be supplied as needed.

Interactive capabilities: Batch processing and easy-to-use input formatting. Can be made interactive on dedicated machines with powerful processors.

Programming language used: FORTRAN IV

Computers and terminals on which available: All IBM mainframes and the "super-minis." Implementable with minor modifications on minicomputers such as VAX, PRIME, and SEL.

Documentation: 1) Complete technical description and 2) User's Manual and Programmer's Manual.

Memory and disk requirements: 550K bytes main memory

State of development: Completed; currently in use. Extensions and updates ongoing. Revisions for particular applications considered on a consulting basis.

Availability of code: By license

Person to contact for details: C. P. (Charles) Lefkowitz, Control System Automation, 26833 Via Desmonde, Lomita, CA 90717, (213) 326-2197.

References

Lefkowitz, C. P., "Automated Synthesis of Control Systems: A Design Approach," *Proc. of IFAC Workshop on Applications of Nonlinear Programming to Optimization and Control*, (1984).

Lefkowitz, C. P., "Computer-Aided Synthesis and Optimal Design of Multivariable Control Systems with Varying Dynamics," to be published in IFAC Journal, *Automatica*.

34

Package or program name: CONTROL.lab

Principal developer: M. Jamshidi and T. C. Yenn.

Software capabilities: CONTROL.lab is a CAD language for control design, system analysis, data analysis, simulation, modern control design, Kalman filtering, state/output estimation, model reduction, matrix analysis, mathematical transformation theory, etc. It is preceeded by *MATLAB*–a mathematical laboratory for linear algebra (see reference 1).

Interactive capabilities: Fully interactive.

Programming language used: FORTRAN/77.

Computers and terminals on which available: DEC 11/780/750 VAX (UNIX 4.2 Operating System), SUN Microsystems Workstation.

Documentation: (1) CONTROL.lab User's Guide is being written, (2) on-line HELP documentation, and (3) see references (2-3).

Memory and disk requirements: Dependent on system and problem. Typical

VAX/UNIX executable code requires a 1/2 mega byte.

State of development: Over eighty percent is complete. Test sites have been established.

Availability of code: It is available, please contact Professor Jamshidi for details.

Person to contact for details: Professor M. Jamshidi, Director Laboratory for CAD of Systems/Networks, EECE Department, University of New Mexico, Albuquerque, NM 87131, USA, Tele. (505) 277-5538, or (505) 277-0300.

References

Moler, C. B., *MATLAB-User's Guide*, Department Computer Science, University of New Mexico, Albuquerque, NM 87131, August 1982.

Jamshidi, M. and Yenn, T. C., "CONTROL.lab–A CAD language for Control Systems," 2nd *IEEE Symposium CACSD*, Santa Barbara, CA, March, 1984.

Yenn, T. C., "On the Computational Aspects of Kalman Filtering" and "CONTROL.lab–Software for System Estimation, Identification, and Control Design," M.S. Thesis, Laboratory for CAD of Systems and Networks, EECE Department, University of New Mexico, Albuquerque, NM, (1985).

35

Package or program name: L-A-S: Linear Algebra and Systems

Principal developer: Stanoje P. Bingulac, N. Gluhajic, and Phillip J. West

Software capabilities: L-A-S is a ComputerAided Control System Design (CACSD) Language. The fundamental concept behind a L-A-S program is the notion of a L-A-S "operator." Operators are divided into 5 groups: input/output, data handling, linear algebra, control systems, and L-A-S program control. The user constructs a L-A-S program by combining one or more of the more than 130 operators into a meaningful algorithm.

In all cases we have tried to implement those algorithms that have shown numerical superiority over the years. Most operators can handle a 100th order system at best although the maximum order of any operator is not explicitly limited. Currently, L-A-S operators include discrete and continuous Lyapunov, Riccati, and Kalman filtering equation solvers. Multiple Operator Statements (MOS), user-defined subroutines (macros), and the "echo" input feature add considerable flexibility to the L-A-S language.

Interactive capabilities: L-A-S is a unified control system design language. The user may interact directly or indirectly with the L-A-S language interpreter. This means that macro and subroutine structures are permitted. Also, L-A-S has its own built-in editor. The user can easily manipulate a L-A-S program to suit his/

her tastes. Finally, there are easy to understand error messages and the accompanying help facility provides the user with various levels of documentation and explanations.

Programming language used: FORTRAN ANSI Standard '66 or '77

Computers and terminals on which available: In general, the most common minis and large computer systems including: PDP-11/34 and 11/70, VAX/UNIX, VAX/VMS, PRIME 750, HP-1000, DEC-10, and DEC-20. Terminals include: VT100, VT101, Hazeltine 1500, Heathkit H19, etc., and graphics provided by Tektronix 4010 or equivalent.

Documentation: User's Manual and commented code exist. Also, an extensive help file provides quick, on-line assistance.

Memory and disk requirements: Virtual memory machines allow double precision computations throughout the whole software. Otherwise, systems of up to 40th order have been studied on a PDP-11 computer with 64K bytes of memory.

State of development: Package development is complete, however, extentions and additions are readily made as the need arises.

Availability of code: Source and/or object code is commericially available through a license agreement.

Person to contact for details: In Europe: Stanoje P. Bingulac, Debarska 21, 11000 Belgrade, Yugoslavia. In America: Phillip J. West, ELAC Systems, P. O. Box 2205, Station A, Champaign, IL 61820.

References:

Bingulac, S. P. and Gluhajic, N., "Computer Aided Design of Control Systems on Mini Computers Using the L-A-S Language," *Proc. of 1982 IFAC Symposium on Computer Aided Design of Multivariable Technological Systems*, Purdue University, Indiana, (1982).

Chow, J. H., Javid, S. H., and Dowse, H. R., "User's Manual for L-A-S Language," System Dynamics and Control Group, GE, Schenectady, NY, (November, 1983).

Bingulac, S. P., "Recent Modifications in the L-A-S (Linear Algebra and Systems) Language and Its Use in CAD of Control Systems," *Proc. of 1983 Allerton Conference*, Monticello, Illinois, (October 1983).

36

Package or program name: POLPAC/MODPAC/IDPAC/SIMNON

Principal developers: Ann-Britt Östberg, Johan Wieslander, Hilding Elmquist

Software capabilities:

POLPAC–Polynomial designs, polynomial operations, analysis design, deadbeat minimum variance, and pole placement designs.

MODPAC–Transformations of models, graphical outputs, matrix operations, polynomial operations, system transformations, Kalman decompositions, controlabilities, observability, transformation from state-space to frequency-domain and polynomial representations, transformation to diagonal and Hessenberger form, and transformation to observability and canonical forms.

IDPAC–Data analysis, spectral analysis, correlation analysis, parameter estimation in linear models with many inputs and many outputs, and model validation.

SIMNON–Simulation and optimization of nonlinear continuous time and discrete time systems. The program permits choices of different integration routines, Runge-Kutta with fixed and variable steplength, predictor corrector and routines for stiff equations. Systems of order 100 have been simulated.

Interactive capabilities: Commands with macro and help facilities.

Programming language used: FORTRAN 66 verified by PFORT in basic routines and FORTRAN 77 and well isolated assembler code.

Computers and terminals on which available: VAX with Tektronix 4010 and 4025 compatible graphics.

Documentation: Tutorials, User's Manuals, reports, and papers describing applications.

Memory and disk requirements:
POLPAC–460K bytes
MODPAC–570K bytes
IDPAC– 470K bytes
SIMNON–360K bytes

State of development: Production code

Availability of code: Licenses available from Department of Automatic Control subject to approval from the Swedish Board for Technical Development, (STU).

Person to contact for details: Karl J. Åström, Department of Automatic Control, Lund Institute of Technology, Box 725, S-220-07 Lund 7, Sweden, Phone: +46-46-108781, Telex: S-33533 LUNIVER.

37

Package or program name: PC-MATLAB

Principal developers: John Little, Cleve Moler, and the MathWorks, Inc.,

Software capabilities: Analytical commands include eigenvalues, eigenvectors, matrix arithmetic, matrix inversion, linear-equation solution, least-squares, regression, determinants, singular-value decomposition, condition estimates, rootfinding, fast-Fourier transforms, digital filtering, convolution, and statistics. Linear control system design and analysis commands are available as an option. Graphics commands include linear, loglog, semilog, polar, and 3-d mesh surface plots.

Interactive capabilities: With a widely acclaimed user interface, PC-MATLAB does scientific and engineering matrix calculations in a natural interpretive matrix environment. For displaying results, two-

and three-dimensional interactive graphics commands provide automatic graph paper. This is the program that originated the user interface employed by MATRIX_X, CTRL-C, and other CACSD packages.

Programming language used: the C Language.

Computers and terminals on which available: The IBM Personal Computer and compatibles. Maybe others in the near future.

Documentation: User's Guide, Reference Guide, and On-line HELP, BROWSE, and demonstrations.

Memory and disk requirements: Requires at least 256K of memory in the PC.

State of development: Commercial quality, production code.

Availability of code: Available by license for $595. per CPU as of December 15, 1984.

Person to contact for details: John N. Little, The MathWorks, Inc., 124 Foxwood Road, Portola Valley, CA 94025, (415) 851-7217.

Software Summaries Index

NO.	PACKAGE NAME	CONTACT PERSON	LOCATION
1	*EASY5*	Robert J. McRae	Boeing Computer Services
2	*SUBOPT*	Peter Fleming	University of North Wales
3	*SSPACK*	Steve Azevedo	Technical Software Systems
4	*DIGICON/APL*	Gene Franklin	Stanford University
5	*DPACS-F*	K. Furuta	Tokyo Institute of Technology
6	*HONEYX*	Steve Pratt	Honeywell
7		G.K.F. Lee	Hewlett-Packard (Colorado State University)
8	*CONCENTRIC*	Neil Munro	UMIST
9	*ISER-CSD*	Cevdet Suleyman	University of Grenoble
10	*LSAP*	Charles Herget	Lawrence Livermore National Laboratory
11	*CAMCSD*	Tahm Sadeghi	Fairchild-Republic (Rensselaer Polytechnic Institute)
12	*CLADP*	Jan Maciejowski	Prime CADCAM Ltd., (Cambridge University)
13	*KEDDC*	Chr. Schmid	Ruhr University
14	*TRIP*	P. P. J. van den Bosch	Delft University of Technology
15	*SUNS*	Derek Atherton	University of Sussex
16		R. Balasubramanian	University of New Brunswick
17	*PSI*	P.P.J. van den Bosch	Delft University
18	*TSPSP*	Michael Schafer	University of Virginia (Notre Dame)
19	*FREDOM/PC, TIMDOM/PC, LSSPAK/PC*	Jamshidi	University of New Mexico
20	$MATRIX_X$ *SYSTEM_BUILD*	Eleanor Vade Bon Coeur	Integrated Systems
21	*AESOP*	Bruce Lehtinen	NASA/Lewis
22	*SLICE*	Michael Denham	Kingston Polytechnic
23		Dwight Aplevich	University of Waterloo
24	*ORACLS*	Ernest Armstrong	NASA/Langley
25		Ray Magdaleno	Systems Technology
26		Louis L. Scharf	Parametrics
27	*SIRENA*	Y. Yem	Societe RSI, France
28	*CYPROS*	Tor Onshus	Norwegian Institute of Technology
29	*CTRL-C*	John Little	Systems Control Technology

NO.	PACKAGE NAME	CONTACT PERSON	LOCATION
30	*MADPAC*	Franco Davoli	University of Genoa
31	*LISPACK*	P. Petkov	Sofia, Bulgaria
32		P. M. Taylor	University of Hull, U.K.
33	*AUTOCON*	Charles Lefkowitz	Control System Automation
34	*CONTROL.lab*	M. Jamshidi	University of New Mexico
35	*L-A-S*	Phillip West	University of Illinois, Urbana-Champaign
36	*POLPAC*	Karl J. Åström	Lund Institute of Technology
	MODPAC		
	IDPAC		
	SIMNON		
37	*PC-MATLAB*	John Little	MathWorks, Inc.,